Rediscovery of Genetic and Genomic Resources for Future Food Security

Romesh Kumar Salgotra
Sajad Majeed Zargar
Editors

Rediscovery of Genetic and Genomic Resources for Future Food Security

Springer

Editors
Romesh Kumar Salgotra
School of Biotechnology
Sher-e-Kashmir University of Agricultural
Sciences & Technology of Jammu
Jammu, Jammu and Kashmir, India

Sajad Majeed Zargar
Division of Plant Biotechnology
Sher-e-Kashmir University of Agricultural
Sciences & Technology of Kashmir
Srinagar, Jammu and Kashmir, India

ISBN 978-981-15-0155-5 ISBN 978-981-15-0156-2 (eBook)
https://doi.org/10.1007/978-981-15-0156-2

This Springer imprint is published by the registered company Springer Nature Singapore Pte Ltd.
The registered company address is: 152 Beach Road, #21-01/04 Gateway East, Singapore 189721, Singapore

Preface

Plant genetic resources have supported humankind for several millennia. These resources are the basis for food security in addition to the sources of energy, animal feed, fibre and other ecosystem services. They are important in addressing the global challenges that are currently facing the human population, particularly the twin challenge of climate change and food scarcity. Owing to their great importance, effective conservation and sustainable utilization of these resources is critically important and has never been more urgent. Plant domestication is an evolutionary process in which humans have used wild species to develop new and altered forms of plants with morphological or physiological traits that meet human needs. Limited number of individuals of progenitor species were used by early farmers and the traits selected usually were related to overall yield, harvesting, and edibility. As a consequence, this strong selection process produced genetic bottlenecks of varying degrees that have resulted in a heterogeneous reduction in the level of genetic variation among annual herbaceous crops. The domestication process has resulted in reduced diversity at both the genome and morphological levels. The domestication has reduced or eliminated genetic diversity at certain loci in modern crops, thus limiting their potential for developing novel varieties with improved traits. Moreover, the selective approach during domestication had left behind various valuable alleles of biotic, abiotic stress resistance, yield and quality traits in the crop wild relative (CWR) species and landraces. There is urgent need to relook at and explore the available genetic resources for future food and nutritional security.

Crop evolution under domestication has led to increased productivity of crop species, but at the same time has narrowed their genetic basis. The potential of the genetic diversity stored in CWRs and landraces for use in crop improvement appears to be much greater than we previously imagined. Recent increases in the use of wild resources have occurred because of the recognition of the usefulness of CWRs along with the availability of various genomic resources. With the advent of modern technologies, such as next-generation sequencing (NGS) and other omics-based high throughput techniques, various genomic resources have promised to revolutionize genetics, plant breeding and biotechnology through molecular characterization, transcript profiling and cloning of whole genomes to understand the structure, function and evolution of genes. A combination of Genome-Wide Association Studies (GWASs) and next-generation-mapping populations have improved our ability to connect phenotypes and genotypes and explore the genetic diversity of wild relatives for crop improvement. The combination of these approaches with the promise of improved genomic technologies provides an opportunity for comparative genomics to apply our understanding of the past to the future for crop improvement. These genomic resources can be deployed to rediscover and explore the new genes/alleles and traits from CWRs for developing novel crop varieties resistance to various biotic and abiotic stresses, and other quality traits.

Genomic approaches have been widely used to identify genes or genomic regions controlling complex traits. High-throughput next-generation sequencing technologies offer opportunities to efficiently discover SNPs associated with important traits in landraces and crop wild relatives of both diploid and polyploid plant species. With recent significant cost reductions, scientists are now able to genotype thousands of individuals by genotyping-by-sequencing (GBS) or resequencing. With the availability of increasing numbers of SNPs and phenotypic data, researchers have been able to validate and fine-map previously identified genes and to discover novel genomic regions underlying valuable agronomic traits in crop wild species by association mapping. The availability of genome-wide data and efficient phenotyping approaches will continue to accelerate the discovery of genes controlling superior traits in CWRs. Other functional omics approaches, including transcriptomics, proteomics and metabolomics, have provided alternative opportunities for global analysis of regulatory genes, expressed proteins or metabolite candidates underlying important traits in CWRs. These omics approaches are also particularly suitable for dissection of the variation in CWRs for further utilization in crop improvement. However, the development of a high-throughput phenotyping pipeline remains challenging, especially in the field conditions. Some of the genomic regions associated with domestication traits have enhanced our understanding of their genetic basis, and will encourage further investigation to see whether allelic variation in those regions in wild relatives can additionally benefit crop improvement.

Rapid progress in advanced biotechnologies that can bridge genotype-phenotype gaps will facilitate the use of CWRs for crop improvement. Thus a number of QTL and SNPs associated with agronomically and ecologically important traits have been identified in wild species by linkage analyses, GWAS and/or combined "omics" approaches. The rapid improvement of biotechnological tools, such as diverse omics approaches, has resulted in promising advances, and no doubt will become routine in plant breeding programmes. Advanced biotechnologies are continuously being developed and will accelerate the conservation and use of genetic diversity retained in CWRs, resulting in agriculture sustainability. Utility of these resources is important in increasing the resilience and productivity of agricultural production systems. However, despite their importance, utility of these resources is poor. This book reviews the real and potential application of the current advances in genomics-based technologies in exploring and utilization of these resources for crop improvement. This book also describes in detail about exploring the untapped genes and traits for crop improvement from wild species which had been ignored during the domestication process. This will also give insight about how to utilize untapped and unexplored genetic diversity of wild species, wild relatives and landraces for crop improvement.

Rediscovery of Genetic and Genomic Resources for Future Food Security is designed to focus on the importance of plant genetic resources in achieving food security in the near future. It describes how the recent genomic resources techniques can be efficiently used in plant breeding programmes to achieve food security in the future. This book describes in detail about exploring the new genes and traits for crop improvement from wild species at the shortest possible time. The book also gives insight about how to utilize untapped and unexplored genetic diversity of wild species, wild relatives and landraces for crop improvement. It breaks the mould, offering an impressive array of balanced analyses, fresh ideas and perspectives, and thoughtful and realistic prescriptions which could help in the sustainable utilization of plant genetic resources with modern biotechnological techniques. The presentation style of the book is easy to follow and comprehend. Professionals, researchers and students are constantly reminded of previous topics of relevance to current topics being discussed. This book is not only an excellent teaching tool, but it is also a suitable reference source for professionals.

Jammu, Jammu and Kashmir, India — Romesh Kumar Salgotra
Srinagar, Jammu and Kashmir, India — Sajad Majeed Zargar

Contents

Contributors

Hasan Muhammad Abdullah GIS and Remote Sensing Lab, Department of Agroforestry and Environment, Bangabandhu Sheikh Mujibur Rahman Agricultural University, Gazipur, Bangladesh

Violeta Andjelkovic Maize Research Institute 'Zemun Polje', Belgrade, Serbia

Parshant Bakshi Division of Fruit Science, Sher-e-Kashmir University of Agricultural Sciences and Technology of Jammu, Chatha, Jammu, Jammu and Kashmir, India

Praveen L. Bennur Department of Plant Bioechnology, College of Agriculture, UAS, GKVK, Bangalore, Karnataka, India

Kaisar A. Bhat Proteomics Laboratory, Division of Plant Biotechnology, SKUAST-K, Srinagar, Jammu and Kashmir, India

Inderpal Singh Bijral Sher-E-Kashmir University of Agricultural Sciences and Technology, Jammu, Jammu and Kashmir, India

Bhagirath Singh Chauhan Queensland Alliance for Agriculture and Food Innovation (QAAFI), The University of Queensland, Gatton, QLD, Australia

Shailendra Singh Chauhan CSB–Central Sericultural Research & Training Institute (CSR&TI), Pampore, Jammu and Kashmir, India

Jahangir Ahmad Dar Proteomics Laboratory, Division of Plant Biotechnology, SKUAST-K, Srinagar, Jammu and Kashmir, India

Rose Mary Francies Department of Seed Science and Technology, Kerala Agricultural University (KAU), Thrissur, Kerala, India

Mudasir Gani CSB–Central Sericultural Research & Training Institute (CSR&TI), Pampore, Jammu and Kashmir, India

Bharat Bhushan Gupta Division of Plant Breeding and Genetics, Sher-e-Kashmir University of Agricultural Sciences and Technology of Jammu, Jammu, Jammu and Kashmir, India

Chainika Gupta School of Biotechnology, Sher-e-Kashmir University of Agricultural Sciences and Technology of Jammu, Jammu, Jammu and Kashmir, India

Mehak Gupta School of Biotechnology, Sher-e-Kashmir University of Agricultural Sciences and Technology of Jammu, Jammu, Jammu and Kashmir, India

Suphla Gupta Plant Biotechnology Division, Council for Scientific and Industrial Research (CSIR), Indian Institute of Integrated Medicine (IIIM), Jammu, India

Mehfuz Hasan Department of Genetics and Plant Breeding, Bangabandhu Sheikh Mujibur Rahman Agricultural University, Gazipur, Bangladesh

Abu Sayeed Md. Hasibuzzaman Department of Genetics and Plant Breeding, Bangabandhu Sheikh Mujibur Rahman Agricultural University, Gazipur, Bangladesh

Ashok Jagtap School of Agricultural Biotechnology, Punjab Agricultural University (PAU), Ludhiana, Punjab, India

Gayatri Jamwal Sher-E-Kashmir University of Agricultural Sciences and Technology, Jammu, Jammu and Kashmir, India

Suzana Jordanovska Institute of Agriculture Skopje, University "Ss Cyrilus and Methodius" in Skopje, Skopje, Republic of Macedonia

Zoran Jovovic Biotechnical Faculty Podgorica, University of Montenegro, Podgorica, Montenegro

Amar Kadam Department of Biotechnology, Lokmangal College of Agricultural Biotechnology, Mahatma Phule Krishi Vidyapeeth (MPKV), Solapur, Maharashtra, India

Md. Mehboob Hasan Kallol Senior Management Counsellor and Head, Computer Services Division, Bangladesh Institute of Management (BIM), Dhaka, Bangladesh

Masrat Kareem Department of Botany, University of Kashmir, Srinagar, Jammu and Kashmir, India

Jashan Jot Kaur Department of Plant Breeding and Genetics, Punjab Agricultural University (PAU), Ludhiana, Punjab, India

Gulshan Mahajan Queensland Alliance for Agriculture and Food Innovation (QAAFI), The University of Queensland, Gatton, QLD, Australia

Reetika Mahajan Division of Plant Biotechnology, Sher-e-Kashmir University of Agricultural Sciences and Technology of Kashmir, Srinagar, Jammu and Kashmir, India

Dragan Mandic Agricultural Institute of Republic of Srpska, Banja Luka, Republic of Srpska, Bosnia and Herzegovina

Rakeeb Ahmad Mir Department of Biotechnology, School of Biosciences and Biotechnology, BGSB University, Rajouri, Jammu and Kashmir, India

Nagesha N Department of Plant Bioechnology, College of Agriculture, UAS, GKVK, Bangalore, Karnataka, India

Narender Negi ICAR–National Bureau of Plant Genetic Resources (NBPGR), Shimla, Himachal Pradesh, India

Pankaj Pandotra Division of Fruit Science, Sher-e-Kashmir University of Agricultural Sciences and Technology of Jammu, Chatha, Jammu, Jammu and Kashmir, India

Novo Przulj Faculty of Agriculture, University of Banja Luka, Banja Luka, Republic of Srpska, Bosnia and Herzegovina

Punya School of Biotechnology, Sher-e-Kashmir University of Agricultural Sciences and Technology of Jammu, Jammu, Jammu and Kashmir, India

Ambavane Ajinkya Rajendra Department of Seed Science and Technology, Kerala Agricultural University (KAU), Thrissur, Kerala, India

Sheezan Rasool Proteomics Laboratory, Division of Plant Biotechnology, SKUAST-K, Srinagar, Jammu and Kashmir, India

Pawan Saini CSB–Central Sericultural Research & Training Institute (CSR&TI), Pampore, Jammu and Kashmir, India

Pooja Saini Department of Biotechnology, Eternal University, Baru Sahib, Himachal Pradesh, India

Romesh Kumar Salgotra School of Biotechnology, Sher-e-Kashmir University of Agricultural Sciences & Technology of Jammu, Jammu, Jammu and Kashmir, India

Arjun Sharma Department of Biotechnology, School of Biosciences and Biotechnology, BGSB University, Rajouri, Jammu and Kashmir, India

Manmohan Sharma School of Biotechnology, Sher-e-Kashmir University of Agricultural Sciences and Technology of Jammu, Jammu, Jammu and Kashmir, India

Anil Kumar Singh School of Biotechnology, Sher-e-Kashmir University of Agricultural Sciences and Technology of Jammu, (SKUAST-J), Chatha, Jammu, Jammu and Kashmir, India

Charan Singh ICAR–Indian Institute of Wheat and Barley Research (IIWBR), Karnal, Haryana, India

Parvaze A. Sofi Dry Land Agricultural Research Station, SKUAST, Kashmir, Jammu and Kashmir, India

Aijaz A. Wani Department of Botany, University of Kashmir, Srinagar, Jammu and Kashmir, India

Sajad Majeed Zargar Division of Plant Biotechnology, Sher-e-Kashmir University of Agricultural Sciences and Technology of Kashmir, Srinagar, Jammu and Kashmir, India

About the Editor and Co-editor

Romesh Kumar Salgotra, PhD, is currently a Professor and Coordinator, School of Biotechnology at Sher-e-Kashmir University of Agricultural Sciences & Technology of Jammu (SKUAST-Jammu) in India. Dr. Salgotra received his PhD in Plant Breeding and Genetics in 1999 from CSK Himachal Pradesh Krishi Vishwavidyalaya, Palampur, India. He was previously an Endeavour Executive Fellow of Queensland Alliance for Agriculture and Food Innovation (QAAFI) at the University of Queensland, Australia. He has worked in different capacities as Junior Scientist and Senior Scientist at SKUAST-Jammu under All India Co-ordinated Rice Improvement Project, National Rice Research Institute (NRRI), Hyderabad, and *Central Research Institute for Dryland Agriculture* (CRIDA), Hyderabad, India. He has also worked as Agriculture Extension Officer in Govt. of Jammu and Kashmir, India. Dr. Salgotra is a recipient of BOYSCAST Fellowship, Tennessee University, USA. He was awarded the prestigious Norman E. Borlaug Fellowship by United States Agriculture Development (USDA) during 2015. He has participated in Genetic Resources and Intellectual Property Rights Programme (GRIP) under Swedish International Development Agency (SIDA), Sweden, and follow-up programme at Pretoria, South Africa. He has received several awards for his work and research. He was the member and representative of Global Forum for Innovative Agriculture (GFIA), Abu Dhabi, UAE. Dr. Salgotra was also one of the members of delegation in the Joint Committee on Science and Technology, Govt. of India, held at Belgrade, Serbia, in 2017. He has chaired a session in International Association of Science & Technology

Development for Computational Bioscience held at Cambridge, Massachusetts, USA. He has more than 50 peer-reviewed publications and has delivered numerous oral and poster presentations in national and international meetings/conferences. He has been affiliated with several internationally reputed journals and is also reviewer of reputed journals *Scientific Reports, PLOS One, Plant Cell Reports, BMC Genetics, Rice Science, Journal of Integrative Agriculture, Australian Journal of Crop Science, SABRAO* and many others. He has guided MSc and PhD students under his supervision. Dr. Salgotra has received several grants for research projects from national and international funding agencies.

Sajad Majeed Zargar, PhD is currently an Assistant Professor at Sher-e-Kashmir University of Agricultural Sciences and Technology of Kashmir (SKUAST Kashmir), India. He was previously a visiting professor at the Nara Institute of Science and Technology, Japan. He has worked as an Assistant Professor at SKUAST Jammu, Baba Ghulam Shah Badshah University, Rajouri (BGSB), India. He has also worked as a scientist at Advanta India Limited, Hyderabad, India, and TERI (The Energy and Resources Institute), New Delhi, India. Dr. Zargar is a recipient of the CREST overseas fellowship from DBT, India, and is also recipient of the Goho grant from the Govt. of Japan. He has received several awards for his work and research. He is also a member and representative of INPPO (International Plant Proteomics Organization). His editorial activities and scientific memberships include publishing research and review articles in international journals and as a reviewer. He has been affiliated with several internationally reputed journals and is also reviewer of reputed journals, *Frontiers in Plant Science, 3 Biotech, Scientia Horticulturae, Methods in Ecology and Evolution, Australian Journal of Crop Science* and many others. Dr. Zargar has been invited to give many lectures at professional meetings and workshops and has received grants for research projects under his supervision.

Abbreviations

ABA	Abscisic acid
AB-QTL	Advanced back crossed QTL
AFLPs	Amplified fragment length polymorphisms
AM	Association mapping
AMPRIL	Arabidopsis multiparent RIL
ATI	Accelerated Trait Introgression
BACs	Bacterial artificial chromosomes
BCP	Biofortification Challenge Program
BILs	Backcross inbred lines
BNF	Biological nitrogen fixation
BGCI	Botanic Gardens Conservation International
BLAST	Basic Local Alignment Search Tool
Bt	*Bacillus thuringiensis*
CAPS	Cleaved amplified polymorphic sequence
CBD	Convention on Biological Diversity
CC	Core collection
cDNA	Complementary deoxy ribose nucleic acid
CGA	Candidate gene approach
CGD	Crop genetic diversity
CGRs	Crop genetic resources
CGIAR	Consultative Group for International Agricultural Research
CIAT	International Center for Tropical Agriculture
CIFOR	Center for International Forestry Research
CIMMYT	International Maize and Wheat Improvement Center
CIP	International Potato Centre
CMS	Cytoplasmic genic male sterility
CRISPR-cas9	Clustered regularly interspaced short palindromic repeats-associated protein 9
CRS	Core reference set
CSSLS	Chromosome segment substitution lines
CMT3	Chromomethylase 3
CWR	Crop wild relatives
CO_2	Carbon dioxide
DArT	Diversity array technology

DH	Doubled haploid
DHPLC	Denaturing high-performance liquid chromatography
DCL	DICER like enzymes
DNA	Deoxy-ribose nucleic acid
DREB	Dehydration responsive element binding
DS	Domestication syndrome
DT	Drought tolerance
EcoTILLING	Eco Type TILLING
EST	Expressed sequence tag
FACE	Free-air carbon dioxide enrichment
FAO	Food and Agriculture Organization
FAnGR	Farm Animal Genetic Resources
GDP	Gross domestic product
GIS	Geographic information system
GAB	Genomic-assisted breeding
GBS	Genotyping-by-sequencing
GC	Gateway cloning
GolS	Galactinol synthase
GenoSIS	Genome Spatial Information System
GEBV	Genomic estimated breeding value
GM	Genetic modification
GMOs	Genetically modified organisms
GPS	Global positioning system
GR	Green revolution
GS	Genome selection
GWAS	Genome-wide association studies
GWE	Genome-wide editing
GWP	Genome-wide prediction
HDB	Homology-directed recombination
HNVs	High natural values
HSPs	Heat shock proteins
HTPs	High-throughput phenotyping
HTPPs	High-throughput phenotyping platforms
HYVs	High yielding varieties
IAEA	International Atomic Energy Agency
IARI	Indian Agricultural Research Institute
IBP	International Biological Program
ICARDA	International Center for Agricultural Research in the Dry Areas
ICRAF	International Council for Research in Agroforestry
ICRISAT	International Crops Research Institute for the Semi-Arid Tropics
IFPRI	International Food Policy Research Institute
IHF	Integration Host Factor
IITA	International Institute of Tropical Agriculture
ILRI	International Livestock Research Institute
IPCC	Intergovernmental Panel on Climate Change

IBPGR	International Board for Plant Genetic Resources
ICC	International Co-ordinating Council
ICN	International Conference on Nutrition
InDels	Insertions or deletions
IFAD	International Fund for Agricultural Development
ILO	International Labour Organisation
IME	Institution of Mechanical Engineers
INIBAP	International Network for the Improvement of Banana and Plantain
IRRI	International Rice Research Institute
ITPGRFA	International Undertaking on Plant Genetic Resources for Food and Agriculture
IUCN	International Union for Conservation of Nature
IWMI	International Water Management Institute
KASP	Kompetitive allele specific PCR
LEA	Late embryogenesis abundant proteins
LPS	Low-pressure storage
LOS	Low-oxygen storage
LN	Liquid nitrogen
LD	Linkage disequilibrium
LR	Landraces
MAB	Marker-assisted breeding
MABC	Marker-assisted backcrossing
MAGIC	Multi-parent advanced generation intercross
MAGP	Marker-assisted gene pyramiding
MARS	Marker-assisted recurrent selection
MAS	Marker-assisted selection
MBC	Map-based cloning
MS	Mass spectrometry
MSBP	Millennium Seed Bank Project
MoEFCC	Ministry of Environment, Forest and Climate Change
MPCA	Medicinal Plant Protection Area
NAM	Nested association mapping
NAFIS	National Bank for Agriculture and Rural Development, All India Rural Financial Inclusion Survey
NAGS	National Active Germplasm Sites
NBPGR	National Bureau of Plant Genetic Resources
NERICA	New Rice for Africa
NGOs	Non-governmental organizations
NHEJ	Non-homologous end joining
NIDM	National Institute of Disaster Management
NILs	Near isogenic lines
NGS	Next-generation sequencing
PAMP	Pathogen-associated molecular pattern
PCR	Polymerase chain reaction
PGR	Plant genetic resources

PGRFA	Plant Genetic Resource for Food and Agriculture
PoU	Prevalence of undernourishment
PROSEA	Plant Resources of Southeast Asia
PPP	Plant protection products
PTM	Post-translational modifications
QTL	Quantitative trait loci
RILs	Recombinant inbred lines
RAD-seq	Restriction-site-associated DNA sequencing
RAPD	Random amplified polymorphic DNA
RFLP	Restriction fragment length polymorphism
RGA	Rapid Generation Advancement
RGPs	Restoration of gene pools
RNA	Ribose nucleic acid
RNAi	RNA interference
RISC	RNA-induced silencing complex
ROS	Reactive oxygen species
SADC	Southern African Development Community
SARP	Sequence-related amplified polymorphism
SCAR	Sequence-characterized amplified region
SDGs	Sustainable development goals
SGS	Second-generation sequencing
siRNA	Small interfering RNA
SNPs	Single nucleotide polymorphism
SMRT	Single-Molecule Real-Time Sequencer
SNVs	Single nucleotide variations
SSD	Single-seed descent
SSLPs	Simple sequence length polymorphisms
SSR	Simple sequence repeats
STRs	Short tandem repeats
STS	Sequence tagged site
TAC	Technical Advisory Committee
TALENs	Transcription activator-like effector nucleases
TCA	Tricarboxylic acid cycle
TFs	Transcription factors
TGS	Third-generation sequencing technologies
TILLING	Targeting-induced local lesions in genomes
UCDP	Uppsala Conflict Data Program
UNCED	United Nations Conference on Environment and Development
UNFAO	United Nations Food and Agriculture Organization
UNEP	United Nations Environment Programme
UNESCO	United Nations Education, Scientific and Cultural Organization
UNHCR	United Nation Higher Commission for Refugees
UNICEF	United Nations International Children's Emergency Fund
UNIX	Uniplexed Information and Computing System
USEPA	United States Environmental Protection Agency

VIR	N. I. Vavilov All-Union Scientific Research Institute of Plant Industry
WGP	Whole genome prediction
WGS	Whole-genome shotgun sequence
WEF	World Economic Forum
WEPs	Wild edible plants
WHO	World Health Organization
ZNFs	Zinc finger nucleases

Potential of Genetic and Genomic Resources for Genetic Improvement of Food Crops

1

Romesh Kumar Salgotra and Sajad Majeed Zargar

Contents

Abstract

Plant genetic resources (PGR) are the major natural resources on which human being has relied on for their livelihood, and their demand will increase in the future due to the continuing growth of population. Scientifically use of PGR is important in increasing the resilience and productivity of agricultural production systems for future food security. However, despite their importance in crop improvement, utility of these resources has been poor. With the advent of new high-throughput technologies like next-generation sequencing (NGS), new genomic resources have been generated. The advancement in genomic tools and

R. K. Salgotra (✉)
School of Biotechnology, Sher-e-Kashmir University of Agricultural Sciences and Technology of Jammu, Chatha, Jammu, Jammu and Kashmir, India

S. M. Zargar
Division of Plant Biotechnology, Sher-e-Kashmir University of Agricultural Sciences and Technology of Kashmir, Srinagar, Jammu and Kashmir, India

R. K. Salgotra, S. M. Zargar (eds.), *Rediscovery of Genetic and Genomic Resources for Future Food Security*,
https://doi.org/10.1007/978-981-15-0156-2_1

reduction in their costs are bringing many more plants within the range of genome- and transcriptome-level analysis. The genomic resources thus generated will be useful for genetic improvement of crops through applications of resources such as marker-assisted breeding (MAB) for gene introgression, mapping quantitative trait loci (QTLs) or identifying new or rare alleles associated with a particular trait. The effective and complementary use of all of genomic resources and available PGR will be required for meeting the challenge posed by the world's expanding demand for food. This chapter focuses on how genomic resources can be used in PGR for crop improvement particularly for major food crops. The real and potential application of the current advances in genomic technologies can be used for efficient utilization of PGR for crop improvement to secure food security.

Acronyms

AB-QTL	Advanced backcrossed QTL
AM	Association mapping
CGA	Candidate gene approach
CGIAR	Consultative Group for International Agricultural Research
CMS	Cytoplasmic genic male sterility
CRS	Core reference set
CWR	Crop wild relatives
DH	Doubled haploid
EcoTILLING	Ecotype TILLING
EST	Expressed sequence tag
GAB	Genomic-assisted breeding
LD	Linkage disequilibrium
MAB	Marker-assisted breeding
MAS	Marker-assisted selection
NGS	Next-generation sequencing
PCR	Polymerase chain reaction
PGRs	Plant genetic resources
QTLs	Quantitative trait loci
RAD-seq	Restriction-site-associated DNA sequencing
RAPD	Random amplified polymorphic DNA
RFLP	Restriction fragment length polymorphism
SCAR	Sequence characterized amplified region
SNPs	Single nucleotide polymorphism
SSR	Simple sequence repeats
STS	Sequence tagged site
TILLING	Targeting induced local lesions in genomes

1.1 Introduction

Genetic resources are sometimes called the "first resource" of the natural resources which can be used for crop improvement. At one level, plant genetic resources (PGRs) include all the individuals of a species, particularly if it is threatened with extinction. PGRs also include populations, gene pools or races of a species which possess important attributes not found uniformly throughout the species. PGR can also be defined as all materials such as wild species, wild relatives, landraces, etc. that are available for improvement of a cultivated plant species. Breeding lines and research materials, such as mutant, genetic or chromosomal stocks, are also genetic resources and are important in animal and plant breeding and in all phases of biologic research. Finally, genetic resources can refer to genes themselves, maintained in selected individuals or cloned and maintained in plasmids. The effort required to utilize these resources in crop improvement is enormous but well justified as the genetic diversity present in these represents a critical component in the world's fight against hunger. These resources are the basis for food security which can be efficiently utilized for crop improvement through various biotechnological interventions.

To feed the ever-increasing population in the scenario of climate change, the demand for the development resilient crop cultivars is imperative. Development of such cultivars through conventional plant breeding methods depends on the availability of natural genetic variations in PGR of a given crop species. Moreover, genetic variability that exists is very low and needs to be widened for further improving the productivity of the crop. Further, there is a need to protect the loss of genetic diversity in several plant species. Efforts have been made since long to collect, conserve and evaluate PGRs, to support the plant breeders with diverse genetic materials, to widen the genetic base and to create new varieties to enhance the crop productivity. Out of 240,000 plant species, only 25–30 are used for human consumption, and of these, rice, wheat and maize together constitute about 75% of global grain production (Cordain 1999). Therefore, conservation, multiplication and sustainable utilization of the existing PGRs, which comprise cultivars, landraces and wild relatives, are essential to combat the food security.

In classical plant breeding, PGR may also include those genetic materials that do not have any immediate use for the plant breeders (Hallauer and Miranda 1981). To satisfy the ever-increasing demands of a growing human population for more food, plant breeders require access to new genetic diversity in plant species (Brozynska et al. 2016). According to the extended gene pool concept, PGR may be divided into primary gene pool, secondary gene pool, tertiary gene pool and isolated genes (Harlan and de Wet 1971). The primary gene pool comprises of the crop species itself and other species that can be easily crossable. The secondary gene pool consists of related species that are more difficult to cross with the target crop. In this case crossing is less successful, and if crossing is there, the progenies are partially sterile. The tertiary gene pool is composed of species which can only be crossed by employing techniques like embryo rescue or protoplast fusion. In fourth class PGR, the isolated genes, may derive from related or unrelated plant species, from animals or microorganisms. The importance of the different classes of PGR for crop improvement depends on the target crop species. The utility of PGR is important in increasing the productivity and production of food crops. However, despite their importance, utility of genetic resources has been poor. The array of PGR together with new genomic resources and other technological tools provides us with a healthy solution to meet the world's future food demands. PGRs or genomic resources alone will not serve the purpose; rather, the complementary utilization of these resources will be required for crop improvement to meet the food requirements.

In plant genome about 50–60,000 genetic loci are present, and in crop improvement, the correct combination of specific alleles is required. The knowledge of where these alleles are best found and how these can be combined in a single species is important for crop improvement. With the emergence of NGS techniques with reduced cost, various genomic resources such as genome sequences, functional genomic resources including microarrays and RNA-seq, sufficient numbers of molecular markers, expressed sequence tags (ESTs) and high-density genetic maps are causing a rapid acceleration of genetics and genomic research of crops. The completion of reference genome sequences of important crops and the ability to perform high-throughput resequencing are providing opportunities for improving our understanding of the PGR for utilization in crop improvement. This is leading to an increase in our knowledge of the genes that are linked to many agronomical and quality traits. These genomic resources have the potential to accelerate gene discovery which are being introgressed/pyramided in crops to enhance crop productivity to ensure food security. This chapter focuses on the potential use of PGR and genomic resources to meet the continually expanding demand for major food crops. This reviews the real and potential application of the current advances in genomic technologies in improving the utilization of these resources. This will also indicate how potentially these resources and technological tools can be used for crop improvement to meet food security in the future.

1.2 Importance of Genetic and Genomic Resources

Genetic and genomic resources play a significant role in crop improvement particularly in the development of crop varieties with desirable characters. To meet the challenge posed by demand for food, the complementary utilization of genetic resources with available genomic will be required. Genetic resources can efficiently be used in crop improvement with the help of the current advances in genomic technologies.

1.2.1 Genetic Resources

Genetic resources play a significant role in agriculture, particularly in crop variety development and improvement programmes. PGR diversity provides the major sources of important genes for diseases and insect pest resistance and yield and quality improvement. Genetic resources form the natural variations that have been utilized to support human kind for several millennia. They include primitive forms of cultivated crop species and landraces, modern cultivars, obsolete cultivars, breeding lines and genetic stocks and related wild species. These resources are the basis for food security in addition to being sources of energy, animal feed, fibre as well as other ecosystem services. They are important in addressing the global challenges such as climate change, global population growth and food scarcity. PGRs are fundamental to our efforts to improve agricultural productivity. These resources, fortunately stored in gene banks around the world and owing to their great importance, effective conservation and sustainable utilization of PGR, are critically important to food security, as evidenced by the huge number of accessions that are conserved in gene banks for various species. It is clear that enormous progress has been made in conserving important germplasm in gene banks, but these genetic resources remain unexploited because of a variety of factors. Genetic approaches have a long history of use in conservation, but the transition to genomic technologies is only just beginning. Earlier very limited biotechnological interventions and genomic resources were available and are the main limitations for exploitation of genetic resources (Supple and Shapiro 2018). With the development of various genomic techniques, the genetic resources can be used efficiently for crop improvement.

PGRs are the most important components of agrobiodiversity. The introgression of genes in wheat crop which provided the foundation for the "Green Revolution" demonstrated the tremendous impact of genetic resources on crop production (Hoisington et al. 1999). Food security mainly depends on the wise use and conservation of agricultural biodiversity and genetic resources (Esquinas-Alcazar 2005). Since importance have been given to relatively small number of crop species for global food security, it is particularly important that their genetic diversity is conserved effectively and managed wisely. So far, only a small part of the total genetic variability has been characterized and used for crop improvement. Owing to their great importance, effective conservation and sustainable utilization of PGR are imperative for food security.

1. *Wild relatives*: The major portion in PGR in plant species is contributed by wild relatives. Wild species possess numerous desirable characters such as genes for biotic and abiotic resistance and nutritional value enhancement for crop improvement. Due to some crossability barriers, wild species are difficult to cross with cultivated genotypes. However, crossability between cultivated varieties and wild species can be overcome using embryo rescue, genetic engineering and protoplast fusion techniques for transferring some useful traits from wild relatives to cultivated genotypes. The wild relatives possess desirable nutritional traits such as protein content in wheat, calcium content and provitamin A in potatoes and tomatoes, respectively.
2. *Landraces*: The indigenous varieties selected by the local people from the traditionally grown germplasm are termed as landraces. And development of varieties from these selected germplasm is called as landrace varieties. These varieties also possess numerous genes for biotic and abiotic resistance. These landraces are grown under species environmental conditions and areas but possess a

number of important traits to be used in crop improvement. Although landrace varieties are low yielding, these are tolerant to major stress factors such as water regime, drought, frost, soil salinity and heat.

1.2.2 Genomic Resources

With the advancement in high-throughput techniques and cost reduction of next-generation sequencing (NGS), many genomic resources such as genome sequences, high-throughput analysis of gene expression, numbers of molecular markers, ESTs and high-density genetic maps have paved the way to the genetic engineering and molecular breeding of plants for crop improvement. The application of these genomic resources to crop species can contribute efficiently to solve the problems of nutritional deficiency and biotic and abiotic stresses. In the last decade, the emphasis on crop improvement using novel genomic tools has shifted toward the identification and functional analysis of miRNAs, one of the hottest research fields in plant sciences (Sun 2012). For the last three decades, globally different genomic resources are being effectively used for identifying the important genes and alleles from the genetic resources to be used in breeding programmes (Fig. 1.1). The following are the various genomic resources which are being used for crop improvement:

1. *Quantitative trait loci*: A number of agronomic traits are controlled by many genes, and these traits are called as quantitative or complex traits. A quantitative trait locus (QTL) represents a region of a genome that contains genes associated with a particular quantitative trait such as yield, grain weight, protein content,

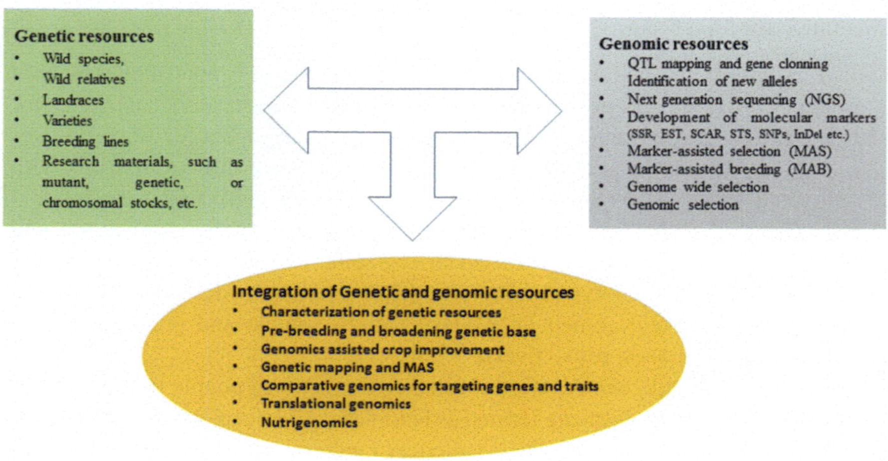

Fig. 1.1 A schematic illustration of the integration of genetic and genomic resources for crop improvement

etc. (Collard et al. 2005). Such regions are associated with the agronomic/phenotypic traits of the plant species. However, their identification in the genome is very difficult because of the interactions between QTL, absolute number of QTL, epistasis and other sources of variation (Doerge 2002). These genes/QTLs are identified and tagged by using molecular markers (Mohan et al. 1997). A tight association between the gene of interest and molecular markers led to marker-assisted breeding (MAB) programme in crop improvement. This will also help in the construction of genetic map based on molecular markers (Francia et al. 2005).

2. *Molecular markers*: With the advances in NGS and molecular marker technology, a number molecular markers such as restriction fragment length polymorphism (RFLP), random amplified polymorphic DNA (RAPD), amplified fragment length polymorphism (AFLP), simple sequence repeats (SSR), sequence tagged site (STS), expressed sequence tag (EST), sequence characterized amplified region (SCAR), single nucleotide polymorphism (SNPs), insertion-deletion (InDel), etc. were developed. These molecular markers have been successfully used in the construction of genetic maps, indirect selection of desired traits in segregating and advanced breeding materials. Today molecular markers become the choice of markers for genetic diversity studies of crop plants. Molecular markers could be used in various crops such as wheat, apple, pear, plum, etc. for identification of desired traits. These markers successfully help in the development of biotic and abiotic stress-resistant varieties (Salgotra et al. 2015).
3. *Marker-assisted selection*: There has been considerable progress during the last two and half decades in mapping and tagging many agronomical desirable traits with molecular markers, which form the basis for marker-assisted selection (MAS). In MAS, molecular markers are used to assist in the selection of target traits of interest such as resistance genes for diseases and insect pests and selection of quality traits of fruit trees (Dirlewanger et al. 2004). MAS has been successfully used in introgression of a number of genes in rice crop varieties which are susceptible to lodging, diseases and insect pests.
4. *Genome-wide association studies (GWAS)*: Association mapping (AM) is an alternative to classical linkage mapping to explicate the genetic basis of complex traits particularly for abiotic stresses (Abdurakhmonov and Abdukarimov 2008; Zhao et al. 2011). Linkage mapping based on biparental progeny has been useful for identification of major genes and QTL mapping (Komatsuda et al. 2007). But linkage mapping based on biparental progeny suffers from several drawbacks (Cosart et al. 2011). The shortcomings of the biparental-based linkage mapping are well addressed in association genetics in several crops (Gupta et al. 2005; Hall et al. 2010; Maccaferri et al. 2011). Further advantages of biparental linkage analysis, along with association mapping in nested association mapping in single unified mapping population, are used for the genome-wide dissection of complex traits (Yu et al. 2008).
5. *Whole-genome* de novo *sequencing*: In several decades "Sanger sequencing" remained predominant for decoding the genomes. The whole-genome sequenc-

ing in less time and low cost is the major landmark discovery in omics. Earlier a small-genome sequencing requires huge funds in multi-institutional efforts. With the advancement of NGS technologies, genome sequencing has become much faster, low cost and efficient by several folds. After the introduction of the first 454 NGS platform, several platforms were introduced such as Illumina, ABI SOLiD, Helicos, PacBio, Ion Torrent and Oxford Nanopore. The whole-genome sequencing of plants has generated huge genomic resources such as development of molecular markers, comparative genomics, gene expression through transcriptome, etc.

6. *Genome resequencing for the discovery of genome-wide variation*: A reference genome is generated once the genome of a plant is sequenced. The reference genome is for studying genetic resources of the same species or related species to detect genetic variations. Thus, whole-genome resequencing of several accessions/genotypes enables to generate useful genomic resources and information. This has also removed bottlenecks of ascertainment bias (i.e. the presence of rare alleles) obtained through biparental mapping population in the estimation of linkage disequilibrium (LD) and genetic relationships between genotypes (Cosart et al. 2011). Genome Project 1001 is one of the largest projects started in 2008 for resequencing of 1001 *Arabidopsis thaliana* to know/discover genome-wide sequence variations. In China genome sequencing project of 3000 rice accessions is ongoing to discover genome-wide sequence variations among the genotypes.

1.3 Genetic and Genomic Resources for Crop Improvement

Conventional research inputs have contributed in solving some of the constraints limiting crop productivity. However, limitations, such as complex genome, narrow genetic base, poor fertility in distant crosses, transferring important genes from wild relatives, susceptibility to biotic and abiotic stresses and long duration to breed elite cultivars, hinder crop improvement programmes (Salgotra et al. 2015). However, the modern biotechnological techniques and genomic resources can overcome the problem of identification of important genes and alleles in the wild and unrelated species. The identified important genes and alleles can be easily transferred in cultivated species embryo rescue techniques, and linkage drag can be minimized with the use of genomic resources such as MAS, MAB, QTL mapping, etc. There are various biotechnological strategies to solve the problems faced by the breeders which can be adopted for PGR to enhance the crop productivity on sustainable basis.

Genetic resources are the basic material for selection and improvement of crop species through breeding to ensure food security for the rapidly increasing population. Different aspects related to PGR such as collection, conservation, evaluation, management and utilization are, however, needed to be done eminently. Biotechnological tools have proved useful in a number of ways to improve the conservation and management of PGR (Hodgkin et al. 2001). Molecular markers help in DNA profiling, identification and verification of accession identity and genetic contamination (Collard et al. 2005; Spooner et al. 2005; Weising et al. 2005) and also have been used to identify eco-geographic races within the domesticated or wild gene pools of crop species (Yu et al. 2003). Different genomic resources have emerged with the availability of high-throughput techniques and sequence of various crop species genomes through NGS techniques. Presently, existing biotechnological approaches also overcome challenges of embryo rescue and somatic hybridization for effective utilization and enhancement of PGR in crop improvement (Rao et al. 2003; Zimnoch-Guzowska et al. 2003). Bringing together plant breeders and biotechnologists is an eminent need for effective utilization of genomic resources techniques in plant breeding programmes. The following are recent genomic resources techniques which can be efficiently used in plant breeding programmes to achieve food security in the future.

1.3.1 Pre-breeding and Broadening Genetic Base

A significant proportion of gene bank collections comprises wild species, which represent the primary, secondary and tertiary gene pool (Harlan and de Wet 1971). These genetic resources have immense value in terms of the useful genes and alleles to improve the gene pool of crop species. Earlier breeders are reluctant to use these valuable resources in their breeding programmes because of linkage drag. For example, breeders in Japan have faced challenges in developing elite varieties with resistance against blast and also possessing good quality traits because of the co-introduction of desirable alleles for blast resistance and the undesirable ones controlling poor grain quality (Fukuoka et al. 2009). While such associations could be because of pleiotropy, they have in most cases been found to be because of tightly linked genes (Fukuoka et al. 2009). Breaking this linkage is usually costly and time-consuming. Most breeders therefore prefer to reuse their usually limited working collections, thereby leading to release of varieties with narrow genetic diversity. This narrow genetic base negatively affects the resilience, productivity and sustainability of agricultural production systems. Pre-breeding is therefore an important activity that helps to improve the genetic value, attractiveness as well as suitability of gene bank materials to breeders. Presently, certain predictive models have been developed which have the capacity to predict those SNP variations that are most likely to lead to deleterious phenotypic effects (Xu 2010). The biotechnological tools can identify the materials with such SNP alleles which can be eliminated from breeding programmes at an early stage. Plant breeders are also reluctant to use wild species due to linkage drag and their unwillingness to disrupt the favourable linkage blocks in their breeding materials.

With the use of high-throughput sequencing and genotyping approaches, it is now possible to obtain cross-specific sequence markers such as SNPs that can be used to saturate the genetic background of both parents (Henry et al. 2010). Using SNP markers, it is possible to monitor the degree of introgression of specific alleles or genomic regions in the offspring (Sharma et al. 2013; McNally et al. 2009).

This monitoring ensures that the genome of the recurrent parent can be efficiently regained, and the tracking of both desired and undesired alien alleles ensures that only narrow segments of the wild species, preferably having only the desired allele, are introgressed (Henry et al. 2010). To minimize linkage drag, it is recommended that the markers to monitor the introgression should be as close as possible to the desired genomic region (Hospital 2001). The use of genic or functional markers linked to the gene of interest is preferable. Deep sequencing of the genomic region controlling a particular trait can help identify the loci/alleles responsible for the undesirable trait and thus select recombinants lacking this undesirable allele (Fukuoka et al. 2009; Varshney et al. 2014). Genomics therefore plays an important role in the identification of both beneficial and deleterious alleles as well as facilitating the transfer of the beneficial ones during crop improvement. This minimizes the challenges associated with wild and unadapted materials, thereby enhancing their utility in crop improvement.

1.3.2 Genomic-Assisted Crop Improvement

Genomic technology has been applied in gene identification laying good foundation for functional genomics research, and to aid us in understanding the gene expression and biological activity, genomics initiatives are focused on fundamental elements of plant biology with regard to growth, development, reproduction, photosynthesis and responses to environmental conditions and pathogens. Cereal genomics carries the strength to shape the future of agriculture and its sustainability (Tuberosa et al. 2002). The better prediction of the phenotype that a particular genotype will produce is a primary goal of genomics-based breeding. Analysis of the crop genome architecture and their expressed components is now possible with the development in crop genomics and subsequently leads to an increase in our knowledge of the genes that are linked to key agronomically important complex traits particularly in major crop species. DNA-based molecular markers including SNPs have played a pivotal role in detecting the genetic variation available in germplasm collections and breeding lines. These DNA markers can be generated in large numbers and can prove to be very useful for a variety of purposes relevant to crop improvement. Their association with genes/QTLs controlling the traits of economic importance has also been utilized in some cases for indirect MAS. Other uses of molecular markers include gene introgression through backcrossing, germplasm characterization, genetic diagnostics, characterization of transformants, study of genome organization and phylogenetic analysis.

Various sets of diverse molecular markers have been developed for many major crop species and are being used extensively for the development of saturated molecular, genetic and physical maps and for the identification of genes or QTL controlling traits of economic importance through MAS (Varshney et al. 2005; Varshney et al. 2006). With the use of genomic resources, various traits of crops have been improved (Table 1.1). Together with MAS other approaches like association

Table 1.1 Potential use of genetic and genomic resources for crop improvement

Crop	Breeding strategies	Objective	References
Maize	AB-QTL	Improved hybrid performance for yield, grain moisture and plant height	Ho et al. (2002)
	MAS	Conversion of normal maize lines into quality protein maize (QPM)	Babu et al. (2004)
	MAS	Improvement of drought adaptation	Ribaut and Ragot (2006)
Barley	Introgression, MAS	Enhancement of tolerance to boron toxicity in two-rowed barley	Emebiri et al. (2009)
	Introgression, MAS	Resistance to barley yellow mosaic virus I–III from donor line "Y4"	Okada et al. (2003)
Rice	Introgression, AB-QTL	Identification of yield-improving QTLs from *O. rufipogon*	Xiao et al. (1998)
	Pyramiding	Bacterial blight (BB) resistance (*X. oryzae* pv. *oryzae*) into elite rice variety PR 106	Singh et al. (2001)
	Pyramiding	Bacterial blight (BB) resistance (*X. oryzae* pv. *oryzae*) into elite rice variety Samba Mahsuri	Sundaram et al. (2010)
	MAS	Introgression of locus conferring submergence tolerance from cultivar "FR13A" into "Swarna"	Xu et al. (2006)
Wheat	Introgression	Stacking of QTL for *Fusarium* head blight (FHB) resistance from non-adapted sources in an elite spring wheat background	Miedaner et al. (2006)
	MAS	Leaf rust resistance (*Puccinia triticina*), stripe rust resistance (*P. striiformis*), leaf, stripe and stem rust resistance gene complex	Chicaiza et al. (2006)
	MAS	Introduction of six *Fusarium* head blight QTLs, orange blossom wheat midge resistance (*Sm1*) and leaf rust resistance (*Lr21*)	Somers et al. (2005)
	MAS	Introduction of three *Fusarium* head blight QTLs into an elite winter wheat breeding population	Wilde et al. (2008)
	Introgression	Leaf rust resistance gene *Lr58* from *A. triuncialis*	Kuraparthy et al. (2011)

Source: Journal of Plant Science & Research

mapping (Ersoz 2007), functional genomics (Schena 1998), genetical genomics (Jansen and Nap 2001), allele mining (Varshney et al. 2005), targeting induced local lesions in genomes (TILLING) and ecotype TILLING (EcoTILLING) (Till 2007) have been available from the past decade.

Development in cereal genomics play a key role in crop improvement for better understanding of the biological mechanisms which can improve the decision-making process for more efficient breeding strategies for screening and selecting superior genotypes (Varshney et al. 2005). These advances and development will provide opportunity for efficient transfer of information systems from model species and major crops to orphan crops (Naylor et al. 2004).

1.3.3 Genetic Mapping and MAS

With the advent of molecular markers and recent advances in genomics research, it has been possible to utilize genomics for enhancing the precision and efficiency of crop breeding termed as genomic-assisted breeding (GAB). GAB includes deployment of structural, functional as well as comparative genomics to identify molecular markers including functional markers, candidate genes and predictive markers for breeding. Several kinds of molecular markers such as SSR and SNP markers have been used for trait dissection and for enhancing precision in selection in plant breeding programs.

MAS can help (1) to select individuals carrying molecular markers that are linked to the trait of interest instead of performing extensive phenotypic tests (foreground selection) and (2) to reduce undesired parts of the donor genome including the linkage drag (background selection). Foreground selection requires a tight linkage between the trait of interest and its flanking markers that are being selected for. Background selection necessitates genotyping with a larger number of markers that cover the whole genome. MAS has proven efficient for the transfer of simply inherited qualitative traits from PGR into elite materials using backcrossing procedures. It is particularly useful for traits that are recessive, that can be assessed only after flowering and that are very difficult and expensive to assess. Introgression of fertility restorer genes in cytoplasmic male sterile (CMS) lines has been successfully done (Geiger et al. 1995; Miedaner et al. 2000; Stracke et al. 2003).

By using a combination of foreground and background selection, the transfer of a monogenic trait from a PGR into a breeding line may be completed within the shortest time (3–4 years) instead of the usual 6–7 generations of backcrossing with the same proportion of the recurrent parent genome (Frisch et al. 1999). Frisch et al. (1999) developed models and strategies for optimal application of MAS to transfer one or two genes from a donor PGR into a recipient genotype. Important parameters that were optimized for given marker intervals around the target gene(s) are the minimum number of individuals to be genotyped, the minimum number of data points in the genotyping and the allocation of marker analyses to different backcross generations.

1.3.4 Comparative Genomics-Based Approach Targeting Genes and Traits

Comparative genomics identify the functional elements in a genome based on the assumption that these elements are conserved through evolutionary time (Morrell et al. 2011). In comparative genomics it is an advantage of evolutionary signature that has been acted on the genome to understand the function and evolution across different crop species. The advent of high-throughput sequencing techniques and availability of whole-genome sequences of crops and plant species, as well as other genomic resources (e.g. microarray methods, EST libraries, high-throughput resequencing technologies), have extended the comparative method to encompass the evolution of genome structure and function (Fredslund et al. 2006; Paterson 2006).

Rapid progress in genomics makes it possible to understand detailed structural and functional comparisons of genes involved in various biological processes among different crop species. Comparisons at varying levels of evolutionary divergence are likely to reveal functional regions characteristic of different plant groups. Moreover intraspecific genomic approaches have been shown to be useful in predicting functional sequence motifs (Boffelli et al. 2004). Genomic collinearity has been reported in various crop species, and the comparative genomics approach using bioinformatic tools might therefore provide an opportunity for efficient transfer of information from model species to another species, i.e. from major crop species to minor crops and orphans (Devos 2005). These crops are regionally or locally important for nutrition and income, particularly in developing and underdeveloped countries (Naylor et al. 2004; Boffelli et al. 2004; Nelson et al. 2004).

1.3.5 Translational Genomics

Upon completion of the maps for several genomes, there are several major post-genomic tasks lying ahead such as the translation of the mapped genomes and the correct interpretation of huge amounts of data that are being rapidly generated. The important task of applying these fundamental results is to derive major benefits in various crop species. Translational genomics is very rapidly advancing through the detection in parallel of mRNA levels for large numbers of molecules, as well as through progress made with miniaturization and high-density synthesis of nucleic acids on microarray solid supports.

Plant translational genomics is a challenge faced by the plant genomics research to develop applications in crop plants which imply the translation of gene functions from a model to a crop species. Candidate gene approach (CGA) renown as a tool for translational genomics has been considered for successful application in crops with determined factors such as the type of crop, the complexity of the trait and the type of genes involved. The CGA is based on the assumption that genes with a proven or predicted function in a "model" species (functional candidate genes) or genes that are co-localized with a trait locus (positional candidate genes) could control a similar function or trait in an arbitrary crop of interest (target crop) (Salentijn et al. 2007). Studying the sequence variation among alleles (paralogs and orthologs) of candidate genes may provide conserved sequence motifs or conserved SNPs associated with a trait (Caicedo and Purugganan 2005). The extrapolations of gene function from a model crop to a more distant species have been well discussed (Gutterson and Zhang 2004; Laurie et al. 2004).

1.3.6 Nutrigenomics

Several decades ago technological developments made it possible to identify mutations in single genes, which interfere dramatically with metabolism and thereby lead to nutrition-related traits and disorders. In addition, the ameliorative and preventive potential of the diet in relation to monogenic disorders was demonstrated, for instance, by the treatment of patients with galactosemia and phenylketonuria (Zlatunich and Packman 2005). In this way, the importance of the interaction between nutrition, metabolism and gene expression for homoeostasis was more and more recognized. However, at the same time, the complexity of this interaction became apparent. Most frequent nutrition-related disorders appeared not to result from the interaction of a single nutrient with a single gene but of a complex mixture of nutrients with multiple genes. In order to understand this interaction, research was undertaken from the angle both of the diet and of the genes. Nutrition was regarded as readily accessible, but our complement of genes, our genome, with the state of technology of that time was regarded as the great unknown.

In recent years crop genomics-based research has been providing the means to uncover the genetic basis of crop characteristics with significance to human life and health. Crop plant genome research integrated with human genome analysis, nutritional science and medicine constitutes a novel discipline of research in support of human welfare (Bouchard and Ordovas 2012; Parnell 2012). This existing new multidisciplinary approach is called nutrigenomics. This approach focuses on the highly complex interplay between human genetic predisposition and nutrition, in regard to both food nutritional quality and disease prevention (Atanassov et al. 2007). Strong priorities are now to focus on the genes that determine characteristics supporting the production of crops in an environmentally friendly and sustainable manner. Millions of poor children in the world particularly in semiarid tropics of Asia and Africa suffer from vitamin A deficiency, which causes blindness and reduces the bioavailability of other important dietary micronutrients, including iron that is important for human health. This serious public health problem is addressed by genetically increasing the levels of provitamin A (primarily green and yellow vegetables) in dietary staples like rice, maize, sorghum and pearl millet. Besides the genes that control the accumulation of bulk nutrients, efforts are taken to uncover the genes that determine the content of valuable compounds such as potential pharmaceuticals, health-promoting probiotics, flavour and fragrance compounds, protectants, biocides, fine chemicals, etc. The rapid development and innovations in crop plant genomics are expected to provide newer knowledge in these areas and will open up ways for targeted crop improvement, both through the direct use of natural genetic diversity and via genetic engineering (Atanassov et al. 2007).

1.4 Important PGR in Crop Improvement

Out of 240,000 plant species, rice, wheat and maize together constitute the major share of global grain production (Cordain 1999). These are the important crops grown over most of the world. “Norin 10”, a cultivar from Japan, provided two very important genes, *Rht1* and *Rht2*, that resulted “Green Revolution” due to reduction in plant height in wheats. Norin 10, in turn, inherited these genes originally from “Shiro Daruma”, a Japanese landrace (Kihara 1983). The incorporation of the *Rht1* and *Rht2* genes into the new varieties enabled to develop and deploy the genes from unadapted materials (Krull and Borlaug 1970). These genes provide tremendous change in wheat plant resulting in higher production due to reduction in lodging via reduced height. It is now clear that they have other direct effects on yield via better nutrient uptake and tillering capacity (Krull and Borlaug 1970).

Besides reduced plant height in wheat crop, the incorporation of host plant resistance genes into modern wheat varieties has significantly increased the yield of wheat crop compared to the fungicides treated wheat varieties. Many of these varieties have incorporated single major genes from different PGR that convey resistance to specific races of the rust pathogen. In wheat there are more than 40 genes for leaf rust resistance that are known; out of these 12 originated in species other than wheat (*T. aestivum*) and durum (*T. turgidum*), while 20 of the 41 known genes for stem rust resistance originated in species other than wheat and durum. Even among the genes originating from wheat, many come from landraces.

In maize unlike wheat, the use of genetic resources in improvement has not been fully explored and exploited or well documented at the global level. The species of *Tripsacum* genus consists of a number of species with varying levels of ploidy and a base chromosome number of 18, yet more distant relatives to maize also offer promising potential. The species, mostly perennials, contain a number of interesting genes; however, only a few studies have referred to the possible use of *Tripsacum* segments for maize. Some maize improvement for increased yield and disease resistance have been observed because of an *Ht* gene derived from *Tripsacum floridanum* (Hooker and Perkins 1980). One more characteristic such as apomixes, asexual reproduction of a plant through the seed, is found in the polyploid species of *Tripsacum* which is still not exploited in the maize-growing world. If polyploid species of *Tripsacum* is explored for apomixis gene and exploited in hybrid seed production, this could provide another revolution in hybrid seed industry. As in developing countries, many farmers cannot take advantage of hybrid technology because high cost of hybrid seed is either unavailable or unaffordable. If hybrid seed could be produced via apomictic hybrids, public and private sector seed companies could produce larger quantities of high-quality seed, at lower cost, to meet the new demand. Exploitation of apomictic technique in hybrid seed hybrid technology to these countries would undoubtedly boost yields, both through heterosis and by providing effective options for introducing resistances and tolerances.

1.5 Role of PGR and Genomic Resources in Industries

PGRs are playing significant role in industries under different sectors. The rapid development of modern biotechnology over the past decades has enabled us to use genetic resources in ways that have not only fundamentally altered our understanding of the living world but also led to the development of new products and practices that contribute to human well-being, ranging from vital medicines to methods that improve the security of our food supplies. It has also improved conservation methods that help safeguard global biodiversity. Genetic resources can be put to commercial or non-commercial use. i) In commercial use, companies can use genetic resources to develop specialty enzymes, enhanced genes or small molecules. These can be used in crop protection, drug development, production of specialized chemicals or industrial processing. It is also possible to insert genes into crops to obtain desirable traits that can enhance their productivity or resilience to disease. ii) In non-commercial use, genetic resources can be used to increase knowledge or understanding of the natural world, with activities ranging from taxonomic research to ecosystem analysis.

In industry sector PGR can be potentially used with the help of biotechnological tools. Biotechnology industries span a wide range of activities including pharmaceutical, industrial and agricultural technology. The use of genetic resources in these industries is extremely varied:

1. *Pharmaceutical industry*: Chemical compounds or substances produced by living organisms found in nature continue to play an important role in the discovery of leads for the development of drugs and contribute significantly to the bottom lines of large pharmaceutical companies. For example, the US National Cancer Institute worked with a small pharmaceutical company to develop compounds called calanolides, derived from a tree in the Malaysian rainforest. Research demonstrated that they have the potential to treat HIV (type 1) and certain types of cancer. Clinical trials are ongoing.
2. *Industrial biotechnology*: Enzymes are used by textile, detergent, food, feed and other industries to improve the efficiency and quality of their products and production processes. Industrial biotechnology companies are particularly interested in genetic resources found in areas with high species diversity, as well as in extreme or unique environments, like salt lakes, deserts, caves and hydrothermal vents.
3. *Agricultural biotechnology*: Seed, crop protection and plant biotechnology industries rely heavily on genetic resources. PGR with traits that improve performance and farming efficiency for major crops are a key focus area for large seed companies. There is considerable growth in the value of the market for plant biotechnology-based products.

1.6 Implication of Genomic Resources for PGR Conservation

There exist two main approaches to genetic resources conservation: off-site (ex situ) conservation, by which is meant the maintenance of the resources in a site or facility which is not their natural or native habitat, and on-site (in situ) conservation, by which is meant the preservation of the resources in their native habitats. Four strategy levels for conservation can be distinguished: (1) conservation of cloned genes, gametes, embryos, seeds, tissues or whole organisms in a quiescent state; (2) conservation of plants, animals or microorganisms in a confined or controlled environment, such as plantations, gardens, zoological parks, reserves or on host organisms in the case of obligate parasites; (3) conservation of plants, animals or microorganisms in their natural habitats where population size and structure are managed; and

(4) conservation of plants, animals or microorganisms in their natural habitats without regard to population size or structure. The successful conservation of any given genetic resource may involve combinations of two or more of these strategy levels, employing both on-site and off-site methods.

Although DNA-based molecular markers are used to identify duplicate samples in the gene bank, yet they suffer from the difficulty to use as a common set of markers for a given set of germplasm of a species. Additionally, often the problem of reproducibility of DNA marker data among the laboratories has been encountered. Importantly, high-throughput sequencing or NGS data do not suffer from such shortcomings and therefore are the most suitable to address the issue of redundancy. However, sequencing of ex situ collections only to eliminate redundancy would be too expensive. Practically, it is impossible to sequence each genotype in a large crop collection. Therefore, there is a need to develop the "core reference set" as an alternative (Glaszmann et al. 2010).

With the advancements of the NGS technologies, sequencing of core reference set is becoming relatively easy with low coverage to develop genome-wide markers for facilitating the rejection of duplicates (Bansal et al. 2010; Davey et al. 2011). Several NGS-based technologies such as reduced representation libraries (Gompert et al. 2010; You et al. 2011), complexity reduction of polymorphic sequences (Mammadov et al. 2010), restriction-site-associated DNA sequencing (RAD-seq) (Baxter et al. 2011) and low-coverage sequencing for genotyping (Huang et al. 2009; Elshire et al. 2011) have been developed recently for genetic analysis of plants including the non-model species, wild species as well as alien species, species with high levels of repetitive DNA or breeding lines with low levels of polymorphism. These methods can be applied to compare SNP or haplotype diversity within and between closely related plant species or within wild natural populations to avoid redundancy in germplasm collections (Ossowski et al. 2010; Pool et al. 2010).

1.7 Impediments to the Use of PGR in Crop Improvement

Lack of environmental adaptation of the PGR is one major reason for the limited use of genetic resources in classical plant breeding. Other reasons impeding the use of PGR in crop improvement are huge performance difference between PGR and actual breeding materials for complex inherited traits; lack of inbreeding tolerance and unknown affiliation to heterotic pools (in the case of hybrid breeding); and genetic problems like pleiotropy, linkage between desired and undesired alleles of the PGR and epistasis or coadaptation of genes within both breeding population and PGR. Pleiotropy is the situation where one gene locus affects several traits. A specific allele from the PGR may be favourable for one trait, but negative for the expression of another trait that is directly or indirectly under control of the same gene locus. Strong linkage between desired and undesired alleles of the PGR makes it difficult to develop overall superior materials. With conventional backcrossing, the linkage drag is reduced only very slowly; about 53 cM remain around the target gene in the third backcrossing generation (BC_3), and in BC_{10}, the average linkage drag is still about 20 cM (Stam and Zeven 1981; Welz and Geiger 2000). If this linkage drag contains undesirable alleles from the PGR, the performance of the backcrossing products can be unsatisfactory.

Epistasis or coadaptation of genes within both breeding population and PGR means that natural or artificial selection has favoured specific combinations of alleles at different gene loci within each type of material. The specific allele combinations are lost after crossing and recombining the two types, leading to so-called recombination losses. It takes several generations to establish new favourable allele combinations through selection.

1.8 Challenges

Though it has been fully endorsed that the crop species can be improved to secure the food production by use of genetic resources through effective utilization of genomic resources. The difference between genetic resources and cultivated varieties and breeding materials for complex inherited traits can be narrowed down with biotechnological techniques such as embryo rescue and doubled haploid (DH) breeding. The major challenges for effective utilization of genomic resources revolve around cost, funding, availability of genomic resources and technical and infrastructural capacity. A large majority of the gene banks, save for those in the CGIAR system and a few selected national gene banks in developed countries, have inadequate infrastructural and bioinformatic capacity. The high-performance computing resources required to store and analyse NGS data are beyond the financial capacity in most of the developing countries. Cloud computing is however becoming popular and provides a ray of hope, as it is increasingly becoming possible to share computing resources between partner institutions. Equally challenging is the lack of genomic resources for a majority of the minor, neglected and underutilized plant species.

1.9 Conclusion

In the rapidly changing field of genomics, there is no telling what the future holds for the application of genomics in crop improvement for future food security. Though it has been suggested that the current genomic revolution has the capacity to dramatically change breeding programmes. This vision is likely to remain a mirage unless there is a paradigm shift in the way genomic resources are used in crop improvement. The scientists should be trained in areas such as genomics, computational biology and population genetics for exploring and identifying more genes/alleles in the genetic resources and use these resources in crop improvement. There is a need to bring a group of plant breeder and biotechnologist together for the use of genomic resource techniques in plant breeding programmes. Linkages and collaborations should be developed which will help leverage on the greater infrastructural, technical and financial capacity to user, which currently, and in the foreseeable future, remains a great constraint in promoting effective conservation and use of genetic resources. Application of genomic resources for effective utilization of PGR has potential of ultimately having an increasing impact in the development of more resilient varieties. This is likely to result in increased agricultural productivity, thereby having a positive impact on global food and nutritional security.

References

Abdurakhmonov IY, Abdukarimov A (2008) Application of association mapping to understanding the genetic diversity of plant germplasm resources. Int J Plant Genomics 5:1–18

Atanassov A, Batchvarova R, Djilianov D (2007) Strategic vision for plant biotechnology and genomics development. Biotech Biotech Equip 21:1–7

Babu ER, Mani VP, Gupta HS (2004) Combining high protein quality and hard endosperm traits through phenotypic and marker assisted selection in maize. Proceedings of the 4th International Crop Science Congress, Brisbane

Bansal V, Harismendy O, Tewhey R, Murray SS, Schork NJ, Topol EJ, Frazer KA (2010) Accurate detection and genotyping of SNPs utilizing population sequencing data. Genome Res 20:537–545

Baxter SW, Davey JW, Johnston JS, Shelton AM, Heckel DG, Jiggins CD, Blaxter ML (2011) Linkage mapping and comparative genomics using next-generation RAD sequencing of a non-model organism. PLoS One 6:19315–19326

Boffelli D, Weer CV, Weng L, Lewis KD, Shoukry MI, Pachter L, Keys DN, Rubin EM (2004) Intra-species sequence comparisons for annotating genomes. Genome Res 14:2406–2411

Bouchard C, Ordovas JM (2012) Fundamentals of nutrigenetics and nutrigenomics. Prog Mol Biol Transl Sci 108:1–15

Brozynska M, Furtado A, Henry RJ (2016) Genomics of crop wild relatives: expanding the gene pool for crop improvement. Plant Biotech J 14:1070–1085

Caicedo AL, Purugganan MD (2005) Comparative plant genomics. Frontiers and prospects. Plant Physiol 2:545–547

Cordain L (1999) Cereal grains: humanity's double-edged sword. Simopoulos AP (ed): evolutionary aspects of nutrition and health. World Rev Nutr Diet 84:19–73

Cosart T, Beja-Pereira A, Chen S, Ng SB, Shendure J, Luikart G (2011) Exome-wide DNA capture and next generation sequencing in domestic and wild species. BMC Genom 12:347–355

Chicaiza O, Khan IA, Zhang X, Brevis JC, Jackson L, Chen X, Dubcovsky J (2006) Registration of five wheat isogenic lines for leaf rust and stripe rust resistance genes. Crop Sci 46:485–487

Collard BCY, Jahufer MZZ, Brouwer JB, Pang ECK (2005) An introduction to markers, quantitative trait loci (QTL) mapping and marker-assisted selection for crop improvement: the basic concepts. Euphytica 142:169–196

Davey JW, Hohenlohe PA, Etter PD, Boone JQ, Catchen JM, Blaxter ML (2011) Genome-wide genetic marker discovery and genotyping using next generation sequencing. Nat Rev Genet 12:499–510

Devos KM (2005) Updating the "crop circle". Curr Opin Plant Biol 8:155–162

Dirlewanger E, Graziano E, Joobeur T, Garriga-Calderé F, Cosson P, Howad W, Arús P (2004) Comparative mapping and marker-assisted selection in Rosaceae fruit crops. Proc Natl Acad Sci U S A 29:9891–9896

Doerge RW (2002) Mapping and analysis of quantitative trait loci in experimental populations. Nat Genet 3:43–52

Elshire RJ, Glaubitz JC, Sun Q, Poland JA, Kawamoto K, Buckler ES, Mitchell SE (2011) A

robust, simple Genotyping by-sequencing (GBS) approach for high diversity species. PLoS One 6:19379–19389
Emebiri L, Michael P, Moody D (2009) Enhanced tolerance to boron toxicity in two-rowed barley by marker-assisted introgression of favourable alleles derived from Sahara 3771. Plant Soil 314:77–85
Ersoz ES (2007) Applications of linkage disequilibrium and association mapping in crop plants. In: Varshney RK, Tuberosa RT (eds) Genomics assisted crop improvement: genomics approaches and platforms. Springer, Berlin, pp 97–120
Esquinas-Alcázar J (2005) Science and society: protecting crop genetic diversity for food security: political, ethical and technical challenges. Nat Rev Genet 6:946–953
Fredslund J, Madsen LH, Hougaard BK, Nielsen AM, Bertioli D, Sandal N, Stougaard J, Schauser L (2006) A general pipeline for the development of anchor markers for comparative genomics in plants. BMC Genom 14:207
Francia E, Tacconi G, Crosatti C et al (2005) Marker-assisted selection in crop plants. Plant Cell Tissue Organ Cult 82:317–342
Frisch M, Bohn M, Melchinger AE (1999) Minimum sample size and optimal positioning of flanking markers in marker-assisted backcrossing for transfer of a target gene. Crop Sci 39:967–975
Fukuoka S, Saka N, Koga H, Ono K, Shimizi T, Ebana K, Hayashi N, Takahashi A, Takahashi A, Hirochika H, Okuno K, Yano M (2009) Loss of function of a proline containing protein confers durable disease resistance in rice. Science 325:998–1001
Geiger HH, Yuan Y, Miedaner T, Wilde P (1995) Environmental sensitivity of cytoplasmic genic male sterility (CMS) in Secale cereale L. In: Kück U, Wricke G (eds) Genetic mechanisms for hybrid breeding. Adv Plant Breed 18: 7–17
Glaszmann JC, Kilian B, Upadhyaya HD, Varshney RK (2010) Accessing genetic diversity for crop improvement. Curr Opin Plant Biol 13:167–173
Gompert Z, Forister ML, Fordyce JA, Nice CC, Williamson RJ, Buerkle CA (2010) Bayesian analysis of molecular variance in pyrosequences quantifies population genetic structure across the genome of Lycaeides butterflies. Mol Ecol 19:2455–2473
Gupta PK, Sachin R, Pawan LK (2005) Linkage disequilibrium and association studies in plants: present status and future prospects. Plant Mol Biol 57:461–485
Gutterson N, Zhang JZ (2004) Genomics applications to biotech traits: a revolution in progress? Curr Opin Plant Biol 7:226–230
Hall D, Tegstro C, Ingvarsson PK (2010) Using association mapping to dissect the genetic basis of complex traits in plants. Brief Funct Genom 9:157–165
Hallauer AR, Miranda JB (1981) Quantitative genetics in maize breeding. Iowa State University Press, Ames
Harlan JR, de JMJ W (1971) Towards a rational classification of cultivated plants. Taxonomy 20(4):509–517
Henry RJ, Rice N, Waters DLE, Kasem S, Ishikawa R, Hao Y, Dillon S, Crayn D, Wing R, Vaugha D (2010) Australian *Oryza*: utility and conservation. Rice 3(4):235–241
Ho C, Mccouch R, Smith E (2002) Improvement of hybrid yield by advanced backcross QTL analysis in elite maize. Theor Appl Genet 105:440–448
Hodgkin T, Roviglioni R, De Vicente MC, Dudnik N (2001) Molecular methods in the conservation and use of plant genetic resources. Acta Horticultura 546:107–118
Hoisington D, Khairallah M, Reeves T, Ribaut JM, Skovmand B, Taba S, Warburtan M (1999) Plant genetic resources: what can they contribute toward increased crop productivity? Proc Natl Acad Sci USA 96:5937–5943
Hooker AL, Perkins JM (1980) Proceedings of the 35th Annual Corn and Sorghum International Research Conference. American Seed Trade Association, Washington, DC, pp 68–87
Hospital F (2001) Size of donor chromosome segments around introgressed loci and reduction of linkage drag in marker-assisted backcross programs. Genetics 158(3):1363–1379
Huang X, Feng Q, Qian Q, Zhao Q, Wang L, Wang A, Guan J, Fan D, Weng Q, Huang T, Dong G, Sang T, Han B (2009) High throughput genotyping by whole-genome re-sequencing. Genome Res 19:1068–1076

Jansen RC, Nap JP (2001) Genetical genomics: the added value from segregation. Trends Genet 17:388–391
Komatsuda T, Pourkeirandish M, He C, Azhaguvel P, Kanamori H, Perovic D, Stein N, Graner A, Wicker T, Tagiri A, Lundqvist U, Fujimura T, Matsuoka M, Matsumoto T, Yano M (2007) Six-rowed barley originated from a mutation in a homeodomainleucine zipper I– class homeobox gene. Proc Natl Aca Sci USA 104:1424–1429
Kihara H (1983) In: Sakamoto S (ed) Proceedings of the 6th international wheat genetic symposium. Plant Germplasm Institute, University of Kyoto, Kyoto
Krull CF, Borlaug NE (1970) In: Frankel OH, Bennett E (eds) Genetic resources in plants: their exploration and conservation. Blackwell Scientific, Oxford
Kuraparthy V, Sood S, Guedira GB, Gill BS (2011) Development of a PCR assay and marker-assisted transfer of leaf rust resistance gene *Lr58* into adapted winter wheats. Euphytica 180:227–234
Laurie DA, Griffiths S, Dunford RP, Christodoulou V, Taylor SA, Cockram J, Beales J, Turner A (2004) Comparative genetic approaches to the identification of flowering time genes in temperate cereals. Field Crops Res 90:87–99
Maccaferri M, Sanguineti MC, Demontis A, El-Ahmed A, Garcia Del Moral L, Maalouf F, Nachit M, Nserallh N, Ouabbou H, Rhouma S, Royo C, Villegas D, Tuberosa R (2011) Association mapping in durum wheat grown across a broad range of water regimes. J Exp Bot 62:409–438
Mammadov JA, Chen W, Ren R, Pai R, Marchione W, Yalc F, Witsenboer H, Greene TW, Thompson SA, Kumpatla SP (2010) Development of highly polymorphic SNP markers from the complexity reduced portion of maize (*Zea mays* L.) genome for use in marker-assisted breeding. Theor Appl Gen 121:577–588
McNally KL, Childs KL, Bohnert R, Davidson RM, Zhao K, Ulat VJ, Zeller G, Clark RM, Hoen DR, Bureau TE, Stokowski R, Ballinger DG, Frazer KA, Cox DR, Padhukasahasram B, Bustamante CD, Weigel D, Mackill DJ, Bruskiewich RM, Rätsch G, Buell CR, Leung H, Leach JE (2009) Genome wide SNP variation reveals relationships among landraces and modern varieties of rice. Proc Natl Acad Sci U S A 106(30):12273–12278
Miedaner T, Glass C, Dreyer F, Wilde P, Wortmann H, Geiger HH (2000) Mapping of genes for male-fertility restoration in 'Pampa' CMS winter rye (*Secale cereale* L.). Theor Appl Genet 101:1226–1233
Miedaner T, Wilde F, Steiner B, Buerstmayr H, Korzun V, Ebmeyer E (2006) Stacking quantitative trait loci (QTL) for *Fusarium* head blight resistance from non-adapted sources in an european elite spring wheat background and assessing their effects on deoxynivalenol (DON) content and disease severity. Theor Appl Genet 112:562–569
Mohan M, Nair S, Bhagwat A, Krishna TG, Yano M, Bhatia CR, Sasaki T (1997) Genome mapping, molecular markers and marker-assisted selection in crop plants. Mol Breed 3:87–103
Morrell P, Buckler E, Ross-Ibarra J (2011) Crop genomics: advances and applications. Nat Rev Genet 13:85–96
Naylor RL, Falcon WP, Goodman RM, Jahn MM, Sengooba T, Tefera H, Nelson RJ (2004) Biotechnology in the developing world: a case for increased investments in orphan crops. Food Policy 29:15–44
Nelson RJ, Naylor RL, Jahn MM (2004) The role of genomics research in improvement of "orphan" crops. Crop Sci 44:1901–1904
Okada Y, Kanatani R, Arai S, Asakura T, Ito K (2003) Production of a novel virus-resistant barley line introgression to the *ryml* locus with high malting quality using DNA marker assisted selection. J Inst Brew 109:99–102
Ossowski S, Schneeberger K, Lucas-Lled_o JI, Warthmann N, Clark RM, Shaw RG, Weigel D, Lynch M (2010) The rate and plant genomic resources for abiotic stress tolerance 9 molecular spectrum of spontaneous mutations in *Arabidopsis thaliana*. Science 327:92–94
Parnell LD (2012) Advances in technologies and study design. Prog Mol Biol Transl Sci 108:17–50
Paterson AH (2006) Leafing through the genomes of our major crop plants: strategies for capturing unique information. Nat Rev Genet 7:174–184
Pool JE, Hellmann I, Jensen JD, Nielsen R (2010) Population genetic inference from genomic

sequence variation. Genome Res 20:291–300
Rao NK, Reddy LJ, Bramel P (2003) Potential of wild species for genetic enhancement of some semi-arid food crops. Genet Res Crop Evol 50:707–721
Ribaut JM, Ragot M (2006) Marker-assisted selection to improve drought adaptation in maize: the backcross approach, perspectives, limitations, and alternatives. J Exp Bot 58:351–360
Salentijn EMJ, Pereira A, Angenent GC, van der Linden CG, Krens F, Smulders MJM, Vosman B (2007) Plant translational genomics: from model species to crops. Mol Breed 20:1–13
Salgotra RK, Gupta BB, Sood M (2015) Biotechnological interventions and their role in sustainable hill agriculture. J Plant Sci Res 2(1):1–8
Schena M (1998) Microarrays: Biotechnology's discovery platform for functional genomics. Trends Biotech 16:301–306
Sharma S, Upadhyaya HD, Varshney RK, Gowda CLL (2013) Pre-breeding for diversification of primary gene pool and genetic enhancement of grain legumes. Front Plant Sci 4:309
Singh S, Sidhu JS, Huang N, Vikal Y, Li Z, Brar DS, Dhaliwal HS, Khush GS (2001) Pyramiding three bacterial blight resistance genes (*xa5, xa13* and *Xa21*) using marker-assisted selection into *indica* rice cultivar PR106. Theor Appl Genet 102:1011–1015
Somers D, Thomas J, Depauw R, Fox S, Humphreys G, Fedak G (2005) Assembling complex genotypes to resist *Fusarium* in wheat (*Triticum aestivum* L.). Theor Appl Genet 111:1623–1631
Spooner D, van Treuren R, de Vicente MC (2005) Molecular markers for genebank management. IPGRI Technical Bulletin No. 10, International Plant Genetic Resources Institute, Rome
Stam P, Zeven AC (1981) The theoretical proportion of the donor genome in near-sogenic lines of self-fertilizers bred by backcrossing. Euphytica 30:227–238
Stracke S, Schilling AG, Forster J, Weiss C, Glass C, Miedaner T, Geiger HH (2003) Development of PCR-based markers linked to dominant genes for male-fertility restoration in pampa CMS of rye (*Secale cereale* L.). Theor Appl Genet 106:1184–1190
Sun G (2012) MicroRNAs and their diverse functions in plants. Plant Mol Biol 80:17–36
Sundaram RM, Sakthivel K, Hariprasad AS, Ramesha MS, Viraktamath BC, Neeraja CN, Balachandran SM (2010) Development and validation of a PCR-based functional marker system for the major wide-compatible gene locus *S5* in rice. Mol Breed 26:719–727
Supple MA, Shapiro B (2018) Conservation of biodiversity in the genomics era. Genome Biol 19:1–12
Till BJ (2007) TILLING and EcoTILLING for crop improvement. In: Varshney RK, Tuberosa RT (eds) Genomics assisted crop improvement: genomics approaches and platforms. Springer, Berlin, pp 333–349
Tuberosa R, Gill BS, Quarrie SA (2002) Cereal genomics: ushering in a brave new world. Plant Mol Biol 48:445–449
Varshney RK, Terauchi R, McCouch SR (2014) Harvesting the promising fruits of genomics: applying genome sequencing technologies to crop breeding. PLoS Biol 12(6):1001883–1001888
Varshney RK, Graner A, Sorrells ME (2005) Genomics-assisted breeding for crop improvement. Trends Plant Sci 10:621–630
Varshney RK, Hoisington DA, Tyagi AK (2006) Advances in cereal genomics and applications in crop breeding. Trends Biotech 24:490–499
Weising K, Nybom H, Wolff K, Kahl G (2005) DNA fingerprinting in plants: principles, methods, and applications. CRC, Boca Raton
Welz HG, Geiger HH (2000) Marker-assisted selection. I Qualitative traits. In: BIG H, Geiger HH, Hess DE, Hash CT, Bramel-Cox P (eds) Application of molecular markers in plant breeding (training manual)., http://www.icrisat.org/text/research/grep/homepage/mol/molecular.htm. ICRISAT, Patancheru
Wilde F, Schon C, Korzun V, Ebmeyer E, Schmolke M, Hartl L, Miedaner T (2008) Marker-based introduction of three quantitative-trait loci conferring resistance to *Fusarium* head blight into an independent elite winter wheat breeding population. Theor Appl Genet 117:29–35
Xiao J, Li J, Grandillo S, Ahn SN, Yuan L, Tanksley SD, Mccouch SR (1998) Identification of trait-improving quantitative trait loci alleles from a wild rice relative, *Oryza rufipogon*. Genetics 150:899–909

Xu K, Xu X, Fukao T, Canlas P, Maghirang-Rodriguez R, Heuer S, Ismail AM, Bailey-Serres J, Ronald PC, Mackill DJ (2006) *Sub1A* is an ethylene response factor like gene that confers submergence tolerance to rice. Nature 442:705–708

Xu Y (2010) Plant genetic resources: management, evaluation and enhancement. Mol Plant Breed. CABI, Wallingford, pp 151–194

You FM, Huo N, Deal KR, Gu YQ, Luo MC, McGuire PE, Dvorak J, Anderson OD (2011) Annotation-based genome-wide SNP discovery in the large and complex Aegilops tauschii genome using next-generation sequencing without a reference genome sequence. BMC Genom 12:59–78

Yu J, Holland JB, McMullen MD, Buckler ES (2008) Genetic design and statistical power of nested association mapping in maize. Genetics 178:539–551

Yu SB, Xu WJ, Vijayakumar CHM, Ali J, Fu BY, Xu JL, Jiang YZ, Marghirang R, Domingo R, Aquino C, Virmani SS, Li ZK (2003) Molecular diversity and multilocus organization of the parental lines used in the international Rice molecular breeding program. Theor Appl Genet 108:131–140

Zhao K, Tung CW, Eizenga CG, Wright MH, Ali ML, Price AH, Norton GJ, Islam MR, Reynolds A, Mezey J, McClung AM, Bustamante CD, McCouch SR (2011) Genome-wide association mapping reveals a rich genetic architecture of complex traits in *Oryza sativa*. Nat Commun 2:467–475

Zimnoch-Guzowska E, Lebecka R, Kryszczuk A, Maciejewska U, Szczerbakowa A, Wielgat B (2003) Resistance to *Phytophthora infestens* in somatic hybrids of *Solanum nigrum* L. and diploid potato. Theor Appl Genet 107:43–48

Zlatunich CO, Packman S, Galactosemia (2005) Early treatment with an elemental formula. J Inherit Metab Dis 28:163–168

Untapped Genetic Diversity of Wild Relatives for Crop Improvement

2

Zoran Jovovic, Violeta Andjelkovic, Novo Przulj, and Dragan Mandic

Contents

Abstract

We live in a time when, thanks to the achievements of a modern civilization, the consumption and destruction are uncontrolled of what ensures human survival on the planet – the biosphere. Special pressure is focused on biodiversity, equally to all its elements – genes, species, and ecosystems. In order to ensure sustainable

Z. Jovovic (✉)
Biotechnical Faculty Podgorica, University of Montenegro, Podgorica, Montenegro

V. Andjelkovic
Maize Research Institute 'Zemun Polje', Belgrade, Serbia
e-mail: avioleta@mrizp.rs

N. Przulj
Faculty of Agriculture, University of Banja Luka,
Banja Luka, Republic of Srpska, Bosnia and Herzegovina

D. Mandic
Agricultural Institute of Republic of Srpska,
Banja Luka, Republic of Srpska, Bosnia and Herzegovina

R. K. Salgotra, S. M. Zargar (eds.), *Rediscovery of Genetic and Genomic Resources for Future Food Security*,
https://doi.org/10.1007/978-981-15-0156-2_2

food supply for increasing human population, new varieties with increased tolerance to environmental stresses need to be created. Due to the richness of genetic diversity, CWR show very high adaptability in a fairly wide range of ecological conditions. Therefore, in order to preserve the adaptability of crops and future food safety, CWR, as a critical component of plant genetic resources for food and agriculture, will be needed more than ever before. There is no doubt that climate change in the future will increasingly affect the survival of wild species. The ever-present loss of their germplasm and habitat requires greater mobility in their protection. It is therefore necessary to take urgent steps to preserve these resources both in their natural habitats (in situ) and in the gene banks (ex situ), while the genetic diversity they contain is still available. Any further postponement of their conservation would lead to further disturbance of biodiversity, and these activities would become even more uncertain in the future. For these reasons it is necessary to develop acceptable conservation agendas as soon as possible that will offer a solution where and how to protect the diversity of wild relatives in situ. Due to their actual or potential value, this should be done as soon as possible in order to ensure continuous crop improvement for food sustainability through the conservation of CWR diversity.

Acronyms

AEGIS	An European Genebank Integrated System
AnGR	Animal genetic resources
CBD	Convention on Biological Diversity
CIAT	International Center for Tropical Agriculture
CT	Crop Trust
ESA	European Seed Association
FAO	Food and Agriculture Organization of the United Nations
ICWRA	Important Crop Wild Relatives Areas
IFPRI	International Food Policy Research Institute
IPCC	Intergovernmental Panel on Climate Change
ITPGRFA	International Treaty on Plant Genetic Resources for Food and Agriculture
IUCN	International Union for Conservation of Nature
MAA	Most Appropriate Accessions
MAWP	Most Appropriate Wild Populations
MEA	Millennium Ecosystem Assessment
NASA	National Aeronautics and Space Administration
PA	Protected area
PGRFA	Plant Genetic Resources for Food and Agriculture
PGR	Plant Genetic Resources
RBG	Royal Botanic Gardens (brand name Kew)
UN	United Nations

2.1 Introduction

Plant genetic resources represent resources of enormous importance because they are basis for world food production. Observed from the aspect of nutrition of a constantly growing population, their importance as the basis for world food security is getting bigger. In addition to playing a major role in increasing production and food safety, they also play an important role in preserving and restoring damaged ecosystems and natural habitats, as well as preserving and protecting endangered plant and animal species. Conservation of agricultural diversity should ensure stable production of quality food and plant products and mitigate the risks associated with intensive plant production systems. Increasing the diversity of crops will also ensure greater food safety. Crop improvement, especially under climatic conditions, will increasingly be based on the diversity of plant genetic resources that have been relatively poorly used and inadequately preserved until now (Jovovic and Kratovalieva 2016).

Agricultural biodiversity (agrobiodiversity), as an important component of total biodiversity, according to the Convention on Biological Diversity (CBD), represents the variability of all animals, plants, and microorganisms that are directly or indirectly used in food production and agriculture. In a broader sense, agrobiodiversity consists of plant genetic resources for food and agriculture (PGRFA), crop wild relatives (CWRs), domestic animals (including fish and other managed aquatic animals) (AnGR), genetic resources of mushrooms and microorganisms, and all other species that contribute to food production (microorganisms in soil, pollinators, predators, etc.). Today it is clear that the Convention on Biological Diversity's 2010 (UN Resolution 65/161 of 22 December 2010) targets to reduce the loss of biodiversity is not met. A large number of facts suggest that under the influence of climate change and uncontrolled exploitation of natural resources, biodiversity is significantly threatened, and the disappearance of plant and animal species is much more intense today than it was in the past. Genetic diversity is rapidly decreasing and therefore the ability of the environment to respond to the growing distortion of biodiversity. The Millennium Ecosystem Assessment (MEA 2005) recognizes climate change as the biggest threat to biodiversity and its conservation as one of the major challenges in the future. PRGFA and CWR face similar threats. An accelerated replacement of traditional varieties with new genetically uniform selection has led to significant genetic erosion. In addition to the fact that many primary sources of diversity have been permanently lost, many of their offspring are also lost. In this way resources of traditional cultivars that were vital to small-scale farmers and the future of plant breeding are significantly decimated. These processes have led to higher population homogenization, which is narrowing genetic variability, and to some extent limit their potential as donors of desirable genes in the development of new varieties with improved properties. Reduced genetic variability reduces the ability of plant species to effectively respond to numerous environmental challenges and thus their chances of survival (Jovovic and Kratovalieva 2016).

Recent research shows that in nature, there are about 50,000–60,000 wild relatives of cultivated crops, of which about 11,000 are used in food production and agriculture. For global food safety, greater importance and need for urgent protection have 700 of them. Although the CWR number in the gene banks is very modest, they are widely distributed in nature. They can be found on all continents except Antarctica (Larson et al. 2014). Their global distribution also points to their enormous diversity. According to the latest data from the world plant gene banks, 29% of the species of wild relatives of cultivated plants have disappeared, while 24% of them are with less than 10 samples. In order to improve their presence in over 70% of species, there is an urgent need for further collection and conservation, while 95% of species are not sufficiently represented in terms of the full range of geographic and ecological variations in their native distributions (Maxted and Kell 2009). Given that the number of inhabitants in the world is increasing rapidly and climate change intensifies, their use in the future will most likely be higher.

Due to the expansion of the human population and many human activities (environmental pollution, urbanization, deforestation, land drainage, etc.), most CVRs are today seriously threatened by real danger of extinction. Recent studies suggest that urgent conservation measures should be undertaken in many geographic regions: Mediterranean, Middle East, Western and Southern Europe, Southeast and East Asia, and South America (Castaneda-Alvarez et al. 2015). Some recent studies predict that CWR can be significantly affected by climatic changes, and by the year 2055, a significant number of species could lose over 50% of their range size. For example, due to climate change and habitat loss by 2055, 61% of peanut species, 12% of potato species, and 8% of cowpea species could be permanently lost (Jarvis et al. 2008). For these reasons, in recent years, a lot of effort has been put into the development of the method for predicting the future geographic distribution of wild species (Rotenberry et al. 2006).

It is estimated that by 2100, the yield of maize and wheat in the valleys will decrease by 40% and rice by 30% if urgent measures to mitigate the consequences of the changing climate do not initiate. Therefore, breeders will be obliged to create such varieties that will be able to suppress all extreme climate manifestations. In these processes, the role of CWR will be invaluable. In spite of their recognizable importance, so far little has been done on their systematic preservation. Since CWRs possess identical biological and ecological characteristics as other wild species, the strategies for their protection would be similar to those used in the in situ conservation of endangered wild species. All the initiatives toward their conservation have met the same challenge – how to preserve so many CWR species and genetic diversity they possess. Only 9% of PGR accessions in the European gene bank collections are CWR, while most of their populations in situ are not actively monitored and managed. Even CWR species of the highest priority for food security are not adequately preserved (Dias et al. 2011). For future crop improvement programs, in situ populations and CWR genetic reserves will be essential because CWR preserved ex situ do not maintain evolutionary adaptation processes encountered in in situ populations. Since CWR conservation is within the jurisdiction of the state in whose territory they are, this quality of national strategies will be fundamental to future global food security.

The primary goal of modern breeding is the creation of high-yielding varieties with tolerance to biotic and abiotic stresses, superior nutritional properties, and the ability to adapt to the changing environment. The main obstacle in the realization of this goal is the narrow genetic basis of modern varieties; therefore the use of wild relatives is a promising approach to improve the genetic diversity of cultivated crops necessary for continued agricultural sustainability (Jovovic et al. 2013). Recognizing the importance of CWR for crop improvement, Russian botanist Vavilov began collecting a large number of wildlife in the early twentieth century. Other scientists around the world wrote about their significance and worked intensively on the development and improvement of breeding programs (Olsen and Wendel 2013a). In addition to the CWR, some other wild species that were never domesticated were studied as well to indicate their importance for improving crops. Although the number of publications on the use of CVR in breeding in recent years has significantly increased, there is still insufficient knowledge of the degree of their diversity, as well as on the ways in which this diversity would be used to improve cultivated plants. In order to their scope of production to be approximate to their real potential, they must work more intensively on their systematic study. With the further development of science and research methods, the need for resources of wild relatives grows, which will result in the stopping of intensive processes of genetic erosion. Bearing in mind the importance they have for food safety and adaptation to climate change, it is necessary to establish effective mechanisms for their conservation as soon as possible, as well as new technologies that will enable their improved use.

2.2 The Importance of Crop Wild Relatives

CWRs are wild plant species that are genetically related to cultivated crops. They include ancestors of cultivated crops and other species that are more or less related to them. Due to relatively close genetic origin, gene donors are indirectly used for crop improvement (Maxted et al. 2006). Neglected by farmers, they have acquired some new features in the long evolution processes, such as improved tolerance to drought and other abiotic stresses, as well as resistance to some plant diseases and pests. Since the beginning of the development of agriculture, they have been an important basis for the creation of new varieties with increased yield and improved nutritional properties. Like all other crops, wild relatives are important for the production of food feed, medicines, textiles, building materials, fuels, etc. Due to their close connection with a certain territory and the local community, they also represent a part of the cultural heritage. Therefore, more attention is paid lately to knowledge related to their use. Many traditional communities around the world have preserved and improved their autochthonous knowledge related to the management of natural resources and agriculture for centuries. This knowledge is a product of the use of these resources and is related to plant species and varieties, production technology, harvesting, storage, use, processing, etc. (Jovovic and Kratovalieva 2016). For their improved deployment and exploitation, detailed analysis of their real value and potential will be required. More active participation of local communities in conservation programs would contribute to more efficient and cost-effective conservation and management of biodiversity. It is therefore important to support the capacity building of the local communities through training, equipping, and other similar actions.

In long domestication processes, man used wildlife to create new plant forms with properties that primarily met his needs. These first selections were mainly focused on yield and quality-related traits (Hua et al. 2015), resulting in a significant reduction in the level of genetic variation within newly created crops (Miller and Gross 2011). For example, in cultivated soybeans, more than half of the genetic variation was lost during the domestication compared to its wild counterpart (Zhou et al. 2015). Genetic diversity has also been significantly reduced in cultivated rice (Xu et al. 2012), while only 2–4% of genes is currently used in maize selection programs (Wright et al. 2005). These examples clearly indicate that modern selection has greatly decreased the genetic diversity of modern cultivars and, to some extent, limit the scope of their use in the development of new varieties. In contrast, CVR retained a high level of genetic diversity at both the population level and the individual level, with the exception of clonally propagated perennial crops that maintain a high level of heterozygosity in domesticated lineages (McKey et al. 2010).

The importance of CWR for modern agriculture is invaluable because it provides a wide range of potentially useful genes for breeders. This precious genetic variation reservoir will be increasingly important for world food security and the improvement of agricultural production in the light of a rapidly growing world population and increasingly intense climate change (Henry and Nevo 2014). The solution to these challenges lies in the creation of tertiary genotypes that will be able to adapt to the new changes. However, this will not be an easy task, and it will be necessary to define new selection programs and strategies and appropriate capacities and approaches for crop biotechnology and breeding (Tester and Langridge 2010). It is certain that agricultural production in the future will increasingly be based on stable genotypes that will be able to deliver stable yields in changed environment conditions. The reaction of certain genotypes to different agroecological conditions is conditioned by their genetic properties and interaction with the external environment. Due to the interactions of the genotype and the environment, approximately balanced yields are provided in relatively different agroecological conditions. Bearing in mind the above said, it is surprising that these basic evolutionary principles have so far been used so little in practice (Jovovic et al. 2012).

The unprecedented growing demand for food and abiotic stresses that crops face due to climate change cannot be solved without the transfer of new genetic variations from CWR into the breeding pools adding to adaptability of crops to changing environmental conditions. Wild relatives and climate change integrate the evolution of crops and production practices by creating acceptable models for sustainable use of biodiversity. Thus, wild relatives appear as a good source of genetic variability for crop improvement, especially for pest and disease resistance and tolerance to abiotic stresses. Therefore, plant breeders must identify the germplasm with features that can be used to create new varieties with greater resistance to abiotic extremes, which will be of great help to producers in regions affected by climate change. There is a high likelihood that many of these features can be found in CWR, which will require the availability of large gene pools. The importance of CWR in adapting to climate change is also emphasized in the FAO report (2015), which, in view of the increased adaptive capacity of wild relatives, recommends the consolidation of their collections. Special emphasis in these collections should be put on stress-adapted genetic material that can contribute to adapting to climate change.

Tolerance to drought and salinity, increased resistance to disease and pests, and some other productive traits of wild relatives have been tested in numerous prebreeding and breeding programs (Olsen and Wendel 2013b). In addition to CWR, other wild plant species such as blueberries and strawberries were also tested in these programs. The results obtained clearly showed that there is a huge untapped potential in nature which, in conditions of growing climate threats, can be an important source of new genes for improving crops. More intensive inclusion of CWR in existing breeding programs (classical breeding or modern biotechnologies) will, on the one hand, create conditions for sustainable farming while, on the other, lead to the establishment of sustainable agriculture and achieving sustainable food security.

CWR, as a source of new genetic variability, have so far been used mostly in the improvement of wheat, barley, rice, and tomato crops but are successfully applied in the cultivation of maize, oats, potatoes, beans, sunflower, chickpea, sugarcane, lettuce, bananas, and cassava as well. Although there are a number of reasons for their significant use in certain areas (cross-compatibilities, the taxonomic relationship between crops and their corresponding wild species, fertility in the F_1 and subsequent progeny, availability of CWR, exploration and utilization of wild germplasms, and financial support possibility), the use of wild relatives for crop improvement is still quite behind their real potential. The Global Trust Crop Diversity (Crop Trust CT), the International Center for Tropical Agriculture (CIAT), and the Royal Botanic Gardens (RBG) have made a significant contribution to the greater use of wild-type germplasm in modern breeding processes in close collaboration with national and international agricultural research institutes. As a result of these activities, numerous information on specific CWR taxa (name of taxon, gene pool categories, and geographical distribution), their potential use in breeding, but also other characteristics were obtained. This information is deposited in different databases and may be of great importance for encouraging greater use of wild relatives for crop improvement.

For the last 30 years, yields of the most important crops in the world have increased significantly. However, the possibilities for the improvement of current crops, that is, the creation of improved varieties with even higher yield potential and high environmental stress tolerance and resistance traits, are, due to narrow genetic variation, quite limited. The situation is further aggravated by the environmental changes caused by climate fluctuations, as well as the rapid evolution of pests and pathogens. Due to all this, the immediate need is to establish effective breeding programs that will use the untapped genetic resources from the wild relatives of cultivated crops. The use of CWR would be a promising approach to improving the genetic diversity of cultivated crops. In order to increase their use in breeding programs, their germplasm must be conserved, characterized, and easily accessible to research communities and plant breeding organizations (Dempewolf et al. 2014). Despite the numerous difficulties encountered in using CWR for crop improvement, especially for certain crops, there is a significant number of successful transfers of superior alleles from CWR to domesticated crops. Advanced biotechnology, such as sequencing of the genes, contributed to faster detection of the gene. As a result, in the CWR genome, a large number of genes of potential carriers of resistance to various types of stress have been discovered. They are currently tested in breeding programs for abiotic and biotic resistance to stress and yield improvement in their cultivated offspring (rice, wheat, barley, soybeans, tomatoes, potatoes, peanuts) (Zhang et al. 2017).

According to the International Food Policy Research Institute (IFPRI), areas under the four main sources of food in the world (wheat, rice, corn, and potatoes) will continue to decline by 2050, which will represent an increasing challenge for agriculture. It is anticipated that during this period, the number of people on the planet could exceed nine billion, which would lead to increased pressure on plant production but also agriculture in general (Dempewolf et al. 2014). In order to meet the food needs of this rapidly growing population by 2050, it will be necessary to increase food production by 60% globally, or 100% in developing countries, which will not be an easy task (Jovovic et al. 2016). The process of adapting to climate change and sustainable production will also mean the abandonment of the cultivation of certain crops and the introduction of some new ones into the production. These challenges can be solved by the development of new varieties and hybrids with increased biotic or abiotic tolerance/resistance that allows their cultivation in poor climatic conditions and marginal lands. Thus, the most probable focus on wild relatives of crops, as a fundamental resource for further crop improvement, will be significantly increased in the coming period (Redden et al. 2015).

2.3 Selection and Prioritization of Species and Area

CWRs represent an important part of total biodiversity, and their role in preserving natural and agricultural ecosystems is enormous. Wild relatives have the ability to cross with the cultivated crops, helping agriculture to adapt to new challenges. Nowadays, man exerts increasing pressure on natural habitats of wild species, which threatens their survival. That is why many CWRs are seriously endangered and significantly suffer from genetic erosion today. Some recent studies indicate that in the next 50 years, 15–37% of wild species (Thomas et al. 2004a, b) and 16–22% of wild relatives of cultivated plants of importance to agriculture could be in great danger of extinction (Jarvis et al. 2008). Although many researchers and plant breeders have recognized the future potential value of CWR diversity as a source of value for the development of new crop varieties, little has been done so far to protect them. The main reason for this lies in the fact that there are not enough human, but also financial, resources in the world to protect all CWRs that are important for agriculture. Often, the problem with these efforts is the strong competition of other components of biodiversity but also the inability to prepare serious management plans and monitoring regimes for all of them (Maxted et al. 2006). Therefore, during the development of national conservation programs, some form of selection should be used so that candidate species can be placed in different priority categories and appropriate forms of genetic conservation. In the event that formal genetic conservation is not possible, then for such CWR species should be created alternative solutions to limit threats to them or their habitats. Given their importance, it is necessary to urgently strengthen the capacity of developing countries, especially those in the centers of origin of biological diversity, to dedicate themselves as best as possible to the preservation, improvement, and management of these resources. Since each state has a sovereign right over the management and use of genetic resources in its territory, the responsibility for their preservation lies at the national level. This means that national strategies for conserving prioritized CWR genetic diversity in situ and ex situ are fundamental for the global conservation of these resources as well. A number of activities aimed at studying and protecting CWR diversity have resulted in significant progress in the development of national CWR conservation strategies, especially in Europe, which has been leading in the development of CWR conservation practices in the last 15 years (Kell et al. 2017). A comprehensive approach to the conservation of CWR includes activities at national, regional, and global levels, both in gene banks and in the wild.

There are a number of criteria for selecting the areas on which CWR would be preserved. If the CWR is a small population or it is geographically limited to small areas, then the selection of the area does not pose a problem; however, if it is a variable species that contains many populations and with extensive geographical distribution within the country, then the situation becomes much more complex. Choosing a conservation area cannot be viewed as a simple choice of that area in relation to another, but that decision must take into account the costs and benefits of alternative conservation strategies, as well as current and future threats to preserve the diversity of species and the functioning of the ecosystem. In recent years, a lot of models for reserve selection have been suggested in the world, most of which emphasize maximum biodiversity coverage. When selecting a priority area, the following criteria should be taken into account:

- The prevalence and abundance of target species
- The level of diversity of the target population
- Number of population and number of individuals within the population
- Current conservation status of the target species
- The presence of the target species in priority areas or centers of plant diversity
- Accessibility
- The quality of the priority area
- State of management of the reserve
- Political and socioeconomic factors

In addition, the following should also be taken into consideration:

- The size of the priority area
- Borders and shape
- The presence of invasive species, etc. (Dulloo et al. 2008)

Criteria used to prioritize species for conservation action are based on scientific, social, economic, and cultural values. The starting point in conservation of CWR is establishing a list of target tax. Regardless of the fact that each tax in a particular gene pool will not be immediately involved and the CWR conservation strategy, a complete list of taxa can be of great use to some future conservation actions (Kell et al. 2012a). There is no universal methodology for selecting species or populations to be given priority for in situ conservation. In practice, selection often depends on the priorities of an institution or agency involved in conservation activities (Ford-Lloyd et al. 2008). Sometimes the priority is given to CWR of economically important crops and sometimes to CWRs that are most at risk. The level of vulnerability is determined based on their conservation status, which represents an assessment of their current state in terms of their distribution, population size and numbers, genetic variation, the availability of habitat, potential threats to current maintenance, and survival prospects in the short, medium and long term. In addition to the above criteria, the choice of tax can also be influenced by some other reasons, such as:

- Their value to society in terms of ensuring food and nutrition security.
- Probability of success of conservation.
- Costs of conservation activities.
- That the taxa is taxonomically well known.
- It is also easy to collect.
- What is actual or potential use.
- Whether the tax is endemic or is widespread.
- Whether the occurrence of the species is in a continuous decline and whether there is evidence of genetic erosion.
- Whether it has a cultural significance.
- Whether it has some kind of protected status.
- Its biological characteristics, etc. (Heywood and Dulloo 2005).

The choice of priority species often depends on the level at which the conservation activities are carried out. Thus, for global food safety, the priority is certainly the conservation of wild relatives of major food crops (bread wheat, maize, and rice), while on a regional or national level, a higher priority is given to conservation of minor crops (cassava, millets, sweet potato, etc.). Generally speaking, due to their fundamental role in sustaining human life, production of food intended for human consumption is given the highest importance. In addition, top priorities include crops of high economic value due to the importance they have for sustainable economic growth (Kell et al. 2012b).

Thanks to the great advancement of gene detection and transfer techniques, it is discovered that almost all wild species are potential donors of genes to crops. Therefore, their potential value for variety improvement is taken as a very important criterion in the prioritization of species for conservation. As the use of biotechnology is a relatively expensive and technically demanding plant breeding tool, the use of conventional breeding techniques for interspecies gene transfer between closely related species will continue to be the main method for improving crops in the future (Maxted and Kell 2009). The reduction of the CWR conservation to a limited number of taxon remains the main concept of conservation strategy at the moment. However, we should not ignore the wild relatives of unknown utilization potential as those which with the discovery of some new breeding techniques may in future become a very important source of desirable genes. That is why all CWRs that are endangered or are expected to be exposed to genetic erosion in the near future should be given priority protection.

The degree of vulnerability of CWR species also falls under the basic criteria for conservation planning. The relative status of their vulnerability is determined according to their Red List status or by taking new estimates. It should always be remembered that even if some CWR is not on the Red List, it does not mean that it is not threatened and that it is not under any form of threat. The classification of the taxon vulnerability can be done according to their comparative distribution which can be considered as an indicator of the relative degree of threat when actual threats to the populations or habitats in which they are found are not known. Thus, CWR with limited distribution level can be assigned higher priority status in relation to widespread types. When species prioritization is carried out on the basis of the Red List, attention should be paid to differences that may arise between national, regional, and global estimates. Red List assessments of 572 native European CWR showed that at least 11.5% of species were endangered, of which 3.3% were critically endangered, 4.4% endangered, and 3.8% vulnerable. Another 4.5% of species is rated as near threatened (http://ec.europa.eu/environment/nature/conservation/species/redlist/index_en.Htm).

CWR conservation issues were also addressed in the Second Global Plan of Action for Plant Genetic Resources for Food and Agriculture (PGRFA 2011), which highlights the importance of preserving populations in situ allowing the continuation of evolutionary processes, that is, the generation of adaptive traits. In addition, this report also emphasizes the need to extend the program to ex situ conservation in order to ensure greater conservation of species diversity, including those adapted to extreme climatic and soil conditions, as well as those that come from areas that can be seriously affected by climate change. Ex situ conservation of CWR in gene banks, field collections, and botanical gardens facilitates their use in agricultural research and breeding. Preservation of CWR germplasm in gene banks greatly facilitates their use, and their use in selection without ex situ preservation would be impossible. On the other hand, in situ conservation is important for the maintenance and recovery of populations in their natural environment, which enables the continued evolution of new traits. Since in situ conservation is necessarily carried out at the national level, the national CWR strategies are essential for their preservation. Because of this, the complementary conservation of CWR both in situ and ex situ is the best way of keeping them safe.

2.4 In Situ Conservation of CWR and Conservation Strategies

Crop wild relatives are wild plant species or populations that are more or less related to domesticated plants. They differ from the crop, having important variation that is missing in cultivated plants, since they have not passed through domestication. Although they represent a vital pool of genetic variation that can be used in breeding as a source of valuable traits, many crop wild relatives are at risk of extinction. Conservation of CWR is a stressed as a priority aim in numerous international documents and organizations (Convention on Biological Diversity, CBD 2010; Food and Agriculture Organization of the United Nations, FAO 2015), because of the threats they are exposed to and their possible utilization in future.

Their geographic origins can be traced to regions known as Vavilov's "centers of origin," e.g., geographic area where crops were domesticated firstly (Hummer and Hancock 2015). The eight "centers of origin" are as follows:

1. East Asian Center (Central and West China, Korea, Japan, and Taiwan); native region for soybeans, millet, many vegetable crops, and fruits.
2. Hindustani (Tropical) Center (India, Indochina, Southern China, and the islands of Southeastern Asia); native land of rice, sugarcane, tropical fruit, and vegetables.
3. Inter-Asiatic center (interior mountains of Asia Minor, Iran, Syria, Palestine, Trans-Jordania, Afghanistan, Inner Asia, and Northwestern India); the native crops include the origin of wheat, rye, and fruit trees.
4. Caucasian (South West Asiatic) center (Turkey, Iran, and Afghanistan) include original species of temperate fruit trees. Additional species of wheat and rye are also found here.
5. Mediterranean center includes countries bordering on the Mediterranean Sea. Ancient civilizations selected many crops including olives and the carob tree.
6. Abyssinian center (Ethiopia) produced teff (*Eragrostis abyssinica* Link.), niger-seed oil plant (*Guizotia abyssiniaca* Cass.), a banana [*Ensete ventricosum* (Welw.) *Cheesman*], and coffee (*Coffea arabica* L.).
7. Central American center (Central America, south Mexico, and the west Indian Islands); native area for maize, cotton, beans, pumpkins, cocoa, avocados, and subtropical fruits.
8. Andean center (Peru, Ecuador, Bolivia, and Chile); this is native land for tuber-bearing crops such as potatoes, quinine tree, and the coca bush.

Nowadays, scientists proposed term "center of diversity" for the area where a high degree of genetic variation for a particular crop exists, instead of "centers of origin" (Hummer and Hancock 2015).

The wild relatives closely related to the cultivated plants within the same gene pool could be of higher importance if they are utilized in crop improvement by conventional breeding than more distant species. However, some distantly related crops are not used as a valuable source of genes in breeding but could be potential donors in future and, therefore, should be conserved. Besides, CWR conservation in long term should emphasize both widespread and rare crops, for their potential usage for crop improvement (Maxted and Kell 2009). Better collecting and conservation could be obtained by application of gap analysis through four ways (Nabhan 1990):

1. Targeting sites were species missing from entire collection could be easily found
2. Determination of "under-collected" or "over-collected" areas, according to the distribution of particular taxon
3. Locating regions with higher richness of species compared to the others
4. Defining ecological area of each species, so ecotypes could be easily sampled for every adapted population

The strategies used for CWR conservation are ex situ (primarily as seed in gene banks, but also in tissue culture or cryopreservation, or as plants in field collections) and in situ (i.e., in natural habitats managed as genetic reserves).

In situ ("on-site," "in place") conservation includes determination, management, and monitoring of biodiversity in the same area where crops were found. The in situ strategy is best explained in comparison to ex situ, namely, ex situ ("off-site") conservation methods are applied out of target's plant natural habitat. According to Heywood and Dulloo (2005), in situ conservation approaches can be applied as species-centered (at populations of targeted plants) or ecosystem-based (at whole ecosystems). The common goal for both approaches is to enable biodiversity conservation in natural habitat, i.e., to enable species's self-replication and self-maintenance. This encompasses conservation of all components in ecological system where plants developed particular traits. It could be a natural habitat or area significantly influenced by human, such as arable fields for agricultural crops. Therefore, the conservation of agricultural biodiversity includes conservation of agroecosystems together with farmer's traditional practices (Myer et al. 2000).

The basis of in situ conservation is protected area (PA). "Protected area is clearly defined geographical space, recognized, dedicated and managed, through legal or other effective means, to achieve the long-term conservation of nature with associated ecosystem services and cultural values" (Dudley 2008). A protected area is necessary for maintenance of particular taxon in its natural habitats and for preservation of the other natural or seminatural ecosystems. However, today is growing importance of broaden in situ conservation out of protected areas, because of different socioeconomic and political reasons (Rodrigues et al. 2004). Protected areas were used for conservation of target species and their natural habitat for future needs as well as to prevent degradation and extinction. At present, there are about 200,000 PAs in 245 countries and territories worldwide, and they are stored in the World Database on Protected Areas (www.protectedplanet.net).

Ex situ strategy represents conservation of biological diversity beyond its natural habitats (CBD 2010), as a safety measure for threaten species against their extinction. In the past, in situ was the preferred approach compared to ex situ conservation. In situ strategy was considered as more comprehensive conserving processes in the natural habitats that are not preserved in ex situ conditions. The CBD (2010) highlighted complementarity of both in situ and ex situ strategies and specifies that "ex situ facilities and techniques should predominantly be implemented for the purpose of complementing and supporting in situ measures," and not as an alternative (http://biodiversitya-z.org/). The in situ conservation of CWR is important for future (Maxted and Kell 2009), whereas the ex situ conservation of germplasm already conserved in situ represents safety duplications of natural plant diversity, threatened by the impacts of human (Thuiller et al. 2005; Van Vuuren et al. 2006). If it is not possible to implement in situ conservation, ex situ methods could be applied with the highest priority in preservation of plant diversity by collecting representative samples.

Estimated number of crop wild relatives of more than 58 000 is represented only by 2–6% of global ex situ collections, and about 6% of the total number of CWR have any accessions conserved ex situ (Maxted and Kell 2009; FAO 2011).

The world's most important crop wild relatives of *Triticum* and *Aegilops* are inadequately represented in ex situ collections, while gene banks worldwide have only 10% of wild rice accessions (Tanksley and McCouch 1997). Similarly, in national Chinese soybean collection, the largest in the world, CWRs are poorly represented (Zhao et al. 2005). Nowadays, gene banks develop ex situ collections with priority to CWR. Good example is Israel Plant Gene Bank which prioritized 323 plant species for ex situ conservation, including many CWR (Barazani et al. 2008).

Ex situ storage of CWS is a good backup solution, but it is not sufficient to keep and maintain their diversity. Protected areas with CWS worldwide are aimed for conserving both natural habitats and populations in in situ conditions, with appropriate monitoring and management (Engels et al. 2006). Also, Convention of Biological Diversity and International Treaty on Plant Genetic Resources for Food and Agriculture highlighted the necessity of in situ conservation of CWS in existing protected areas as their natural ecosystems (Iriondo et al. 2008). Bioversity International (https://www.bioversityinternational.org/) supports local, national, and global in situ conservation through active involvement of local institutions in the following tasks:

- Determination of the status and threats of CWR
- Management approaches for conservation of CWR with priority and cost-effectiveness
- Development of long-term threshold levels for sustainable utilization of CWR without endangering their preservation

A great variability of CWR, their natural habitats, utilization, threats, and priorities for preservation induce that there is no general or unique protocol or general approach for conservation of wild plants. Bioversity International has published *Crop Wild Relatives: A Manual of in Situ Conservation* in 2011 (http://www.cropwildrelatives.org), based on experience and practices from countries, important centers of diversity of CWR: Armenia, Bolivia, Madagascar, Sri Lanka, and Uzbekistan.

There are numerous potential approaches for conservation of global CWR diversity. Maxted et al. (2007) proposed three complementary approaches, distinguished as individual, national, and global:

- Individual approach – involves an individual protected area or activity of gene bank manager for CWR conservation.
- National approach – involves activity of an individual country on CWR conservation for systematic protection of CWR diversity in in situ conditions, as well as ex situ storage of samples in gene banks, as safety backup. The aim is to protect CWR in that area or state and to promote their potential utilization.
- Global approach – is a worldwide strategy which focuses on crop wild relatives with high priority for conservation. It enables systematic in situ conservation of CWR of the highest importance for mankind through global network and also in gene bank collections. Vavilov's "centers of diversity" are with preferences for CWR conservation, compared to the other areas in the world.

Another insight was based on wideness of applied CWR conservation strategy and encompasses two complementary approaches, floristic and monographic (Maxted et al. 2011). Floristic approach refers to CWR conservation in particular geographic area (protected area, country, particular region, or even whole globe); monographic approach is related to certain plant gene pool, no matter how big is the given geographic area. For example, in Europe a combination of the two approaches is applied through the EU project PGR Secure (www.pgrsecure.org, Kell et al. 2004).

Necessity for all of proposed distinct and complementary approaches is to include ex situ collections for all in situ conserved CWR, giving priority to plants that could be utilized for food production within changed and increased demands of mankind. Last century was characterized by growing genetically uniform modern cultivars and loosing genetic variability, e.g., according to FAO (2011), 75% of genetic diversity of agricultural crops was lost during that period.

There are many examples of negative effect of growing genetically uniform crops: the potato hunger in 1840 in Ireland, Europe, and North America, caused by growing potato varieties susceptible to the potato blight that were spread, and in 1970 the southern corn blight resulted in serious yield losses of growing few genetically uniform hybrids, with Texas male-sterile cytoplasm (Tcms). Importance of CWR is in preventing such losses by maintaining genetic diversity and obtaining secure food production.

In the twentieth century, in 1970s and 1980s (Hoyt and Brown 1988), utilization of CWR in breeding of new varieties started, as a source for valuable traits, predominantly resistance to biotic stress (Prescott-Allen and Prescott-Allen 1986). Landmark example is introduction of late blight [caused by *Phytophthora infestans*

(Mont.) de Bary] resistance from the wild potato *Solanum demissum* Lindl. (Pavek and Corsini 2001) and stem rust (caused by *Puccinia graminis* ssp. *graminis* Pers.:Pers.) resistances from the wild wheat *Aegilops tauschii* Coss. (Prescott-Allen and Prescott-Allen 1986; Kilian et al. 2010). Also, CWR have been used as a source for wheat curl mite resistance (Malik et al. 2003) and grassy stunt disease in rice (Brar and Khush 1997).

Considering abiotic stress tolerance, wild relatives are used for improvement of drought tolerance in wheat (Farooq and Azam 2001) and tolerance to high temperatures in rice (Sheehy et al. 2005). They represent a valuable source of nutritional content like protein content in durum wheat (Kovacs et al. 1998), potato calcium content (Bamberg and Hanneman 2003), and increased carotene content in tomato (Pan et al. 2000).

The main criteria in choosing which CWR should be conserved are based on priority of certain crop, potential utilization, and threatened status (Kell et al. 2012). Besides, monitoring and management of protecting area for CWR is of high importance. In situ (in nature) conservation includes genetic reserve management that encompasses maintenance of the farming system. Nowadays there are only a few active in situ conserved CWR:

- *Beta patula* in Madeira, Portugal
- *Triticum* species in Ammiad, Eastern Galilee, Israel
- *Aegilops* species in Ceylanpinar, southeast Turkey
- *Zea perennis* in the Sierra de Manantlan, Mexico
- *Citrus*, *Oryza*, and *Alocasia* species in Ngoc Hoi, Vietnam
- *Solanum* species in Pisac Cusco, Peru
- Coffee species in Yayu Forest Biosphere Reserve, Ethiopia

According to Kell et al. (2012), 16% of native European CWR are threatened, and 4% are critically endangered. The estimated number of CWR in the EU countries is 17.495, and a half of them are endemic. They are represented in ex situ collections only by 6% that include 1.095 species.

In the last decade, interest about CWR conservation in Europe and their preservation for the future has increased. Under the guidance of the European Cooperative Programme for Plant Genetic Resources (ECPGR), a group of authors wrote "ECPGR Concept for in situ conservation of crop wild relatives in Europe," with the aim to offer it to the EU Commission for future policy on in situ conservation of CWR diversity in Europe (Maxted et al. 2015).

Systematic conservation of CWR includes national and regional (European) level of conservation strategy. Despite the importance of CWR conservation, it is necessary to define criteria and mark off species and population with the highest importance and priority for conservation. Within An European Genebank Integrated System (AEGIS, http://www.ecpgr.cgiar.org/aegis/aegis-homepage/), member countries with ex situ collections nominate unique and important Most Appropriate Accessions (MAA) to be included in the system. Alike, a concept for in situ conservation proposed a definition of Most Appropriate Wild Populations (MAWP). If particular species is with priority as MAWP, it is not MAA for ex situ collections, but it has to be present in the collection as safety duplicate. Another difference between MAWP and a MAA is that MAWP is dynamic and could be changed and developed over time, while MAA is static and genetically stable.

An MAWP is included in the European CWR conservation strategy if it fulfills the following criteria:

- The population is autochthonous or was growing in that area for more than ten generations.
- The population possesses particular genetic diversity or traits of interest.
- The population has to be maintained and manage in accordance with long-term in situ conservation standards (Iriondo et al. 2012).
- The population is not particularly endangered in a long term (100 years), and possible threats are insignificant and could be controlled.
- The nomination of potential MAWP has to be done via national agencies; samples have to be available to public and included in AEGIS ex situ collections.

Theoretically, MAWPs existed in protected areas, but in practice, they could be found out of them. In both cases integrated conservation strategies for CWR in Europe have to be applied, according to proposed instructions by Maxted et al. (2015):

- *National CWR conservation strategies* – refer to every European country with national CWR conservation strategy that encompasses both in situ and ex situ activities. Besides, each country has to propose national MAWPs. National agencies in each country are responsible for conservation activities for MAWP at regional level.
- *Regional (European) CWR conservation strategy* – include in situ CWR conserved populations (MAWPs), with their safety duplication in ex situ collections. MAWPs will be proposed by the ECPGR in situ and on-farm conservation network.
- *Integrated CWR conservation strategy for Europe* – both national and regional strategies are integrated in two ways:
  - (a) *Bottom-up integration* – national MAWPs are nominated for official identification in the European network of priority in situ CWR populations.
  - (b) *Top-down integration* – regional MAWPs nominated at European level are included in the European network of national and regional MAWPs and integrated in appropriate national CWR conservation strategies.

At both national and regional levels, hotspots of CWR have to be identified and included in integrated CWR conservation strategy. Hotspots are areas where “exceptional concentrations of endemic species experience exceptional loss of habitat” (Myer et al. 2000). CWR hotspots are defined based on their richness, uniqueness, or utilization and may be designated as Important Crop Wild Relatives Areas (ICWRA). After determination of European ICWRA, they will attract particular attention and importance for CWR conservation and will contribute to public awareness of the importance of CWR for European food security. Also, establishing of the European network of MAWPs is important contribution for global network of CWR important for mankind (FAO 2013).

In general, it is of high importance to determinate species and areas with high priority for conservation and to apply integrated strategies to protect genetic diversity of crop wild relatives for future.

2.5 CWR and Climate Changes

Global warming, as a part of climate change, includes increase in average temperature and accompanying effect like changes in rainfall patterns, and in the frequency of extreme weather events, in the last two centuries. The terms "global warming" and "climate change" are mutually used. In 2008 NASA distinguished global warming as "the increase in Earth's average surface temperature due to rising levels of greenhouse gases," from climate change – "a long-term change in the Earth's climate, or of a region on Earth" (https://www.nasa.gov/topics/earth/features/climate_by_any_other_name.html). In 2013, the Intergovernmental Panel on Climate Change (IPCC) reported that the global temperature increase between 1880 and 2012 was 0.9 °C, and in the second part of the twentieth century, it was even 1.1 °C, predominantly due to human activities (e.g., increased emissions of greenhouse gases and dust). Predictions are that by the end of the twenty-first century, global average temperature will increase by 0.3 to 5.4 °C, compared to period 1986–2005 (http://www.ipcc.ch/report/ar5/wg1/).

Effects of climate change differ between regions, estimated to be greater on the land than on the oceans. They include ice melting, rising sea levels, changing rainfalls, and widening of deserts (Zeng and Yoon 2009). *The other changes are more frequent occurrence of severe weather events, like heat waves, droughts, heavy rainfalls and snowfalls, floods, increased presence of pests and diseases, and ocean acidification* (ESA 2014). Many of these changes will last for centuries, seriously threaten global food security, by decreasing crop yields and arable land, alongside with increase human population (Clark 2016).

In general, increase of every degree of global temperature is expected to induce (http://dels.nas.edu/resources/static-assets/materials-based-on-reports/booklets/warming_world_final.pdf):

- 5–10% changes in precipitation
- 3–10% increases in the amount of rain during rainfall season
- 5–10% changes in stream flow in river basins
- 15% reductions in the average temperature over the Arctic Ocean, resulting in long-term increasing global sea level by 4–7.5 m
- 5–15% decreasing in crops yields

As example, *the new study of* Scheelbeek et al. (2018) estimated global vegetable and legume yields decrease about 35% by 2100 due to climatic changes. Another research (Tigchelaar et al. 2018) estimated maize yield reduction by 50% with temperature increase for 4 °C that is predicted by the end of the twenty-first century. Even limiting temperature increase to less than 2 °C will reduce maize yield about 7–10% in the countries which are the most important producers and exporters in the world.

Climate change will also influence rainfall distribution worldwide. About 80% of cultivated crops are growing in rainfed conditions, and extreme rainfalls and flooding could significantly affect yield.

Changed weather conditions will be in favor to expansion of weeds, pests, and insects, modifying their number and distribution area, together with earlier springs and mild winters. Widespread of new diseases and pests will affect crops which had no time to adapt and became resistant to them. For instance, new mutant of wheat rust, not present for more than 50 years, was spread from Africa to Asia, the Middle East, and Europe, destroying crops.

Moderate climatic changes in the last century already affected prevalence, number, growth, and development of a large number of species. Some of them shifted toward poles or to higher altitude or to earlier planting time to avoid negative effect of drought during sensitive vegetative and reproductive phases (Root et al. 2003; Parmesan 2006).

Different models (Thomas et al. 2004a, b) indicate wide range of extinctions as a consequence of changing climate. It implicates particular attention for conservation, determination of protected areas, and coordinated activities on preservation of threatened crops at national and global level.

The world's population is predicted to reach over 9.3 billion by the year 2050, which is going to increase food demands, but due to climate changes, reduction in global food production is expected. Implications of changing climate will be on the quality, quantity, and availability of the crop and animal production, as well as in agriculture ecosystem, at the global, regional, and local level.

Agriculture and climate change are in complex interrelationship. Agricultural production releases great amount of CO_2 that influence climate. Higher levels of CO_2 contribute to increased plan growth and crop yields, but changing of the other climatic factors (extreme weather, heat waves, or drought) at the same time negatively affects potential yield increase. For instance, optimal temperatures over a crop reduce available water and nutrients and negatively affect protein, vitamin and mineral content, and final yield. Besides stimulating of plant growth, increased level of CO_2 reduces protein and nutrient content in crops, too. There are estimations that protein content in some crops decreases 6–15% if CO_2 level reaches 540–960 ppm that is predicted to be by the end of the twenty-first century (at present it is 409 ppm). The content of some minerals (iron, zinc, calcium, magnesium, copper, sulfur, phosphorus, and nitrogen) and vitamin B is predicted to decrease, as well. For example, with increased CO_2 concentrations, rice content of vitamin B1 (thiamine) and B2 (riboflavin) decreased for 17%, vitamin B5 (pantothenic acid) for 13%, and vitamin B9 (folate) even 30% (https://blogs.ei.columbia.edu/2018/07/25/climate-change-food-agriculture/).

Optimal temperature for each crop differs, likewise, the influence of increased temperature on growth, development, and reproduction. Higher temperatures enable farmers to plant crops that are typical for warmer environments but could harm some crops that are growing nowadays and lower their yield. For example, in 2010 and 2012, high temperatures during summer reduced maize yield in the USA; mild winter in 2012 induced premature budding of cherries and cause more than 200,000 $ losses in Michigan (USGCRP 2014). In Australia, wheat, barley, and canola yields were reduced by drought over 40% (Lobell et al. 2008).

An increase in temperature and decrease in precipitation resulted in dry soils. Somewhere it could be possible to apply irrigation, but in some regions, there is a lack of available water when necessary. Widening cultivated land to increase food production is limited and could not be a solution for increased food demands, but higher food production could be achieved by increasing crop yields in a sustainable way (Garnett et al. 2013).

Agricultural production at the beginning of the second half of the twentieth century increased worldwide due to "Green Revolution," i.e., breeding of crops' modern varieties by discovering of new genes in genetically diverse landraces on national and international level (Evenson 2005). Later, this "revolution" was replaced by "Gene Revolution" with new high-yielding crops developed by application of new technologies and greater investment of private companies in agricultural research and development (Breithaupt 2008). Development of new cultivars for changing climate needs new genes and new sources. Traditional landraces hold a wide genetic diversity to be utilized (Vancetovic et al. 2010; Mladenovic Drinic et al. 2011). However, within predicted climate change, agricultural production has to be broadened to marginal regions and extreme environments, with new crops. Genes for cultivars in different and new environments in future could be found in crop wild relatives, rather than in landraces (Breithaupt 2008). Using wild species as sources of sets of genes for quantitative traits was also proposed by Zamir (2001). They are recommended as an important source of new genes and alleles, and preservation of CWR was particularly highlighted by FAO (2015), with the aim to adapt agriculture production to future climatic changes. Appropriate conservation of CWR, as a source for broadening genetic pool for breeding of modern varieties, is essential for future agricultural systems. Planned conservation is necessary, since all wild plants are exposed to numerous threats: climate change, changing in land usage, biotic exchange including invasive species, urbanization, and pollution.

Different projections of climate change in Europe by 2080 showed that half of wild species, including CWR, could be seriously affected or threatened (Thuiller et al. 2005). Those crops could be very susceptible to global weather changes, like it was presented in the research on peanut (*Arachis*) in South America, wild cowpea (*Vigna*) in Africa, and wild potato (*Solanum*) in South America (Jarvis et al. 2008). Namely, model projections by 2055 with different scenarios of crops shifts (unlimited, limited, and no migration) predicted that 16–22% of the three species are going to be extinct and most of them will lose half of the size. The greatest extinction risk is predicted for wild peanuts (about 48–62%), although it is not easy to accurately predict which CWR will be most endangered. Another studies on wild cowpea (*Vigna* spp.) predicted that about 50% will be extinct by 2050 due to climate change (Anonymous 2007).

Vegetation distribution is expected to change toward poles and to higher altitudes, but no prediction model is precise in details related to local area, such as changes in soil moisture and texture as a consequence of changing climate. Expansion of pests, diseases, and insect (Diffenbaugh et al. 2008), as well as occurrence of new races, is going to affect flowering time and fertilization.

Wild species, involving aromatic and medicinal plants and CWR locally grow in mountains, could be at risk from changing climate, if they have to move to new habitats (Heywood 2011). Numerous examples indicate that novel plant distributions due to temperature or rainfall changes could negatively influence their existence by exposing to nonoptimal climatic conditions and to competition with other plants (Cavaliere 2009). Increased sea level as a consequence of global warming and ice melting will influence all species in certain costal area. Forage species *Vigna marina* typical for the African coast is predicted to be particularly affected with increase of sea levels (Padulosi et al. 2011).

Bioclimatic model was applied for distribution prediction of eight wild *Cucurbitaceae* species in Mexico: *Cucurbita argyrosperma* subsp. *sororia*, *Cucurbita lundelliana*, *Cucurbita pepo* subsp. *fraterna*, *Cucurbita okeechobeensis* subsp. *martinezii*, *Sechium chinantlense*, *Sechium compositum*, *Sechium edule* subsp. *sylvestre*, and *Sechium hintonii* (Lira et al. 2009). Most of them showed resistance to some diseases, and prevalence is limited to particular area. In the analysis of the importance of their protected areas in future, under severe climatic change scenario, all species will be maintained in 29 PAs out of 69 existing nowadays. Probably, most of eight examined *Cucurbitaceae* species are going to be extinct in such weather conditions.

Estimation of number of CWR that could be lost as a consequence of climate change is very difficult, since they are numerous and diverse, although some information about number of species at potential risk could be obtained from national Red Lists. However, species that are not determinate as threatened now, could become in future, since criteria of International Union for Conservation of Nature (IUCN) are not containing influence of climate changes.

Wild relatives of crops are important factors for improved and sustainable agriculture (FAO 2008). With the anthropogenic climate change, they are becoming very important for food security and stability in future. CWR are related to socioeconomically important cultivated crops and represent valuable source of beneficial traits for them. They are already used for drought, cold, and salinity tolerance improvement in crops, but their importance is remarkable as a source of resistance to biotic stress, since climate change will induce migration of plant pathogens, shifting to different area and appearance of new one (Garrett et al. 2006).

Increased temperature within global climate change will affect flowering, fertilization, and final grain yield. Wild rice (*Oryza officinalis*) is successfully used to shift flowering time in cultivar Koshihikari (*Oryza sativa*, Ishimaru et al. 2010). Another wild source of drought tolerance is *Oryza glaberrima*, likewise sources of plant height and tillering, which were found in wild rice species *Oryza barthii*, *Oryza australiensis*, and *Oryza meridionalis*. These wild species are valuable sources for tolerance to high temperatures and drought, as well (Sanchez et al. 2013).

Advances in new technologies could increase utilization of CWR in breeding and increase rate of successful introgression of important traits into crops (Hajjar and Hodgkin 2007). Application of marker-assisted selection allows screening a large number of examined plants, after mapping and validation of important trait (van de Wiel et al. 2010). Quantitative trait loci (QTL) and genome-wide association studies (GWAS) are most commonly used mapping approaches (Takeda and Matsuoka 2008; Morrell et al. 2011). As example, in wild bean identified QTLs were associated with seed size and yield, and further genetic analysis was performed on cultivated varieties related to heat and drought tolerance (Wright and Kelly 2011). Application of QTL in morphological analysis and drought-related traits is used in numerous studies (Nikolic et al. 2012, 2013; Gahlaut et al. 2017). Application of genomics allowed development of single-nucleotide polymorphism (SNP) platforms (Comadran et al. 2012; Wang et al. 2014) for numerous species *Hordeum* species (Bayer et al. 2017), *Brassica napus* L. (Clarke et al. 2016), *Capsicum annuum* L. (Hulse-Kemp et al. 2016), and *Helianthus annuus* L. (Livaja et al. 2016). For diverse genepool of some crops, as well as for CWR, it is possible to use high-throughput sequencing for mapping analysis (Kilian and Graner 2012; Assenov et al. 2013).

Genome resequencing has started after completing genome sequencing for some crops such as soybean (Lam et al. 2010). Crop wild relatives of soya have been sequenced, and 17 whole-genome sequences were obtained, with wider diversity compared to cultivated soybean crops. New biotechnological tools enabled screening of large number of CWR, and by next-generation sequencing, it is possible to focus on specific gene by targeted resequencing and provide them to conventional breeders. Large-scale resequencing allowed identification of genes related to drought tolerance in crops, as well as genes related to response to stress (www.generationcp.org).

Application of new biotechnological tools is necessary for crop improvement and for utilization of potential from CWR. Wild relatives are still underutilized and underestimated resource in breeding and food production, but potential future use means appropriate and planned conservation. Coordinated activities on national, regional, and global level in defining priority taxa for preservation and establishment of genetic reserves are of the highest importance. Maximum exploitation of new technologies is required for a discovery of a new gene within wide natural diversity and a successful incorporation in new varieties in order to achieve a higher level of genetic diversity compared to domesticated crops. The development of new varieties will be accelerated by the application of new genomic tools for the dissection of genetic traits within the untapped diversity of CWR. Efficient conservation of CWR, together with their sustainable utilization in traditional and molecular plant breeding, will ensure food security for future demand.

2.6 The Use of Wild Relatives in Crop Improvement

Plant breeding can be described as a process of successive selection cycles, which eventually lead to a reduction in genetic divergence in relation to crop wild relatives. The process of continuous selection has resulted in crops on which production is based on the nutrition of the population today. However, in this way there was a decrease in cultivated crop genetic variability in relation to wild ancestors and other crop wild relatives from which they were domesticated (van Heerwaarden et al. 2011). Two phases in the domestication process can be defined when genetic diversity has been reduced. The first, initial domestication bottleneck occurs when wild populations of a single plant species are isolated and their cultivation begins. After this initial bottleneck, loss of divergence occurs in the process of breeding for desirable traits during crop improvement, and this is an improvement bottleneck. The study of the effect of domestication bottlenecks has been difficult for many species for three reasons:

Domestication was initiated probably over thousands of years.

There was an introgression of genetic material through the hybridization process with wild relatives.

The change in the distribution and extinction of natural populations, cultivation, and interactions with wild relatives (Flint-Garcia 2013).

Reducing genetic variability during domestication is the consequence of the creation of high-yielding varieties and hybrids for which the market and regulatory requirements have been required to be phenotypically uniform. As a result, many favorable genetic variants have been lost during the selection. However, in spite of that, many of the traits on which the selection was made, such as reactions to biotic and abiotic factors, are kept in CWR. Crossbreeding varieties with WCR, breeders have recently revealed the genetic variability lost during the domestication and the crop improvement process.

CWR have been used as sources for widening genetic diversity during domestication process and in crosses with cultivated crops for their improvement by breeders. Pre-breeding is a concept that provides breeders with more "usable" form of wild plant diversity for introgression in modern cultivars (Sharma et al. 2013). It encompasses basic and applied research with the aim to create genetic pool of useful traits that will be used for breeding new varieties, improving the trait of interest. The application and success of using pre-breeding concept, i.e., exotic germplasm or crop wild relatives, are difficult to estimate. Public institutions rarely reported pedigree of released varieties, rather some steps of pre-breeding material under development, but from private seed company, it is almost impossible to get any pedigree data of new varieties. Introgression of wild plants for widening of genetic diversity is possible by two approaches: (1) "Choose first," wild material is chosen according to phenotypic and genotypic data for targeting crosses, and the progeny is evaluated afterward, or (2) "cross first," numerous crosses are made between domestic and wild plants, and offsprings are estimated for introgressed traits in domestic material (Dempewolf et al. 2014). The first approach, also known as predictive characterization (Thormann et al. 2014), comprehend phenotyping (targeting a particular trait), genotyping, and their combination through statistical analysis. After identification of specific trait, additional crosses and backcrosses are made for incorporation in the crop. The latter approach includes wide range of crosses between wild and domesticated crops and screening their progeny and later generations for beneficial traits. Advantage of the second approach is appearance of potentially new sources of diversity and important traits (Hajjar and Hodgkin 2007; Moore 2015). After incorporation of important traits into modern cultivars, few generations of backcrossing are necessary. Disadvantage is the presence of nondesirable characteristics from wild crops, such as low yield, logging, susceptibility to diseases and abiotic stresses, and small fruit size (Salamini et al. 2002). Regardless of all difficulties, scientists agree that wild relatives and populations are valuable source for crop improvement for future needs. Potentially useful CWR include 4157 possible "usage," within 970 taxa and 127 different crops (http://www.cwrdiversity.org/checklist/). They are classified into seven "breeding use classes": biotic stress (2427), abiotic stress (700), agronomic trait (485), fertility trait (272), morphological trait (20), phenological trait (54), and quality trait (199).

CWR is a source of gene tolerance to pests and diseases; improves tolerance to environmental conditions such as extreme temperatures, drought, and flooding; and improves nutrition properties and handling qualities. CWRs have strong contribution to agriculture, crop processing, and generally world economy (Stolto et al. 2006). CWR are species that are closely related to cultivated plant species and represent an important source of genetic diversity for crop improvement. In the process of domestication and crop breeding, the genetic variability of cultivated plants decreased, resulting in the cultivated plants having a significantly narrower genetic base than CWR, which is the basis of their sensitivity to biotic and abiotic stresses. CWRs have been used in breeding programs over a century, especially in programs to increase the tolerance of cultivated plants to pests and diseases.

CWRs have a genetic basis that can be used to increase crop plant resistance to biotic and abiotic stress, which ultimately results in increased yields and production stability (Guarino and Lobell 2011). Since CWR developed in nature in different ecological conditions, under various abiotic and biotic stresses and pressures, but did not go through the genetic bottlenecks of domestication, through various genetic changes, they adapted to the stresses under which they developed and grew (Vollbrech and Sigmo 2005). Today, some modern varieties are replaced with stress-tolerant varieties to ensure the agricultural viability of the crop in the same locations (Yadav et al. 2011). The question arises as to whether it is possible to maintain the achieved yield and quality of cultivated plants under conditions of abiotic and biotic stress without the greater use of exotic germplasm in the process of breeding (Feuillet et al. 2008). However, due to the different activities of man, besides climatic changes, genetic erosion occurs in CWR as well as in other wild plant species today (Bilz et al. 2011).

CWR genepool of some crops (alfalfa, cassava, chickpea, cowpea, finger millet, maize, sweet potato) has potential source of resistance and genetic variability for traits of interest (Smýkal et al. 2018). For example, maize wild species, *Teosinte* and *Tripsacum*, are not explored and utilized like the other wild relatives of crops with similar importance for humans. Possible explanation is that there is enough genetic variability for different traits, not completely explored and used yet, in local landraces and populations (Vancetovic et al. 2013; Babic et al. 2015; Andjelkovic et al. 2016). Another factor is that private breeding companies are not interested in investment in long-term processes of introgression of CWR that could result in minor yield increase, especially in maize and soybean. Besides, wild plants could bring unknown and unwanted traits that could seriously affect high yield of modern cultivars.

Some species, such as barley, bread and durum wheat, sorghum, rice, potato, soybean, etc., do not have genetic variability for some traits within the domesticated genepool, which is why new alleles have to be searched for in CWR. In some cultivated crops, large genetic bottleneck effects are created, and inadequate genetic variability has become very sensitive to abiotic and especially biotic stresses such as various pathogens. Therefore, the importance of CWR as the source of the resistance genes is becoming more and more important.

In order to create varieties and hybrids tolerant to biotic stress, over the last few decades, great progress has been made in introducing traits from CWR in cultivated crops. Although the main goal of the breeding today is to create modern varieties that will bring profit, breeders are increasingly studying CWR in the search for resources that will enable the extension of the genetic basis of cultivated crops (Moore 2015). Jha et al. (2014) consider that wild populations that have not undergone domestication bottleneck are a significant resource of an additional source of alleles. CWR are adapted to a wide variety of habitats and a range of environmental conditions and represent an important source of resistance to biotic factors and tolerance to abiotic factors (Dempewolf et al. 2014).

Potato suffers from many pests and diseases among which late blight, caused by the oomycete *Phytophthora infestans*, is the worst. The long-term and worldwide effort to breed for resistance so far has had little effect. The high-level late blight resistance in a wild potato relative, *S. bulbocastanum* Dunal subsp. *bulbocastanum*, is mainly controlled by a single resistance gene *RB*. Transgenic potato lines containing the *RB* gene have showed strong late blight resistance, comparable to the backcrossed progenies derived from the somatic hybrids between potato and S. *bulbocastanum* (Colton et al. 2006).

The genus *Oryza* has 24 species, out of which 2 are cultivated (*O. sativa* and *O. glaberrima*) and 22 are wild species. All the 22 wild species of *Oryza* are a vast reservoir of genes for biotic and abiotic stress resistance. Some of the yield enhancing traits/genes from AA genome wild species have been identified and mapped with molecular markers for their integration into *O. sativa* genome. A broad-spectrum resistance gene for bacterial blight resistance (*Xa21*) has been identified in *O. longistaminata* and introduced into many rice cultivars. Some important genes *Pi40* and *Bph18* for resistance to blast and brown plant hopper, respectively, have been successfully transferred into elite cultivars from *O. australiensis*, and the function of one blast resistance gene (*Pi9*) derived from *O. minuta* is elucidated. Many important genes from the most distantly related wild species such as *O. alta*, *O. granulata*, *O. longiglumis*, and *O. coarctata* are expected to be transferred into cultivated rice in the future using the latest tools of molecular genetics and biotechnology (Jena 2010).

Wheat (*Triticum* spp.) is the major staple food crop in many parts of the world in terms of cultivated area and food source, and the nutritional quality of wheat grains has a significant impact on human health and well-being worldwide. Although global cereal grain yields have increased due to breeding, a cereal-based diet falls short in providing sufficient nutrients for a balanced human diet (Cakmak et al. 2010). Biofortification of wheat grain, as the major staple food crop in many parts of the world, through genetic strategies is a powerful approach for changing the nutrient balance in the human diet. Since genetic diversity of crop plants has been significantly decreased by domestication and breeding a major objective of modern breeding is to identify valuable alleles in the wild ancestors of crop plants and to reintroduce them into cultivated crops (Smýkal et al. 2018). Wild emmer wheat (*Triticum turgidum* ssp. *dicoccoides*) is a progenitor of cultivated wheat, and it is fully compatible with the tetraploid durum wheat and can be crossed with the hexaploid bread wheat. Due to that the gene pool of the wild emmer wheat offers a rich allelic stock for the improvement of numerous economically important traits, e.g., grain protein, micronutrients, etc. (Peleg et al. 2009).

No resistance against wheat curl mite in common wheat had been reported until Harvey and Martin (1992) found several wheat accessions with strong resistance. Prior to this, several sources of resistance against wheat curl mite had been transferred into wheat from related species. The first of these was the 1B–1R chromosome from rye (Martin et al. 1984). These findings documented the value of the rye gene and lead to the development of the cultivar "TAM 107," which became widely used in the Great Plains region of the USA. This gene was later mapped and named *Cmc3* (Malik et al. 2003). Additional sources of resistance to mite colonization have been identified from partial amphiploids of wheat and related species.

Maize (*Zea mays* L.) is one of the most important field crops in the world grown for livestock feed and for biofuel. *Teosinte* (*Z. mays* ssp. *parviglumis* Iltis & Doebley) and *Tripsacum* are two CWRs that are donors of desirable genes in maize breeding for economically important traits (tolerances to insects, dishes, weeds, abiotic stresses). Due to the morphological similarities, including highly specialized cupulate fruitcase and the possibility of crossing with Zea and obtaining viable but generally infertile hybrids, *Tripsacum* is considered a close relative to Zea (Eubanks 2006). Nine perennial grasses that include habitats from southern Canada to Southern Chile belong to the genus *Tripsacum* (Eubanks 2006). One of them, *T. dactyloides* or Eastern gamagrass, is used to create intergeneric hybrids with corn (Mammadov et al. 2018). The *Teosinte* and *Tripsacum* are less used than wild species most likely due to the large amount of genetic diversity in maize landraces. After almost one century, it was found that *Balsas teosinte* (*Z. mays ssp. parviglumis* Iltis & Doebley) is an ancestor of corn (Matsuoka et al. 2002). *Teosinte* is a wild grass that grows in Mexico and some states of the central part of the American continent, such as Nicaragua, Guatemala, and Honduras (Standley 2015). The *Zea* genus encompasses a 1-year *Zea luxurians*, a perennial diploid species *Z. diploperennis*, perennial tetraploid species *Z. perennis*, and polytypic annual species *Z. mays* (Fukunaga et al. 2005). *Z. mays* encompasses four subspecies – ssp. *mays* (maize), ssp. *mexicana*, ssp. *parviglumis*, and ssp. *huehuetanangensis* (Fukunaga et al. 2005).

CWR is also used to improve the nutritional value of some crops, such as protein content in durum wheat, calcium content in potatoes, and the content of provitamin A in tomatoes. Mineral nutrient malnutrition, and particularly deficiency in zinc and iron, afflicts over three billion people worldwide. Wild emmer wheat, *Triticum turgidum* ssp. *dicoccoides*, genepool harbors a rich allelic repertoire for mineral nutrients in the grain (Peleg et al. 2009). Increased calcium in potatoes may increase the production rate by enhancing tuber quality and storability. Increased calcium levels in cultivated crops may help ameliorate the incidence of osteoporosis. Potato tubers expressing the *Arabidopsis* H+/Ca2+ transporter *sCAX1* contain up to threefold more calcium than wild-type tubers. In view of the importance of potato consumption worldwide, these transgenic plants may be a means of marginally increasing calcium intake levels in the population (Park et al. 2005).

In the gene banks are collectibles and wild relatives of crop species originating from the primary, secondary, and tertiary gene pool (Harlan and de Wet 1971). CWRs have a huge relevance because they contain useful genes and alleles that can be used to improve the genetic basis of crop species. However, breeders are reluctant to use CWR in their breeding programs because of the linkage drag. In order to make CWR more effective in the process of breeding, it is necessary to include and pass the pre-breeding process. Pre-breeding involves all activities of identifying desirable properties and genes in unadapted materials, i.e., CWR that cannot be directly used in breeding and creating superior varieties. Intermediate stock of materials after transferring preferred properties/genes possesses mostly positive traits and introduced genes from CWR. After that breeders can use the intermediate stock in producing new varieties for commercial production. Pre-breeding is a necessary first step in the use of CWR diversity. Through the introgression of desirable genes from CWR into genetic backgrounds readily used by the breeders, pre-breeding provides a unique possibility to overcome linkage drag. Modern molecular genetic technologies are expected to contribute to a more efficient determination of preferred tolerance genes on abiotic factors and their transfer from CWR to cultivated plants (Hajjar and Hodgkin 2007).

The precisely defined relationships between cultivated plants and their CWRs are of particular importance in the use of the CWR germplasm. More efficient use of CWR requires accurate identification and biosystematics of the accessions. Identification based on morphological traits often leads to incorrect systematization, especially in the case of morphologically similar and related species. A large portion of the accessions in the gene banks was identified only up to the level of the genus. Accurate identification is important for the determination of genes and defining gene transfer strategies during plant breeding.

2.7 Barriers to Increased Use of CWR for Crop Improvement

Social and scientific awareness of the necessity of preserving genetic resources has resulted in the formation of gene banks. Seed storage is the most common way of ex situ conservation of genetic resources. Preserving the seed in the gene banks has significant advantages over ex situ conservation, such as the ability to store a large number of samples in a small space, price, and low labor requirements. Since the DNA degrades in the seed over time, it must be periodically renewed, and fresh seeds can be preserved for the next period. The use of these resources in breeding and improving the resilience and productivity of agricultural production systems is very important. However, regardless of their importance, very low results have been achieved so far in the use of these resources (Wambugu et al. 2018). Today, 1625 gene banks are conserved in about 7.4 million accessions of the world's PGR (FAO 2010). There are no precise data on the use of CWR and genetic resources for breeding, although this is considered to be <1% (Sharma et al. 2013). The lack of data on the genetic value and the content of useful alleles in CWR is the main reason for the inadequate use of CWR in the breeding of cultivated plants (Khoury et al. 2016).

Large size of germplasm collections in gene banks and meaningful multi-locational evaluation of a small portion of conserved accessions is a factor that causes low use of CWR. Generally, there is a lack of information on characteristics of an economic character, which often shows a high genotype × environment interaction. All this poses a problem for breeders to choose the appropriate genetic diversity and sources of variability for their breeding programs. Plant genetic banks have greatly facilitated the use of CWR, but their disadvantage is that they are conserving only a part of the total genetic variability that exists in CWR. Ex situ conservation accessions are not exposed to natural selection processes that affect natural populations, which is the conservation of the evolutionary process. The advantage of in situ conservation is that it allows the creation of a new variability that arises as a result of population adaptation to environmental conditions and biotic interactions, which is particularly significant in the present climate change conditions. This means that besides ex situ collections, it is also necessary to have in situ conservation in order to maintain a much larger pool of genetic diversity and to ensure that habitats where CWR occur are protected and wild species continue to evolve in the wild (Jarvis et al. 2008).

In today's breeding programs, hybridization is the basic way of introducing new genes into cultivated crops. A large number of cultivated crops have wild relatives that are cross-compatible, and it is possible by simple hybridization to use their genes for improving cultivated crops. Harlan and de Wet (1971) based on crop genepool concept categorize plant species into three genepools – primary, secondary, and tertiary – and on the basis of this define the relationship between cultivated crops and wild relatives. Species in primary genepools, as a rule, do not have crossing barriers with the crop and, when crossed, give fertile offspring, while crossing with species from secondary and tertiary genepool produces sterile offspring. Poor germination of the pollen, inability to form embryos, sterility of hybrids, poor vigor, and hybrid fertility are the main obstacles to the use of CWR (Dempewolf et al. 2012).

There are several techniques that are used to overcome cross barriers. One of the widely used techniques is based on a large number of crossings with a large number of accessions and in the descendants of selecting a small number of lines with desirable properties (Porch et al. 2013). Van de Wiel et al. (2010) suggest intermediaries, an intermediary species, which can be crossed with both parents and thus transfer the CWR genes into a recurrent parent. In cases where it is difficult to obtain viable seeds or healthy embryos, the embryo rescue method is used (Shen et al. 2011). By this method, the embryo is separated from ovules and grown on artificial medium. In the case of pollen deficiency, a method of somatic hybridization or protoplast fusion (Holmes 2018) is used. If the crossing is done with CWRs with different ploidy levels, it is necessary to double the number of chromosomes, for which special techniques are used.

Finding new methods and mastering hybridization of cultivated crops with CWR are permanent tasks in order to use the genetic variability of CWR. After hybridization with which CWR was undertaken, a sufficiently large population of progeny should be developed with the purpose to study the effect of the introduced gene, its mapping in the genome, and its exploitation using MAS (Przulj and Perovic 2005). Genome-wide association studies or quantitative trait locus mapping approaches are commonly used to tagging desirable alleles with molecular markers. Xu et al. (2010) cited in rice (*Oryza sativa* L.) specific types of recombinants – chromosome segment substitution lines (CSSLs). CSSLs represent a series of introgression lines (ILs) obtained by backcrossing the donor and recipient parents using marker-assisted selection (MAS). Each CSSL carries one or more donor chromosome segments in the genetic background of the recurrent parent. CSSLs allow mapping QTLs throughout the genome and studying the interaction between QTLs and multi-environments (Wan et al. 2005). Combined use of CSSLs and MAS is a valuable strategy of introgression of novel genes or alleles from wild relatives in cultivated crops. CSSLs enable precise identification of QTL that govern complex agronomic and quality traits (Wang et al. 2006). In addition to CSSLs with the introduced chromosome segment, another type of introgression lines is used to isolate the single CWR gene in cultivated crops.

Methods of molecular genetics enable the determination of preferred loci in the CWR as well as the assessment of the success rate of post-domestication introgression with wild relatives (Russell et al. 2016). Genotyping platforms such as single-nucleotide polymorphism (SNP), whole-genome shotgun sequence (WGS), restriction site-associated DNA sequencing (RAD-seq) (Baxter et al. 2011), genotyping-by-sequencing (GBS) (Elshire et al. 2011), and NimbleGen Exome capture (Warr et al. 2015) have been used to characterize CWR germplasm. These methods allow for rapid fine mapping and can saturate mapping populations in terms of detecting all of the recombination events. Desirable genes in CWR can be determined by landscape genomics and environmental association analysis (Storfer et al. 2018). Also genomics supports the collecting, conservation, and use of CWR.

The lack of data on the phenotype and genotype for CWR is the biggest barrier to their more intensive use in the improvement of cultivated crops. Standard systems for identifying accessions and correct taxonomy are basic measures for improving the CWR documentation. Collections in gene banks are often incomplete and do not include CWR's existing variability. Collecting CWR should focus specifically on areas with extreme climates where it is possible to find genes/properties that are not in normal ecological conditions. Accessions should follow passport data, data on environmental factors, soil type, and diseases present.

The biological barriers to crossing, lack of characterization of wild species, linkage drag, lack of knowledge on genotype of complex features, poor management and maintenance of wild species in ex situ collections, poor institutional support, and funding of genetic resources and perceptions of wild species' inferiority relative to elite material are also constraints which prevent the intensified use of WCR (Dempewolf et al. 2017). Identification of traits of interest is complex because of the uncertainty in predicting phenotypes of CWR alleles when they are transferred to cultivated crops under intensive conditions. Preferred CWR alleles do not directly produce the desired phenotype, and this hidden variation can sometimes be discovered in the process of transgressive segregation (Gibson and Dworkin 2004). In the crossing between WCR and cultivated crops, hidden variations can often be obtained by unexpected and superior offspring. However, it is difficult to predict the potential benefit of CWR with many other undesired agronomic properties or to identify preferred alleles when their effect is masked with superior wild background traits. Breeders often avoid crossing with CWR due to the large number of unfavorable agronomic properties they possess. It is necessary to do a detailed evaluation and screening of the variability of properties in WCR in ex situ collections. A common and coordinated phenotypic evaluation of the pre-breeding material by more researchers is a significant and useful assessment of genotype – environment interaction. Due to limited resources and different cultivation methods, phenotypic evaluation of some CWR of clonal crops, such as long-lived perennials, may be particularly demanding.

The exchange of introgression lines with CWR alleles and their testing and cultivation in specific agroecological conditions can be a way of creating a locally adapted material containing CWR alleles. However, most of these genetic resources that carry beneficial traits introgressed from CWR are not adequately maintained and are not available for future use. The adoption of the International Treaty on Plant Genetic Resources for Food and Agriculture (ITPGRFA) by the Conference of the Food and Agriculture Organization of the UN completes one of the tasks set by the Earth Summit in Rio in 1992. The objectives of the ITPGRFA are the conservation and sustainable use of all plant genetic resources for food and agriculture and the fair and equitable sharing of the benefits arising out of their use, in harmony with the Convention on Biological Diversity, for sustainable agriculture and food security. In general, it can be said that there are no adequate human resources – experts in botany, cytogenetics, taxonomy, and curators of gene bank collections of wild material – necessary for the proper use of CWR stored in gene banks.

However, since CWR is at the same time significant for food security, climate adaptation, and biodiversity conservation, financing of one of these problems is financing in a great percentage of the other two, and donors can present the possibility of allocating resources in a way that simultaneously contributes to all these goals. Establishing long-term financing of progressive programs through public-private partnerships (PPPs) is one of the potential strategies for financing wheat, LOLA-VISP (http://vvv.vheatisp.org/) and VHEALBI (http://vvv.vhealbi.eu/), and apples (https://sites.google.com/a/nordgen.org/ppp-apples/).

The process of introducing genetic diversity from CWR into cultivated crops is a work that requires genetic resources, human capacities, and time. It is a long-lasting process that begins in nature, usually beyond human influence, where botanists and taxa and genetic resource experts collect CWRs. Gene bank curators perform conservation and characterization; geneticists, agronomists, phytopathologists, entomologists, and physiologists further characterize and evaluate, and then pre-breeders and breeders incorporate desirable alleles into new varieties. Providing breeders with wild genetic diversity in a more immediate useable pre-breeding form is a key process that allows the use of desirable properties from CWR in the creation of modern varieties (Sharma et al. 2013). In the process of pre-breeding, no cultivars can be obtained that can be cultivated but genotypes that are carriers of desirable alleles, which will be incorporated in crossbreeds into modern varieties. Even a more detailed study of the literature and genetic basis of modern varieties today cannot determine the extent to which CWR is used in the pre-breeding process (Hunter and Heywood 2011). When a preferred allele is introgressed in a genotype, it is rapidly expanding in breeding material, although breeding companies, especially private, do not make pedigrees public. Catalogs of varieties sometimes contain a shortened genealogy where data on the use of CWRs used in crossings are not included. Since it is generally difficult to determine the genetic contribution of CWR to a variety, it is difficult to estimate the economic effect of CWR use.

Although there are several ways to use wild diversity, it is possible to distinguish two most frequently used ones – choose first and cross first (Dempewolf et al. 2017). When choosing a first method, based on locality, phenotype or genotype CWR that possesses desirable properties crosses with a recurrent parent and, in subsequent generations, evaluates and selects progeny. In the cross first method, there is a large number of crossings between CWR and recurrent parents, and in the progeny, it is analyzed whether the preferred trait is in a recurrent parent. When crossing the CWR with a recurrent parent, in addition to the transfer of the desired allele, a large number of unwanted alleles and properties (linkage drag) were transferred. In general, CWR possesses a large number of genes/alleles that cause poor agronomic properties such as low yield, small fruits, sensitivity to lodging, dispersal of seeds, etc. In essence, linkage drag is a major problem in the use of CWR alleles, which is why many breeders prefer to use advanced lines in crossing that have the highest number of desirable properties. Linkage drag is trying to be solved using molecular marker methods (Neeraja et al. 2007). In order to exclude the deleterious diversity, the backcrossings of the obtained progenies with the recurrent parent are carried out.

Rice breeders in Japan are trying to create varieties with resistance against blast and also possessing good quality traits; however, by introducing desirable alleles for blast (*Pyricularia oryzae*) resistance, due to tightly linked genes, undesirable gene controlling poor grain quality is always introduced (Fukuoka et al. 2009). Since the interruption of this linkage requires a lot of time and money, breeders prefer to use available advanced lines and elite varieties than CWR, which usually have narrow genetic diversity. In general, the lack of genetic variability has a negative effect on the resilience, productivity, and sustainability of agricultural production.

Regardless of the current difficulties in using genetic diversity of CWR as a source of desirable genes, most breeders support the concept of introducing alleles from CWR to a modern variety (Prescott-Allen and Prescott-Allen 1986; Maxted and Kell 2009). Dempewolf et al. (2017) have done a detailed analysis of the literature data on the use of CWR published as an online resource http://www.cwrdiversity.org/checklist/. From this database they find that 4157 CWR from 970 CWR taxa were used in improvement of 127 different crops. CWR alleles were used for the improvement of agronomic traits (485), abiotic stress tolerance (700), biotic stress resistance (2427), fertility traits (272), morphological traits (20), phenological traits (54), and quality traits (199). CWR are mostly used in breeding sunflower, bread wheat, and potato. In sunflowers, CWR were most used in breeding for biotic stresses and as sources of cytoplasmic male sterility (Skoric and Jocic 2004). In wheat and potato, CWRs were mostly used for abiotic stress resistances. CWRs were mostly used in breeding of rice (*Oryza sativa* L.), tomato (*Solanum lycopersicum* L.), and wheat (*Triticum aestivum* L.) (Hajjar and Hodgkin 2007). In other crops, pre-breeding programs with CWR are in the early stages of developing or in the phase of wild species study as sources of traits of interest.

The use of wild relatives in the breeding of cultivated plants on salt stress tolerance is controversial, since only a few varieties of salt-tolerant varieties were obtained using this method (Farooq 2004). Although it has long endeavored with the method of wide crossing with halophytes to create wheat tolerant on salt in wide production, no varieties of wheat tolerating this abiotic stress were developed (Flowers 2004). Low yields of salt-tolerant lines were the main limitation of the introduction of these genotypes into production. Generally, the inclusion of bad characteristics from wild relatives in cultivated plant species was the main obstacle to using this approach for wheat breeders. In addition, the inclusion of tolerance genes to slaughtered soil has led to waterlogging sensitivity (Fedak 1999). Therefore, the use of CWR is much more effective in disease resistance breeding and other important traits under the control of a smaller number of genes.

Using throughput sequencing and genotyping, it is possible to obtain cross-specific sequence markers such as SNPs that can be used to saturate the genetic background of both parents (Wambugu et al. 2018). These markers allow to follow the intensity of introgression of specific alleles or genomic regions in the offspring (McNally et al. 2009). This technology enables successful regaining of the recurrent parent genome and introgression across only narrow segments of the wild species possessing only the desired gene.

The availability of molecular markers will greatly assist in reducing linkage drags and increasing the efficiency of introgression in pre-breeding programs. There are predictive models that are able to predict those SNP variations that are most likely to lead to undesirable phenotype (Xu 2010). Due to that breeders are able to eliminate genotypes with such SNP alleles from breeding material at the early stage. In addition to linkage drag, breeders are reluctant to use CWR because by including CWR there is a disruption of the favorable linkage blocks that have been created in the process of breeding.

Recently DNA barcoding has used for the correct identification of a plant (de Vere et al. 2015). DNA barcoding uses specific short sequences of DNA between 400 and 800 base pairs long which are easily isolated and characterized for all species (Hebert et al. 2003). DNA barcodes provide the ability to identify previously discovered and described plant species as well as the possibility of discovery of previously unknown plant species (Cowan et al. 2006). Four primary gene regions (*rbcL*, *matK*, *trnH-psbA*, and *ITS*) have generally been agreed upon as the standard DNA barcodes of choice in most applications for plants (Li et al. 2015).

Of all the above limitations, essentially funding is the main factor for more efficient use of CWR in breeding programs for cultivated crops. The use of CWR in breeding of cultivated crops is a long-term job that requires significant human and financial resources. Identification of the desired alleles in CWR, their isolation and transfer into the genetic basis of cultivated crops requires skilled professionals and intense commitment of the business whose results can sometimes be debatable. Since the use of CWR is not just a matter of food security and climate change adaptation but also the identification and preservation of biodiversity, funds can contribute to all of these ends at once. One of the possible financing of the pre-breeding programs can be through public-private partnerships, for example, in wheat, LOLA-WISP LOLA-WISP (http://www.wheatisp.org/) and WHEALBI (http://www.whealbi.eu/), and apples (https://sites.google.com/a/nordgen.org/ppp-apples/).

2.8 Conclusion

Lately, the flora of the planet, and thus the wild relatives of cultivated plants, is increasingly exposed to various types of natural hazards, of which climate change and habitat loss are the biggest threat. From the current point of view, it will be one of the most important obstacles for global food security in the coming decades. Due to climate change and other challenges (diseases and pests, water scarcity, land degradation, etc.), many regions around the world will face reductions in crop yields. In order to mitigate these effects, our crops should be genetically diversified. Such diversity is deposited in wild relatives of cultivated crops that are relatively closely related to domesticated crops and with different traits that can be transferred to crops. CWR can often be found in a wide range of habitats and in different habitat conditions and as such tend to possess higher levels of genetic diversity. High genomic diversity and expanded traits make CWR an excellent source for crop improvement in order to meet future food requirements. Therefore, the need to improve the adaptive capacity of crop tax for resistance to acute biotic and abiotic stress has never been greater. More intensive use of these resources would lead to an increase in agricultural productivity worldwide, and at the same time, it would be a very effective response to the increase in global demand for agricultural products. The use of this untapped genetic diversity for crop improvement acts as a truly appealing option that can be greatly accelerated with the use of modern molecular and genomic tools to identify the genetic properties of the CWR.

References

Andjelkovic V, Ristic D, Babic V, Dumanovic Z, Kravic N (2016) Maize landraces as a source for adaptation to climate change. Ratar Povrt 53(1):24–29

Anonymous (2007) The conservation of global crop genetic resources in the face of climate change. Summary Statement from a Bellagio Meeting organized by the Global Conservation Trust, held on September 3–7, 2007. Available from: http://www.croptrust.org/documents/WebPDF/Bellagio_final1.pdf

Assenov B, Andjelkovic V, Ignjatovic-Micic D, Vancetovic J, Nikolic A, Christov NK, Tsonev S, Abu-Mhadi N, Vassilev D, Muhovski Y, Ilchovska M, Todorovska E (2013) Identification of SNP mutations in *MYBE-1* gene involved in drought stress tolerance in maize. Bulgarian J Agric Sci 19:181–185

Babic V, Vancetovic J, Prodanovic S, Kravic N, Babic M, Andelkovic V (2015) Numerical classification of Western Balkan drought tolerant maize (*Zea mays* L.) landraces. J Agric Sci Technol 17:455–468

Bamberg JB, Hanneman RE (2003) Calcium rich potatoes: it's in their genes. Agricultural Research Magazine, March 2003

Barazani O, Perevolotsky A, Hadas R (2008) A problem of the rich, prioritizing local plant genetic resources for ex situ conservation in Israel. Biol Conserv 141:596–600

Baxter SW, Davey JW, Johnston JS, Shelton AM, Heckel DG, Jiggins CD, Blaxter ML (2011) Linkage mapping and comparative genomics using next-generation RAD sequencing of a non-model organism. PLoS One 6(4):e19315

Bayer MM, Rapazote-Flores P, Ganal M, Hedley PE, Macaulay M, Plieske J, Ramsay L, Russell J, Shaw PD, Thomas W, Waugh R (2017) Development and evaluation of a barley 50k iSelect SNP Array. Front Plant Sci 8:1792

Bilz M, Kell SP, Maxted N, Lansdown RV (2011) European red list of vascular plants. Publications Office of the European Union, Luxembourg

Brar DS, Khush GS (1997) Wide hybridization for rice improvements: Alien gene transfer and molecular characterization of introgression. In: Jones MP, Dingkhun M, Johnson DE, Fagade SO (eds) Interspecific hybridization: progress and prospect. Bouake, WARDA, pp 21–29

Breithaupt H (2008) Up to the challenge? Rising prices for food and oil could herald a renaissance of plant science. EMBO Reports 9:832–834

Cakmak I, Pfeiffer WH, McClafferty B (2010) Biofortification of durum wheat with zinc and iron. Cereal Chem 87:10–20

Castaneda-Alvarez NP, de Haan S, Juarez H, Khoury CK, Achicanoy HA and Sosa CC (2015) Ex situ conservation priorities for the wild relatives of potato (*Solanum* L. section petota). PLoS One 10: e0122599

Cavaliere C (2009) The effects of climate change on medicinal and aromatic plants. Herbal Gram (American Botanical Council) 81:44–57

CBD (2010) Global strategy for plant conservation. Secretariat of the Convention on Biological Diversity, Montreal. https://www.cbd.int/gspc/

Clark PU (2016) Consequences of twenty-first-century policy for multi-millennial climate and sea-level change. Nat Clim Chang 6(4):360–369

Clarke WE, Higgins EE, Plieske J, Wieseke R, Sidebottom C, Khedikar Y et al (2016) A high-density SNP genotyping array for *Brassica napus* and its ancestral diploid species based on optimised selection of single-locus markers in the allotetraploid genome. Theor Appl Genet 129:1887–1899

Colton ML, Groza IH, Wielgus MS, Jiang J (2006) Marker-assisted selection for the broad-spectrum potato late blight resistance conferred by gene RB derived from a wild potato species. Crop Sci 46:589–594
Comadran J, Kilian B, Russell J, Ramsay L, Stein N, Ganal M et al (2012) Natural variation in a homolog of Antirrhinum CENTRORADIALIS contributed to spring growth habit and environmental adaptation in cultivated barley. Nat Genet 44:1388–1392
Conway E (2008) What's in a name? Global warming vs. climate change. NASA. 5 December 2008
Cowan RS, Chase MW, Kress WJ, Savolainen V (2006) 300,000 species to identify: problems, progress, and prospects in DNA barcoding of land plants. Taxon 55:611–616
De Vere N, Rich TC, Trinder SA, Long C (2015) DNA barcoding for plants. Methods Mol Biol 1245:101–118
Dempewolf H, Hodgins KA, Rummell SE, Ellstrand NC, Rieseberg LH (2012) Reproductive isolation during domestication. Plant Cell 24:2710–2717
Dempewolf H, Eastwood RJ, Guarino L, Khoury CK, Muller JV, Toll J (2014) Adapting agriculture to climate change: a global initiative to collect, conserve, and use crop wild relatives. Agroecol Sustain Food Syst 38:369–377
Dempewolf H, Baute G, Anderson J, Kilian B, Smith C, Guarino L (2017) Past and future use of wild relatives in crop breeding. Crop Sci 57:1070–1082
Dias S, Dulloo ME, Arnaud E (2011) The role of EURISCO in promoting use of agricultural biodiversity. In: Maxted N, Dulloo ME, Ford-Lloyd BV, Frese L, Iriondo J, de Carvalho MAA P (eds) Agrobiodiversity conservation – Securing the diversity of crop wild relatives and landraces. CABI, Wallingford, pp 270–277
Diffenbaugh N, Krupke CH, White MA, Alexander CA (2008) Global warming presents new challenges for maize pest management. Environ Res Lett 3(4):1–9
Dudley N (2008) Guidelines for applying protected area management categories. IUCN, Gland
Dulloo ME, Labokas J, Iriondo JM, Maxted N, Lane A, Laguna E, Jarvis A, Kell SP (2008) Genetic Reserve Location and Design. In: Iriondo JM, Maxted N, Dulloo ME (eds) Conserving plant genetic diversity in protected area, CAB International, Wallingford, pp 23–64
Elshire RJ, Glaubitz JC, Sun Q, Poland JA, Kawamoto K, Buckler ES, Mitchell SE (2011) A robust, simple genotyping-by-sequencing (GBS) approach for high diversity species. PLoS One 6(5):e19379
Engels JMM, Ebert AW, Thormann I, de Vicente MC (2006) Centres of crop diversity and/or origin, genetically modified crops and implications for plant genetic resources conservation. Genetic Resour Crop Evol 53:1675–1688
ESA (2014) ESA report. www.seedworld.com
Eubanks WM (2006) A genetic bridge to utilize *Tripsacum* germplasm in maize improvement. Maydica 51:315–327
Evenson RE (2005) Besting Malthus, the Green Revolution. Proc Am Philos Soc 149:469–486
FAO (2008) Establishment of a global network for the in situ conservation of crop wild relatives: status and needs. FAO, Rome
FAO (2010) Second report on the world's plant genetic resources for food and agriculture. FAO, Rome, p 299
FAO (2011) The second global plan of action for plant genetic resources for food and agriculture https://goo.gl/uucTlL
FAO (2013) Report from Technical Workshop. Towards the establishment of a global network for in situ conservation and on-farm management of PGRFA. Rome, 13 November 2012. Food and Agriculture Organization of the United Nations, Rome Italy www.fao.org/agriculture/crops/thematic-sitemap/theme/seeds-pgr/onfarm-network/en/
FAO (2015) The state of food insecurity in the world. Meeting the 2015 international hunger targets: taking stock of uneven progress. FAO, Rome
Farooq S (2004) Salt tolerance in Aegilops species: a success story from research and production to large-scale utilization of salt-tolerant wheats. In: Taha FS, Ismaial S, Jaradat A (eds) Prospects

of saline agriculture in the Arabian peninsula. Amheerst Scientific Publishers, Massachusetts, pp 121–134

Farooq S, Azam F (2001) Production of low input and stress tolerant wheat germplasm through the use of biodiversity residing in the wild relatives. Hereditas 135:211–215

Fedak G (1999) Molecular aids for integration of alien chromatin through wide crosses. Genome 42:584–591

Feuillet C, Langridge P, Waugh R (2008) Cereal breeding takes a walk on the wild side. Trends Genet 24:24–32

Flint-Garcia AS (2013) Genetics and consequences of crop domestication. J Agric Food Chem 61:8267–8276

Flowers TJ (2004) Improving crop salt tolerance. J Exp Bot 55:307–319

Ford-Lloyd B, Kell SP, Maxted N (2008) Establishing conservation priorities for crop wild relatives. In: Maxted N, Ford-Lloyd BV, Kell SP, Iriondo JM, Dulloo ME, Turok J (eds) Crop wild relative conservation and use. CAB International, Wallingford, pp 110–119

Fukunaga K, Hill J, Vigouroux Y, Matsuoka Y, Sanchez J, Liu K, Edward S, Buckler SE, Doebley J (2005) Genetic diversity and population structure of teosinte. Genet 169:2241–2254

Fukuoka S, Saka N, Koga H, Ono K, Shimizu T, Ebana K, Hayashi N, Takahashi A, Hirochika H, Okuno K, Yano M (2009) Loss of function of a proline-containing protein confers durable disease resistance in rice. Science 21:998–1001

Gahlaut V, Jaiswal V, Tyagi BS, Singh G, Sareen S, Balyan HS, Gupta PH (2017) QTL mapping for nine drought-responsive agronomic traits in bread wheat under irrigated and rain-fed environments. PloS One 12(8):e0182857

Garnett T, Appleby MC, Balmford A, Bateman IJ, Benton TG, Bloomer P et al (2013) Sustainable intensification in agriculture: premises and policies. Sciences 341:33–34

Garrett KA, Dendy SP, Frank EE, Rouse MN, Travers SE (2006) Climate change effects on plant disease: genomes to ecosystems. Annu Rev Phytopathol 44:489–509

Gibson G, Dworkin I (2004) Uncovering cryptic genetic variation. Nat Rev Genet 5(5):681–690

Guarino L, Lobell DB (2011) A walk on the wild side. Nat Clim Chang 1:374–375

Hajjar R, Hodgkin T (2007) The use of wild relatives in crop improvement: a survey of developments over the last 20 years. Euphytica 156:1–13

Harlan JR, de Wet JMJ (1971) Toward a rational classification of cultivated plants. Taxon 20: 509-517

Harvey LT, Martin JT (1992) Resistance to the wheat curl mite (Acari: Eriophyidae) in common wheat. Cereal Res Commun 20:63–66

Hebert PDN, Cywinska A, Ball SL, de Waard JR (2003) Biological identifications through DNA barcodes. Proc R Soc B: Biol Sci 270: 313-321

Henry RJ, Nevo E (2014) Exploring natural selection to guide breeding for agriculture. Plant Biotechnol 12:655–662

Heywood VH (2011) Selection and prioritization of species/populations and areas. In: Hunter D, Heywood V (eds) Crop wild relatives, A manual of in situ conservation. Issues in Agricultural biodiversity. Earthscan, London, pp 129–168

Heywood VH, Dulloo ME (2005) In situ conservation of wild plant species: a critical global review of good practices. Food and Agriculture Organisation, Rome

Holmes M (2018) Somatic hybridization. The rise and fall of a mid-twentieth-century biotechnology. Hist Stud Nat Sci 48:1–23

Hoyt E, Brown S (1988) Conserving the wild relatives of crops. IBPGR, Rome

Hua L, Wang DR, Tan LB, Fu YC, Liu FX, Xiao LT et al (2015) LABA1, a domestication gene associated with long, barbed awns in wild rice. Plant Cell 27:1875–1888

Hulse-Kemp AM, Ashrafi H, Plieske J, Lemmj SK, Hill T et al (2016) A HapMap leads to a *Capsicum annuum* SNP infinium array: a new tool for pepper breeding. Hortic Res 3:16036

Hummer KE, Hancock JF (2015) Vavilovian centers of plant diversity: implications and impacts. Hortic Sci 50(6):780–783

Hunter D, Heywood V (eds) (2011) Crop wild relatives: a manual of in situ conservation. Biodiversity International, Rome

Iriondo JM, Dulloo E, Maxted N (2008) Conserving plant genetic diversity in protected areas: population management of crop wild relatives. CAB International Publishing, Wallingford

Iriondo JM, Maxted N, Kell SP, Ford-Lloyd BV, Lara-Romero C, Labokas J, Magos Brehm J (2012) Quality standards for genetic reserve conservation of crop wild relatives. In: Maxted N, Dulloo ME, Ford-Lloyd BV, Frese L, Iriondo JM, de Carvalho MAA P (eds) Agrobiodiversity conservation: securing the diversity of crop wild relatives and landraces. CAB International, Wallingford, pp 72–77

Ishimaru T, Hirabayashi H, Ida M, Takai T, San-Oh YA, Yoshinaga S, Ando I, Ogawa T, Kondo M (2010) A genetic resource for early-morning flowering trait of wild rice *Oryza officinalis* to mitigate high temperature-induced spikelet sterility at anthesis. Ann Bot 106(3):515–520

Jarvis A, Lane A, Hijmans RJ (2008) The effect of climate change on crop wild relatives. Agric Ecosyst Environ 26:13–23

Jena KK (2010) The species of the genus Oryza and transfer of useful genes from wild species into cultivated rice, *O. Sativa*. Breed Sci 6:518–523

Jha UC, Bohra A, Singh NP (2014) Heat stress in crop plants: Its nature, impacts and integrated breeding strategies to improve heat tolerance. Plant Breed 133:679–701

Jovovic Z, Kratovalieva S (2016) Global Strategies for sustainable use of agricultural genetic and indigenous traditional knowledge. In Salgotra, R.K. and Gupta, B.B. (Eds): Plant genetic resources and traditional knowledge for food security. Springer New York, 39-72

Jovovic Z, Dolijanovic Z, Kovacevic D, Velimirović A, Biberdzic M (2012) The productive traits of different potato genotypes in mountainous region of Montenegro. Genetika 44(2):389–397

Jovovic Z, Stesevic D, Meglic V and Dolnicar P (2013) Old potato varieties in Montenegro. Monograph, University of Montenegro, Biotechnical faculty Podgorica

Jovovic Z, Micev B, Velimirovic A (2016) Impact of climate change on potato production in Montenegro and options to mitigate the adverse effects. Acad J Environ Sci 4(3):047–054

Kell S, Maxted N, Frese L, Iriondo JM, Ford-Lloyd B, Kristiansen K, Katsiosis A, Teeling C, Branca F (2004) In situ conservation of crop wild relatives: a methodology for identifying priority genetic reserve sites http://aegro.jki.bund.de/

Kell SP, Maxted N, Bilz M (2012a) European crop wild relative threat assessment: knowledge gained and lessons learnt. In: Maxted N, Dulloo ME, Ford-Lloyd BV, Frese L, Iriondo JM, Pinheiro de Carvalho MAA (eds) Agrobiodiversity conservation: securing the diversity of crop wild relatives and landraces. CAB Int, Wallingford, pp 218–242

Kell SP, Maxted N, Frese L, Iriondo JM (2012b) In situ conservation of crop wild relatives: a strategy for identifying priority genetic reserve sites. In: Maxted N, Dulloo ME, Ford-Lloyd BV, Frese L, Iriondo JM, de Carvalho MAA (eds) Agrobiodiversity conservation: securing the diversity of crop wild relatives and landraces. CAB International, Wallingford, pp 7–19

Kell SP, Ford-Lloyd BV, Brehm JM, Iriondo JM, Maxted N (2017) Broadening the base, narrowing the task: prioritizing crop wild relative taxa for conservation action. Crop Sci 57:1042–1058

Khoury CK, Castañeda-Álvarez NP, Dempewolf H, Eastwood RJ, Guarino L, Jarvis A, and Struik PC (2016) Measuring the state of conservation of crop diversity: a baseline for marking progress toward biodiversity conservation and sustainable development goals. Crop Wild Relatives project policy brief, 6 p. http://hdl.handle.net/10568/7448

Kilian B, Graner A (2012) NGS technologies for analyzing germplasm diversity in genebanks. Brief Funct Genom 11:38–50

Kilian B, Martin W, Salamini F (2010) Genetic diversity, evolution and domestication of wheat and barley in the fertile crescent. In: Glaubrecht M (ed) Evolution in action. Springer, Berlin/Heidelberg/Berlin/Heidelberg, pp 137–166

Kovacs MIP, Howes NK, Clarke JM, Leisle D (1998) Quality characteristics of durum wheat lines deriving high protein from Triticum dicoccoides (6b) substitution. J Cereal Sci 27:47–51

Lam HM, Xu X, Liu X, Chen W, Yang G, Wong FL, Li MW, He W, Qin N, Wang B, Li J, Jian M, Wang J, Shao G, Wang J, Sun SS, Zhang G (2010) Resequencing of 31 wild and cultivated soybean genomes identifies patterns of genetic diversity and selection. Nat Genet 42:1053–1059

Larson G, Piperno DR, Allaby RG, Purugganan MD, Andersson L, Arroyo-Kalin M et al (2014) Current perspectives and the future of domestication studies. Proc Natl Acad Sci USA 111:6139–6146

Li XW, Yang Y, Hentry RJ, Rossetto M, Wang Y, Chen S (2015) Plant DNA barcoding: from gene to genome. Biol Rev Cambridge Philos Soc 90:157–166

Lira R, Téllez O, Dávila P (2009) The effects of climate change on the geographic distribution of Mexican wild relatives of domesticated *Cucurbitaceae*. Gen Res Crop Evol 56(5):691–703

Livaja M, Steinemann S, Schön C-C (2016) Sunflower polygalacturonase-inhibiting proteins (HaPGIP) are genetically conserved in cultivated sunflower (*Helianthus annuus* L.) but diverse in wild species. Mol Breed 36:17

Lobell DB, Burke MB, Tebaldi C, Mastrandrea MD, Falcon WP, Naylor RL (2008) Prioritizing climate change adaptation needs for food security in 2030. Science 319:607–610

Malik R et al (2003) Assessment of Aegilops tauschii for resistance to biotypes of wheat curl mite (Acari: Eriophyidae). J Econ Entomol 96:1329–1333

Mammadov J, Buyyarapu R, Guttikonda KS, Parliament K, Ibrokhim Y Abdurakhmonov YI, Siva P, Kumpatla PS (2018) Wild relatives of maize, rice, cotton, and soybean: treasure troves for tolerance to biotic and abiotic stresses. Front Plant Sci https://doi.org/10.3389/fpls.2018.00886

Martin TJ, Harvey TL, Bender CG, Seifers DL (1984) Control of Wheat streak mosaic virus with vector resistance in wheat. Phytopathology 72:963–964

Matsuoka Y, Vigouroux Y, Goodman MM, Sanchez J, Buckler E, Doebley J (2002) A single domestication for maize shown by multilocus microsatellite genotyping. Proc Natl Acad Sci USA 99:6080–6084

Maxted N, Kell SP (2009) Establishment of a global network for the in situ conservation of crop wild relatives: status and needs. Food and Agriculture Organization of the United Nations Commission on Genetic Resources for Food and Agriculture, Rome. Background study paper no 39

Maxted N, Ford-Lloyd BV, Jury SL, Kell SP, Scholten MA (2006) Towards a definition of a crop wild relative. Biodiv Conserv 15:2673–2685

Maxted N, Scholten MA, Codd R, Ford-Lloyd BV (2007) Creation and use of a national inventory of crop wild relatives. Biol Conserv 140:142–159

Maxted N, Kell S, Magos Brehm J (2011) Options to promote food security: on farm management and in situ conservation of plant genetic resources for food and agriculture. Food and Agriculture Organization of the United Nations, Rome

Maxted N, Avagyan A, Frese L, Iriondo JM, Magos Brehm J, Singer A, Kell SP (2015) ECPGR Concept for in situ conservation of crop wild relatives in Europe. Wild Species Conservation in Genetic Reserves Working Group. European Cooperative Programme for Plant Genetic Resources, Rome

McKey D, Elias M, Pujol B, Duputie A (2010) The evolutionary ecology of clonally propagated domesticated plants. New Phytol 186:318–332

McNally KL, Childs KL, Bohnert R, Davidson RM, Zhao K, Ulat VJ, Zeller G, Clark RM, Hoen DR, Bureau TE, Stokowski R, Ballinger DG, Frazer KA, Cox DR, Padhukasahasram B, Bustamante CD, Weigel D, Mackill DJ, Bruskiewich RM, Rätsch G, Buell CR, Leung H, Leach JE (2009) Genomewide SNP variation reveals relationships among landraces and modern varieties of rice. Proc Natl Acad Sci USA 106:12273–12278

MEA (2005) Ecosystems and human well-being: biodiversity synthesis. World Resources Institute, Washington, DC

Miller AJ, Gross BL (2011) From forest to field: Perennial fruit crop domestication. Am J Bot 98:1389–1414

Mladenović Drinić S, Anđelković V, Ignjatović-Micić D (2011) Genetic diversity of maize landraces as a source of favorable traits. In: Genetic diversity, Book 2. Caliskan M (eds) published by InTech, pp 89–112

Moore G (2015) Strategic pre-breeding for wheat improvement. Nat Plants 1(3):15018

Morrell PL, Buckler ES, Ross-Ibarra J (2011) Crop genomics: Advances and applications. Nat Rev Genet 13(2):85

Myer L, Klemick H, Guarino L, Smale M, Br AHD, Sadiki M, Sthapit B (2000) A training guide for in situ conservation on-farm. Version 1. International Plan Genetic Resources Institute, Rome

Nabhan GP (1990) Wild phaseolus ecogeography in the Sierra Madre Occidental, Mexico: Areographic techniques for targeting and conserving species diversity. Systematic and Ecogeographic Studies on Crop Genepools 5. International Board of Plant Genetic Resources (IBPGR)

Neeraja NC, Maghirang-Rodriguez R, Pamplona A, Heuer S, Collard CYB, Septiningsih ME, Vergara G, Sanchez D, Xu K, Ismail MA, Mackill JD (2007) A marker-assisted backcross approach for developing submergence-tolerant rice cultivars. Theor Appl Genet 115:767–778

Nikolic A, Ignjatovic-Micic D, Dodig D, Andjelkovic V, Lazic-Jancic V (2012) Identification of QTLs for yield and drought-related traits in maize: assessment of their causal relationship. Biotechnol Biotechnol Equip 26(3):2952–2960

Nikolic A, Andjelkovic V, Dodig D, Mladenovic Drinić S, Kravic N, Ignjatovic-Micic D (2013) Identification of QTL-s for drought tolerance in maize, II: Yield and yield components. Genetika 45(2):341–350

Olsen KM, Wendel JF (2013a) Crop plants as models for understanding plant adaptation and diversification. Front Plant Sci 4:290

Olsen KM, Wendel JF (2013b) A bountiful harvest: Genomic insights into crop domestication phenotypes. Annu Rev Plant Biol 64:47–70

Padulosi S, Heywood V, Hunter D, Jarvis A (2011) Underutilized species and climate change: current status and outlook. In: Yadav SS, Redden RJ, Hatfield JL, Lotze-Campen H, Hall AE (eds) Crop adaptation to climate change. Wiley, Published by Blackwell Publishing Ltd., Oxford

Pan Q, Liu YS, Budai-Hadrian O, Sela M, Carmel-Goren L, Zamir D, Fluhr R (2000) Comparative genetics of nucleotide binding site leucine rich repeat resistance gene homologues in the genomes of two dicotyledons: tomato and *Arabidopsis*. Genet 155:309–322

Park S, Kang T-S, Chang-Kil K, Han J-S, Kim S, Smith HR, Pike ML, Hirschi DK (2005) Genetic manipulation for enhancing calcium content in potato tuber. J Agric Food Chem 53:5598–5603

Parmesan C (2006) Ecological and evolutionary responses to recent climate change. Ecol Evol 37:637–669

Pavek J, Corsini D (2001) Utilization of potato genetic resources in variety development. Am J Potato Res 78:433–441

Peleg Z, Cakmak I, Ozturk L, Yazici A, Jun Y, Budak H, Korol BA, Fahima T, Saranga Y (2009) Quantitative trait loci conferring grain mineral nutrient concentrations in durum wheat × wild emmer wheat RIL population. Theor Appl Genet 119:353–369

Porch T, Beaver J, Debouck D, Jackson S, Kelly J, Dempewolf H (2013) Use of wild relatives and closely related species to adapt common bean to climate change. Agronomy 3:433–461

Prescott-Allen C, Prescott-Allen R (1986) The first resource: wild species in the North American economy. Yale University Press, New Haven

Przulj N, Perovic D (2005) Molecular markers. II Microsatellites. A Periodical of Scientific Research on field and vegetable crops 41: 299–312

Redden R, Yaday SS, Maxted N, Dullo ME, Guarino L, Smith P (2015) Crop wild relatives and climate change. Wiley-Blackwell, Hoboken

Rodrigues ASL, Andelman SJ, Bakarr MI et al (2004) Effectiveness of the global protected area network in representing species diversity. Nature 428:640–643

Root BA, Price JT, Hall K (2003) Fingerprints of global warming on wild animals and plants. Nature 421:47–60

Rotenberry JT, Preston KL, Knick ST (2006) GIS-based niche modeling for mapping species habitat. Ecology 87:1458–1464

Russell J, Mascher M, Dawson IK, Kyriakidis S, Calixto C, Freund F, Bayer M, Milne I, Marshall-Griffiths T, Heinen S, Hofstad A, Sharma R, Himmelbach A, Knauft M, van Zonneveld M, Brown JW, Schmid K, Kilian B, Muehlbauer GJ, Stein N, Waugh R (2016) Exome sequencing of geographically diverse barley landraces and wild relatives gives insights into environmental adaptation. Nat Genet 48:1024–1030

Salamini F, Özkan H, Brandolini A, Schäfer-Pregl R, Martin W (2002) Genetics and geography of wild cereal domestication in the near east. Nat Rev Genet 3:429–441

Sanchez PL, Wing RA, Brar DS (2013) The wild relative of rice: genomes and genomics. In: Zhang Q, Wing R (eds) Genetics and genomics of rice. Springer, Berlin, pp 9–25

Scheelbeek PFD, Bird FA, Tuomisto HL, Green R, Harris FB, Joy EJM, Chalabi Z, Allen E, Haines A, Dangour AD (2018) Effect of environmental changes on vegetable and legume yields and nutritional quality. Proc Natl Acad Sci U S A 115(26):6804–6809

Sharma S, Upadhyaya HD, Varshney RK, Gowda CLL (2013) Pre-breeding for diversification of primary gene pool and genetic enhancement of grain legumes. Front Plant Sci 4:309

Sheehy JE, Elmido A, Centeno G, Pablico P (2005) Searching for new plant for climate change. J Agric Meteorol 60:463–468

Shen X, FGJr G, Grosser JW (2011) Immature embryo rescue and culture. Methods Mol Biol 710:75–92

Skoric D, Jocic S (2004) Achievements of sunflower breeding at the IFVC in Novi Sad. In: Seiler GJ (ed). Proc. 16th Intl. Sunflower Conf., Fargo, ND, USA, 29 August–4 September 2004. Intl. Sunflower Assoc., Paris, France. II:441–448

Smýkal P, Nelson NM, Berger DJ, von Wettberg JBE (2018) The Impact of Genetic Changes during Crop Domestication. Agronomy 8:119. https://doi.org/10.3390/agronomy8070119

Standley PC (2015) Teosinte in Honduras. Revista Ceiba 1:58–61

Stolto S, Maxted N, Ford-Lloyd B, Kell SP, Dudley N (2006) Food stores: using protected areas to secure crop genetic diversity. The arguments for protection series. WWF, Gland

Storfer A, Patton A, Fraik KA (2018) Navigating the interface between landscape genetics and landscape genomics. Front Gen 13(9):68. https://doi.org/10.3389/fgene.2018.00068

Takeda S, Matsuoka M (2008) Genetic approaches to crop improvement: responding to environmental and population changes. Nat Rev Genet 9:444–457

Tanksley SD, McCouch SR (1997) Seed banks and molecular maps: unlocking genetic potential from the wild. Science 277:1063–1066

Tester M, Langridge P (2010) Breeding technologies to increase crop production in a changing world. Science 327(5967):818–822

Thomas CD, Cameron A, Green RE, Bakkenes M, Beaumont LJ, Collingham YC, Erasmus BFN, Ferreira De Siqeira M, Grainger A, Hannah L, Hughes L, Huntley B, Van Jaarsveld AS, Midgley GF, Miles L, Ortega-Huertas MA, Peterson AT, Phillips OL, Williams SE (2004a) Extinction risk from climate change. Nature 427:145–148

Thomas CD, Cameroon A, Green RE et al (2004b) Extinction risks from climate change. Nature 427:145–148

Thormann I, Parra-Quijano M, Endresen D (2014) Predictive characterization of crop wild relatives and landraces: technical guidelines version 1. Biodiversity International, Rome

Thuiller W, Lavorel S, Araújo MB, Sykes MT, Prentice IC (2005) Climate change threats to plant diversity in Europe. Proc Natl Acad Sci U S A 102:8245–8250

Tigchelaar M, Battisti DS, Naylor RL, Ray DK (2018) Future warming increases probability of globally synchronized maize production shocks. Proc Natl Acad Sci U S A 115(26):6644–6649

USGCRP, Hatfield J, Takle G, Grotjahn R, Holden P, Izaurralde RC, Mader T, Marshall E, Liverman D (2014) Ch. 6: Agriculture. Climate change impacts in the United States: the third national climate assessment. In: Melillo JM, Richmond T, Yohe GW (eds) U.S. Global Change Research Program, pp 150–174

van de Wiel C, Schaart J, Niks R, Visser R (2010) Traditional plant breeding methods. Rep. 338. Wageningen UR Plant Breeding, Wageningen, the Netherlands

van Heerwaarden J, Doebley J, Briggs WH, Glaubitz JC, Goodman MM, de Jesus Sanchez Gonzalez J, Ross-Ibarra J (2011) Genetic signals of origin, spread, and introgression in a large sample of maize landraces. Proc Natl Acad Sci U S A 108: 1088-1092

Vancetovic J, Mladenovic Drinic S, Babic M, Ignjatović-Micic D, Andjelkovic V (2010) Maize genebank collections as potentially valuable breeding material. Genetika 42(1):9–21

Vancetovic J, Ignjatovic-Micic D, Bozinovic S, Babic M, Filipovic M, Grcic N, Andjelkovic V (2013) Grain quality of drought tolerant accessions within a maize germplasm collection. Spanish J Agric Res 12(1):186–194
Vollbrech E, Sigmo B (2005) Amazing grass: developmental genetics of maize domestication. Biochem Soc Trans 33:1502–1506
van Vuuren DP, Sala OE, Pereira HM (2006) The future of vascular plant diversity under four global scenarios. Ecol Soc 11: 25
Wambugu WP, Ndjiondjop M-N, Henry JR (2018) Role of genomics in promoting the utilization of plant genetic resources in genebanks. Brief Funct Genom 17:198–206
Wan XY, Wan JM, Weng JF, Jiang L, Bi JC, Wang CM, Zhai HQ (2005) Stability of QTLs for rice grain dimension and endosperm chalkiness characteristics across eight environments. Theor Appl Genet 110:1334–1346
Wang J, Wan X, Crossa J, Crouch J, Weng J, Zhai H, Wan J (2006) QTL mapping of grain length in rice (*Oryza sativa* L.) using chromosome segment substitution lines. Genetic Res 8:93–104
Wang S, Wong D, Forrest K, Allen A, Chao S, Huang BE et al (2014) Characterization of polyploid wheat genomic diversity using a high-density 90,000 single nucleotide polymorphism array. Plant Biotech J 12:787–796
Warr A, Robert C, Hume D, Archibald A, Deeb N, Watson M (2015) Exome sequencing: current and future perspectives. G3: Genes Genomes Genet 5:1543–1550
Wright EM, Kelly JD (2011) Mapping QTL for seed yield and canning quality following processing of black bean (*Phaseolus vulgaris* L.). Euphytica 179:471–484
Wright S, Bi IV, Schroeder SG, Yamasaki M, Doebley JF, McMullen MD, Gaut BS (2005) The effects of artificial selection of the maize genome. Science 308:1310–1314
Xu Y (2010) Plant genetic resources: management, evaluation and enhancement. Mol Plant Breed (Wallingford, UK CABI):151–194
Xu J, Zhao Q, Du P, Xu C, Wang B, Feng Q, Liu Q, Tang S, Gu M, Han B, Liang G (2010) Developing high throughput genotyped chromosome segment substitution lines based on population whole-genome re-sequencing in rice (*Oryza sativa* L.). BMC Genom 11:656
Xu X, Liu X, Ge S, Jensen JD, Hu FY, Li X et al (2012) Resequencing 50 accessions of cultivated and wild rice yields markers for identifying agronomically important genes. Nat Biotech 30:105–111
Yadav SS, Redden R, Hatfield JL, Lotze-Campen H, Hall A (eds) (2011) Crop adaptation to climate change. Wiley-Blackwell, Chichester
Zamir D (2001) Improving plant breeding with exotic genetic libraries. Nat Rev Genet 2:983–989
Zeng N, Yoon J (2009) Expansion of the world's deserts due to vegetation-albedo feedback under global warming. Geophys Res Lett 36(17):17401
Zhang H, Mittal N, Leamy LJ, Barazani O, Song BH (2017) Back into the wild—apply untapped genetic diversity of wild relatives for crop improvement. Evol Appl 10(1):5–24
Zhao L, Dong Y, Liu B, Hao S, Wang K, Li X (2005) Establishment of a core collection for the Chinese annual wild soybean (*Glycine soja*). Chin Sci Bull 50:989–996
Zhou ZK, Jiang Y, Wang Z, Gou ZH, Lyu J, Li WY et al (2015) Resequencing 302 wild and cultivated accessions identifies genes related to domestication and improvement in soybean. Nat Biotech 33:408–414

Molecular Approaches for Harvesting Natural Diversity for Crop Improvement

3

Pawan Saini, Pooja Saini, Jashan Jot Kaur, Rose Mary Francies, Mudasir Gani, Ambavane Ajinkya Rajendra, Narender Negi, Ashok Jagtap, Amar Kadam, Charan Singh, and Shailendra Singh Chauhan

Contents

P. Saini (✉) · M. Gani · S. S. Chauhan
CSB–Central Sericultural Research & Training Institute (CSR&TI),
Pampore, Jammu and Kashmir, India
e-mail: pawansaini-coapbg@pau.edu

P. Saini
Department of Biotechnology, Eternal University, Baru Sahib, Himachal Pradesh, India

J. J. Kaur
Department of Plant Breeding and Genetics, Punjab Agricultural University (PAU),
Ludhiana, Punjab, India

R. M. Francies · A. A. Rajendra
Department of Seed Science and Technology, Kerala Agricultural University (KAU),
Thrissur, Kerala, India
e-mail: rose.mary@kau.in

N. Negi
ICAR–National Bureau of Plant Genetic Resources (NBPGR),
Shimla, Himachal Pradesh, India

R. K. Salgotra, S. M. Zargar (eds.), *Rediscovery of Genetic and Genomic Resources for Future Food Security*,
https://doi.org/10.1007/978-981-15-0156-2_3

Abstract

The evolution of new species revolves around the natural as well as induced genetic variation created through revolutionary and conservative forces. Evolution and domestication are the two important phenomenon which occurred in nature, while domestication acts as a signature of evolution for harnessing the important diversification traits in crop plants which plays an important role in global food security target through crop improvement. Conservation of natural genetic diversity is the utmost importance for tackling future biotic and abiotic threats for achieving global food and nutritional security. To feed the approximately 10 billion world population by 2050, it is necessary to increase the yield of staple crops up to 70–110%. Therefore, utilization of natural genetic diversity present in the form of wild relatives (CWR), landraces, and modern cultivars is indispensable for achieving food security which is of prime importance for crop improvement programs. Crop genetic resources with a broad genetic base are the valuable assets in crop plants which can be utilized through base-broadening

A. Jagtap
School of Agricultural Biotechnology, Punjab Agricultural University (PAU), Ludhiana, Punjab, India
e-mail: abj-coasab@pau.edu

A. Kadam
Department of Biotechnology, Lokmangal College of Agricultural Biotechnology, Mahatma Phule Krishi Vidyapeeth (MPKV), Solapur, Maharashtra, India

C. Singh
ICAR–Indian Institute of Wheat and Barley Research (IIWBR), Karnal, Haryana, India
e-mail: charan.singh1@icar.gov.in

approaches such as pre-breeding and core and mini-core collection. Besides these, a wide array of molecular approaches is available for exploitation and harnessing of natural diversity for achieving higher genetic gains for future food and nutritional security. Through this chapter, we have tried to discuss the domestication events with respect to genetic architecture in crop plants during the course of evolution, crop improvement for food and nutritional security, role of crop genetic resources and their importance, genetic bottlenecks responsible for narrowing down of genetic diversity, significance of conservation of natural variation, and base-broadening conventional approaches. Various molecular approaches (forward and reverse genetics) with genetic modification, genome editing, and sequencing methodologies for harnessing the natural variation for maximization of genetic gain and future scope of natural diversity from plant to crop with three Bs (biotechnology/biodiversity/biomimicry) and optimistic approach of conservation of diversity for future sustainable crop improvement are also discussed.

Acronyms

AFLPs	Amplified fragment length polymorphisms
AM	Association mapping
AMPRIL	Arabidopsis multiparent RIL
ATI	Accelerated Trait Introgression
BACs	Bacterial artificial chromosomes
BCP	Biofortification Challenge Program
BILs	Backcross inbred lines
Bt	*Bacillus thuringiensis*
CBD	Convention on Biological Diversity
CC	Core collection
CGD	Crop genetic diversity
CGRs	Crop genetic resources
CGIAR	Consultative Group for International Agricultural Research
CIAT	International Center for Tropical Agriculture
CIFOR	Center for International Forestry Research
CIMMYT	International Maize and Wheat Improvement Center
CIP	International Potato Center
CO_2	Carbon dioxide
CRISPR-cas9	Clustered regularly interspaced short palindromic repeats-associated protein 9
CSSLS	Chromosome segment substitution lines
CWR	Crop wild relatives
DArT	Diversity array technology
DH	Doubled haploid
DHPLC	Denaturing high-performance liquid chromatography

DNA	Deoxyribonucleic acid
DS	Domestication syndrome
EST	Expressed sequence tag
FAO	Food and agriculture organization
GDP	Gross domestic product
GBS	Genotyping by sequencing
GC	Gateway cloning
GEBV	Genomic estimated breeding value
GM	Genetic modification
GMOs	Genetically modified organisms
GS	Genome selection
GWAS	Genome-wide association studies
GWE	Genome-wide editing
GWP	Genome-wide prediction
HDB	Homology-directed recombination
HYVs	High-yielding varieties
IAEA	International Atomic Energy Agency
IARI	Indian Agricultural Research Institute
IBP	International Biological Program
ICARDA	International Center for Agricultural Research in the Dry Areas
ICRAF	International Council for Research in Agroforestry
ICRISAT	International Crops Research Institute for the Semi-Arid Tropics
IFPRI	International Food Policy Research Institute
IHF	Integration Host Factor
IITA	International Institute of Tropical Agriculture
ILRI	International Livestock Research Institute
IPCC	Intergovernmental Panel on Climate Change
IBPGR	International Board for Plant Genetic Resources
ICN	International Conference on Nutrition
ILO	International Labour Organization
IME	Institution of Mechanical Engineers
INIBAP	International Network for the Improvement of Banana and Plantain
IRRI	International Rice Research Institute
ITPGRFA	International Undertaking on Plant Genetic Resources for Food and Agriculture
IUCN	International Union for Conservation of Nature
IWMI	International Water Management Institute
LD	Linkage disequilibrium
MAB	Marker-assisted breeding
MABC	Marker-assisted backcrossing
MAGIC	Multiparent advanced generation intercross
MAGP	Marker-assisted gene pyramiding
MARS	Marker-assisted recurrent selection

MAS	Marker-assisted selection
MBC	Map-based cloning
NAFIS	National Bank for Agriculture and Rural Development All India Rural Financial Inclusion Survey
NAGS	National Active Germplasm Sites
NBPGR	National Bureau of Plant Genetic Resources
NERICA	New Rice for Africa
NHEJ	Nonhomologous end joining
NIDM	National Institute of Disaster Management
NILs	Near isogenic lines
NGS	Next-generation sequencing
PCR	Polymerase chain reaction
PGR	Plant genetic resources
PoU	Prevalence of undernourishment
QTL	Quantitative trait loci
RILs	Recombinant inbred lines
RAPD	Random amplified polymorphic DNA
RFLP	Restriction fragment length polymorphism
RGA	Rapid Generation Advancement
SADC	Southern African Development Community
SARP	Sequence-related amplified polymorphism
SCAR	Sequence-characterized amplified region
SNPs	Single nucleotide polymorphism
SSLPs	Simple sequence length polymorphisms
SSR	Simple sequence repeats
STRs	Short tandem repeats
STS	Sequence tagged site
TAC	Technical Advisory Committee
TALENs	Transcription activator-like effector nucleases
TILLING	Targeting induced local lesions in genomes
UCDP	Uppsala Conflict Data Program
UNCED	United Nations Conference on Environment and Development
UNEP	United Nations Environment Programme
UNHCR	United Nation Higher Commission for Refugees
USEPA	United States Environmental Protection Agency
VIR	N. I. Vavilov All-Union Scientific Research Institute of Plant Industry
WGP	Whole-genome prediction
WGS	Whole-genome shotgun sequence
WHO	World Health Organization
ZNFs	Zinc-finger nucleases

3.1 Introduction

Evolution refers to the origin of new species or organisms through a long-drawn and continuous process. The initial step involves the creation of variation followed by selection of those genotypes which tend to satisfy the conditions of maximum fitness. Usually, intermediate forms are most favored and extremes are promptly eliminated by natural selection – the prime agency of evolution. Those intermediate types establish themselves in diverse ecological niches and form new interbreeding groups called geographical races or subspecies, and these subspecies may be diverge and further develop into new species. Thus, the complex process of evolution revolves around the genetic variation created by evolutionary forces operating regularly on every organism. The evolutionary forces responsible for the creation of genetic variability in sexual organisms are revolutionary and conservative forces. Revolutionary forces produce genetic novelty by creating Mendelian variation due to mutation and recombination after hybridization. On the other hand, conservative forces preserve certain constellations of genes over a large number of generations by selection pressure, genotypic factors, and phenotypic factors. There are three main processes generating genetic variability in nature: (i) recombination of genes in the process of sexual reproduction, (ii) spontaneous mutation and polyploidy, and (iii) spontaneous hybridization between related plant taxa.

Mutation including polyploidy gives rise to new species at a single stroke; genetic drift and selection, and hybridization followed by recombination, which alter the gene frequency of the populations, are slow processes and speciation proceeds gradually. Among the conservative forces, the impact of natural selection on evolutionary processes is most profound in that it favors balanced genotypes with favorable gene combinations and preserves them. Selection plays a major role for accumulation of small genetic changes generated by mutation and also has stabilizing effect in polymorphic species. From this, it is clear that natural selection eliminates the unfit types and increases the frequency of desirable gene combinations which are essentials for progressive process of evolution.

Thousands of years ago, plant domestication and continued crop improvement via artificial selection generated the novel phenotypes that sustain human populations. Approximately 2500 plant species worldwide have undergone domestication with over 160 families contributing 1 or more crop species (Zeven and de Wit 1982; Dirzo and Raven 2003). Well-studied crops of major economic importance come from a subset of well-studied crops particularly model crops (i.e., crops that have had their genomes analyzed and are transformable) that have been critical for developing our fundamental understanding of domestication as a continuum of ongoing processes. Domestication episode took place in a short time window that separated two long-term historical processes – the preceding buildup of the perceptual and technological background and the later plant evolution under domestication processes. Domestication can be loosely defined as the bringing of a wild species under the management of human. It is a method of plant breeding which provides domestic types that are superior to ones previously available. Before the beginning of agriculture which itself is a multi-episode process, humans were hunter-gatherers and depend on wild species for fulfilling their requirements. The beginning of agriculture dates back 12,000 years ago, and till date, humans have domesticated more than hundreds of plant and animal species as sources of food, fiber, forage, and fuel to changeover from nomadic lifestyle of hunter-gatherers to sedentary lifestyle. As pointed out by Stabbins (1957), "… in some way plant breeding is merely a continuation of natural evolution of the crop species, changing its course in the direction of greater use of mankind." Evolution refers to the origin of new species or organisms through a long-drawn and continuous process which paralleled to domestication through crop improvement and crop diversification. Progress in understanding the crop evolution began with morphological studies and archeological findings of early domesticates. In the early 1900s, chromosome homology and after 1950s allelic variants of enzymes were used to investigate the origin of crop plants, but with the advancement in the molecular genetics and discovery of DNA-based molecular markers in 1980, DNA markers are widely used to distinguish origin of species at genetic level. These advances not only allow the investigation of the overall genetic architecture of the wild-crop transition but also make possible the identification of genomic regions and genes that were subjected to election during the evolution of various crops. In addition to this, recent genetic advances help evolutionary biologists to study on the issue of single or multiple domestication of crops.

3.2 Plant Domestication with Respect to Genetic Changes That Occurred in Crop Plants

Domestication has been seen as the major transition from gathering wild plants to cultivation involving increasing interaction between humans and the plants they used. The earlier archeological records represent the earliest cultivated types of plants such as barley and wheat which are Asian in origin. According to Harlan (1957), barley types have changed little in Egypt during the past 5000 years, and he believes the ancient Egyptians and Europeans obtained their cultivated plants from earlier plant breeders. Insights into the single or multiple origin of crop plants provide information whether there are single or multiple genetic paths leading to a phenotype as this will enable the evolutionary biologists to confirm the identity of crop progenitors and the number of times a particular crop was domesticated and to disentangle complex genetic consequences of domestication and crop improvement (Ballini et al. 2007). Nikolai Vavilov (1926) considered domestication of crop plants as single event. To date, maize (*Zea mays* L.) is a best characterized example of single domestication event from its wild relative teosinte under human influences in Mexico beginning around 9000 years before the present (year B.P.), traversed Central America by ~7500 year B.P., and spread into South America by ~6500 year B.P. (Kistler et al. 2018). Similarly, the most primitive domesticated wheat – einkorn (*Triticum monococcum* L.) (Heun et al. 1997) and two-rowed barley (*Hordeum vulgare* L.) (Badr et al. 2000) – exhibited single domestication event within its native range in the Northern and Western Fertile Crescent. There are several crop species which support multiple origins of domestication events such as rice (*Oryza sativa* L.)

(Londo et al. 2006), barley (Takahashi and Hayashi 1964), rajma (*Phaseolus vulgaris* L.) (Chacón et al. 2005), and *Cucurbita pepo* L. (Sanjur et al. 2002). The overall time required to domesticate a species has decreased since the earliest domestication events. The frequencies of some domestication syndrome traits like non-shattering have decreased over time, while others like secondary metabolites changes have increased.

During the process of domestication, domesticated plants dramatically underwent subsequent genetic changes in several morphological, physiological, developmental, or biochemical traits such as loss of seed dormancy, reduced seed dispersal rate, seed shattering, reduced branching, plant habit, color and shape of harvesting organ, flavor, palatability, and cooking attributes (Harlan 1992; Ladizinsky 1998; Doebley et al. 2006; Miller 2007; Purugganan and Fuller 2009; Gross and Olsen 2010; Sakuma et al. 2011; Olsen and Wendel 2013) that differ from their wild progenitors. Such types of traits are designated as domestication traits, and these will underlay the concept of domestication syndrome (DS) which is defined as the set of characters that distinguishes the crop plants from its wild progenitors. These traits arise at least in part from human selection and hence relate to ways in which the plants are cultivated and harvested (Brown et al. 2008). The morphophysiological and biochemical changes between domesticated and wild progenitors cannot be attributed to ancient domestication episode as crop plants are dynamic genetic entities in nature (Evans 1993; Harlan 1995; Ladizinsky 1998; Abbo et al. 2012). From this, it is clear that differentiation between domestication traits is associated with the domestication process and further changes resulted from crop diversification and domestication (Doebley et al. 2006; Bruke et al. 2007; Pickersgill 2009; Hufford et al. 2012; Abbo et al. 2012; Olsen and Wendel 2013; Larson and Burger 2013; Ronfort and Glemin 2013). The subsequent genetic changes in these plants resulting in the domestication of some of these cultivated species reflect the genius of early farmers, who were the first plant breeders. Although we may not know when man becomes a plant breeder, we can be sure that nature has always been one.

For major cereal crops, domestication syndrome can be divided into seven components, viz., loss of seed dispersal, loss of grain dispersal, increase in grain size, loss of sensitivity to environmental cues for germination and flowering, synchronous flowering, compact growth habit, and enhanced culinary chemistry. In understanding the genetic changes involved in domestication, it would be particularly helpful to determine the relative roles played by natural as well as artificial selection. Heritable variations certainly occurred both before and after the beginning of cultivation and domestication. When the objective is to keep the progeny of some plants while culling out of the progeny of others, on the other hand, natural selection helps in improving the adaptedness and survivability in heterogeneous populations in the environmental conditions in which successive generations are grown (Allard 2013). In any case, natural and artificial selection acting together have provided the modern plant breeder with a liberal heritage of plant materials under the conditions of cultivation.

In general, domestication occurs in response to selection by artificial predominantly and by natural selection up to some extent in different regions with seasonal climates (Pickersgill 2018). There is a clearly defined and well-accepted conceptual framework between domestication and improvement-related traits which depend on the genetic changes associated with crop domestication and crop evolution (Abbo et al. 2014). Improvement-related traits are also associated with genetic changes. The population genetics to find agronomically important genes independently of phenotype is popularly known as the "bottoms-up" approach (Ross-Ibarra et al. 2007) or alternately as "sweep-mapping." Improvement traits are defined as any plant trait that was not imperative for domestication and did not limit the adoption and management of the respective species by humans. Such traits are evolved at any time after the ancient domestication episode. These traits are the result of crop evolutionary processes under domestication. Crop improvement traits include growth habit, higher yield, disease and pest resistance, grain quality, and adaptations to new growing environments. Both domestication and improvement traits are widely accepted by the breeders as there is no such classification criteria (Sakuma et al. 2011; Dempewolf et al. 2012; Olsen and Wendel 2013; Ronfort and Glemin 2013; Sang and Ge 2013; Meyer and Purugganan 2013; Lenser and Theißen 2013). To distinguish between domestication and improvement traits, the concept of crucial domestication traits is proposed. Domestication syndrome (DS) traits are very useful for crop domestication which differentiates between extant domesticated germplasm and wild gene pool of respective crop plant; therefore, crucial domestication traits, without which the adoption of a species for domestication would be impossible, are important for our understanding of plant domestication and crop evolution (Bruke et al. 2007; Pickersgill 2009). Perennial crop domestication increased 2000–3000 ya, and, correspondingly, domestication traits related to aerial vegetative parts, fruit morphology, and secondary metabolites also increased sharply. Few traits include loss of seed dormancy in grain legumes (Ladizinsky 1987, 1993; Abbo et al. 2011) and loss of bitterness in cucurbitaceae and almond (Heppner 1923; Paris and Brown 2005; Ladizinsky 1999; Zhang et al. 2013).

3.3 Parallel vs Convergent Evolution in Crop Plants with Respect to Genetic Architecture in Domestication Process

The new species or organisms evolved are the result of selective pressure which led to the independent evolution of similar or distantly adaptive traits through parallel or convergence evolution phenomenons. Pickersgill (2018) clearly explained both parallel and convergent evolution phenomenon at phenotypic, genetic, and nucleotide sequence. Parallel evolution refers to be independent development of similar phenotypic traits in closely related taxa at phenotypic level. If genes were selected in parallel, the number of genetic solutions to the challenge of domestication would be under constraint. However, considering an experimental evolution assay, considering it an experimental evolution assay, there is no evidence for parallel selection events either between species (maize and rice) or within species (two domestication events within beans) (Gaut 2014). At genetic level, it may be viewed as similar types of phenotypes are produced by orthologous genes (homologous genes) which diverged from a common ancestor and similar phenotypes are produced by identical changes in the same gene at nucleotide sequence. Similarly, convergent evolution is also defined as similar phenotypes occur in distantly related taxa (phenotypic level); similar phenotypes are produced by different, nonhomologous genes (genetic level); and similar phenotypes are produced by different changes in the same gene at nucleotide level. There was so much argument by the several researchers whether parallel or convergent evolution occurred with domestication and parallelism between parallel and convergent evolution in crop domestication. Glémin and Bataillon (2009) and Martinez-Ainsworth and Tenaillon (2016) reached the same conclusion that very little information is available for the parallelism among the DS traits at genetic level of evolution. Poncet et al. (2004) and Lenser and Theißen (2013) considered that the changes occurring in the DS traits are governed by orthologous genes when domesticated species are belonging to the same taxa. On the other hand, Sang (2009) argued that the pleiotropic genes are behind the control of DS traits. But, Gaut (2015) finally kept the questions open for all that up to what extent the parallel evolution occur during domestication of different crops. There are some traits which upon domestication were later treated as diversification traits such as loss of bitter compounds (Abbo et al. 2014).

Qualitative traits governed by few genes (regulatory genes) domesticate from their wild progenitors (Doebley et al. 2006; Sang 2009; Martinez-Ainsworth and Tenaillon 2016). On the other hand, quantitative traits such as crop yield are governed by a number of polygenes (QTLs) which rapidly respond to selection (Poncet et al. 2004). Most of the domestication traits of cereal crops are conditioned by monogenic/digenic, recessive, or loss-of-function alleles (Ladizinsky 1987; Lester 1989; Burger et al. 2002; Kaga et al. 2008; Zhang et al. 2013). According to Pickersgill (2018), there are few traits which are associated with domestication process in one crop and on the other hand the same trait associated with diversification process in other crop which means development of variants within a crop that are adapted to different uses by humans or to different agricultural environments. Within Solanaceae family, there is a parallel loss of abscission zone which is evident from mutation in *JOINTLESS* gene which suppresses development of abscission zone which leads to loss of dispersal which acts as a domestication syndrome trait in chilies as well as a diversification trait in tomato (Mao et al. 2000). There are few examples which distinguish between parallel and convergent evolution under domestication process. Increased size is a well-documented domestication syndrome trait in vegetable crops. Increased size depends upon increased number of cells in primordium, increased cell division in the organ under selection, and increased cell expansion. Increased size depends upon the interaction between genes responsible for proliferation, i.e., *WUSCHEL (WUS)* and *CLAVATA3 (CLV3)*, which promotes cell differentiation. The *WUS-CLV3* signaling pathway regulates the increased size in various crops (Somssich et al. 2016). In tomato, an orthologue gene of *WUS (SlWUS)* is responsible for increased locule number gene (*LOCULE NUMBER*) that increases the fruit size (Van der Knaap et al. 2014). Generally, selection under domestication resulted in larger fruits with more locules in *Capsicum*. Similar to tomato, an orthologue QTL is reported by Barchi et al. (2009) underlying the parallel evolution of increased fruit size in capsicum. In maize, ZmCNR1, an orthologue gene of *FW2.2*, acts in a similar way by controlling the number of cells acting as negative regulator of cell division in tomato as *FW2.2* is a member of *CELL NUMBER REGULATOR (CNR)* gene family. Like maize in cherry (Blanca et al. 2015) and avocado (Dahan et al. 2010), also CNR orthologues are reported for increase in cell number which reveals the parallel evolution in convergent species. Reduced branching is another domestication trait in maize which principally involved *TEOSINTE BRANCHED1 (TB1)* and *GRASSY TILLERS 1 (GT1)* genes. In other crop plants like rice, pea and Arabidopsis targeted strigolactone signaling but not in maize. From this, it is concluded that selection under domestication for reduced branching in other cereals can be achieved by other pathways (Guan et al. 2012). From the abovementioned examples, we can draw inference that phenotypes evolved parallel among the different crop plants from convergent evolution at genetic level under domestication.

3.4 Crop Improvement for Global Food and Nutritional Security in the Twenty-First Century

Advancements in technology have been taking place since the beginning of mankind. From basic stone implements like hammerstones, stone cores, and sharp stone flakes in the early Stone Age to Acheulean hand axes, cleavers, and large cutting tools roughly 1.8 million years ago, mankind is witness to an array of technological advancements. Somewhere down the line, the pace of advancement in technology has been exceptionally rapid, so to say, in the twentieth and twenty-first century, technology advanced in leaps and bounds in the field of science, medicine, and agriculture. Despite these advancements, hunger is on the rise. Enough food for all is but a far cry in today's world. Exploding population, incessant instability in conflict-ridden regions, adverse climate events, global warming, increasing pollution, diminishing resources and cultivable land, nonavailability of labor, and economic slowdowns are but a few challenges that are threatening agriculture production and have worsened food and nutritional security today.

The term food security was used in the 1960s and 1970s to refer to the ability of a country or region to assure adequate food supply for its current and projected population. During this period, international and national efforts were focused on growing more food and reducing population growth rates to sustainable levels. Food grain production in sufficiency to ward off famine, improving availability and access to food at affordable cost, to meet the energy requirements, and to prevent chronic undernutrition among the ever-growing population was considered an ordered reference standard of food security. However, over the decades, although there has been a reduction in severe acute food insecurity, it became evident that dietary intake in large segments of the population does not meet energy (hunger) and micronutrient (hidden hunger) requirements, and consequently, undernutrition and micronutrient deficiencies are widespread (Ramachandran 2013).

Taking these into account, the International Conference on Nutrition (ICN) held in Rome in 1992 and later the World Food Summit in 1996 redefined food security in its most basic form as "physical, social and economic access by all people at all times to sufficient, safe and nutritious food which meets their dietary needs and food preferences for an active and healthy life" (FAO 1996c). Therefore, food insecurity arises when access to safe and nutritious food in sufficient amounts required for normal growth and development for leading an active and healthy life is lacking.

Food security and nutritional security are different yet they are intertwined. Availability, accessibility, and utilization of food are the key to food security. The emphasis on "utilization" in the definition of "food security," however, underlines the importance of nutritional security. The reasons for food insecurity may be nonavailability of food, lack of purchasing power, unequal distribution, and poor utilization at the household level. However, in addition to food insecurity, poor health and sanitation and inappropriate socioeconomic status may lead to poor nutritional security.

The number of undernourished people, i.e., those facing chronic food deprivation, has increased to nearly 821 million in 2017 with the share of undernourished people in the world population reaching 10.9% in 2017, from around 804 million in 2016 (SOFI 2018). India has been self-sufficient in food production since the seventies and has low household hunger rates. However, seasonal food insecurity is seen in different pockets of the country even to this day. There have been a substantial reduction in severe grades of undernutrition and micronutrient deficiencies and some improvement in the nutritional status of all the segments of the population. Although, in the last five decades, the rate of decline in undernutrition has been slow, the mortality rate has come down by 50% and the fertility rate by 40%, but the reduction in underweight rates is only 20% (Ramachandran 2013).

3.5 Challenges Faced by Agriculture Sector in the Twenty-First Century

3.5.1 Exploding Population

Population explosion is imposing a lot of pressure on already limited resources and on agriculture. Uncontrolled growth of population and increased life expectancy are keys to the rising population. In 1900, the life expectancy of human being was around 47 years, while it was 58 years in 1950. However, with the advancement in health and medical care today, the world average life expectancy touches almost 73 years (World Bank 2018a, Fig. 3.1). Scientists project that the average age could reach 100 years before the end of the century.

Between 1900 and 2017, the world population increased from 1.5 to 7.53 billion, i.e., three times greater than that during the entire previous history of humankind (World Bank 2018b, Fig. 3.2). It is estimated that more than 108 billion humans were born between this period (Kaneda and Haub 2018), indicating that world's population size today makes up 6.5% of the total number of people ever born.

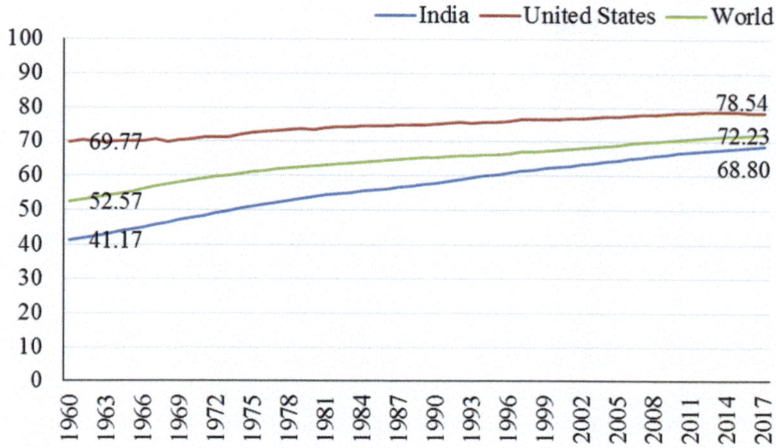

Fig. 3.1 Life expectancy. (Source: World Bank 2018a)

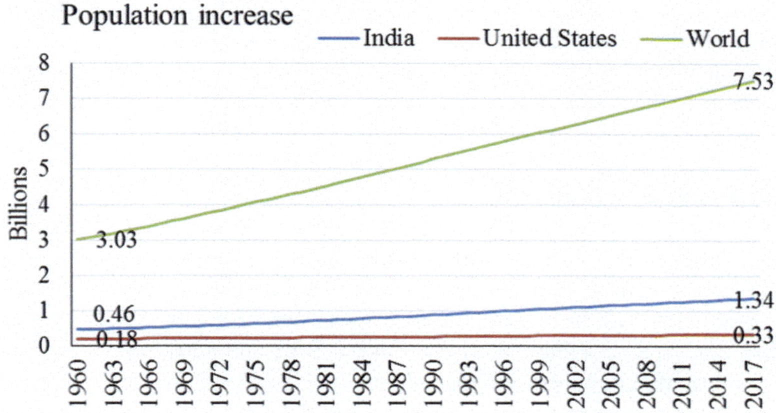

Fig. 3.2 Population growth from 1960 to 2017. (Source: World Bank 2018b)

Projections estimate that the population will reach 9 billion in 2038 and by 2056 the number will cross the 10 billion mark.

This scenario of population explosion warrants more than doubling of world food production. The main hurdle faced by the agriculture sector in achieving this goal is that the population growth rate is faster than the rate of increase in food productivity. The rate of increase in grain production is 0.5% per year, which is way below the rate of population growth which is estimated at 1.5% per year (Heszkyl 2008). In the backdrop of increasing population and diminishing resources, improvement in crop productivity to feed the increasing hungry mouths and the production of nutritious and healthy food are the needs of the hour world over.

3.5.2 Climate Change and Extremities

3.5.2.1 Global Scenario

Climate change is no more a prediction but a reality. Emission of toxic gases in the atmosphere and global warming, deforestation, and other man-made activities have led to a change in climate on a global scale. The forest cover helps intercept rainfall allowing water infiltration in the soil. However, rampant deforestation around the globe has caused precipitation to reach across the land eroding topsoil often leading to floods. The indiscriminate felling of trees has ironically worsened the drought in dry years too owing to quick drying off of soil in the absence of tree cover.

Climate variability and extremities are negatively undermining all dimensions of food security, viz., food availability, access, utilization, and food safety. The occurrence of disastrous climate-related events such as extreme heat, droughts, floods, storms, etc. is reported to have doubled since the 1990s, with an estimated 213 disasters occurring every year between 1990 and 2016 (SOFI 2018). These cripple food security by drastically affecting agricultural productivity leading to shortfalls in food availability, which in turn cause income losses and hikes in food price reducing people's access to food. Hence, addressing the challenges of agriculture and food production even under extreme climate events is critical to ensure food security and nutrition indicators. The Indian scenario is no exception. Indian agriculture is vulnerable to the extremities of fluctuating weather conditions, and the forthcoming menace of climate change will aggravate this vulnerability even further.

3.5.2.2 The Indian Scenario

In many regions of the country, climate extremes have increased in number and intensity, particularly where average temperatures are shifting upward: very hot days are becoming more frequent, and the hottest days are becoming hotter. Extreme heat is associated with increased mortality, lower labor capacity, lower crop yields, and other consequences that undermine food security and nutrition.

Production of major crops without adaptation is expected to worsen as temperatures increase and become more extreme. From December 2002 to January 2003, northern states of India had experienced severe cold wave. This cold wave caused considerable damage to all crops grown in those regions. Crops like brinjal, guava, mustard, papaya, potato, tomato, etc. suffered a severe yield loss that varied between 10 and 100%. The damage was more extreme in low-altitude areas where cold air settled and remained for a longer time (Samra and Singh 2003).

In addition to variable temperatures, the country is experiencing changes in the amount of rainfall received, the rainfall intensity, the nature of rainy seasons, and specifically the timing of precipitation events. In the Rabi season of 2007, 15,000 hectares (Ha.) of the wheat crop was destroyed over Haryana, Punjab, and UP due to heavy rains and hailstorms. Heavy rains again in September in Andhra Pradesh, Karnataka, and Kerala led to floods. A huge crop loss was noticed in several states of the country due to floods in Kharif, 2007, and thus the year 2007 was declared as the flood year in India.

Floods are reported to lead to more climate-related disasters globally than any other extreme climate event. According to SOFI (2018), the increase in the occurrence of flood-related disasters was the highest reaching 65% over the last 25 years. Heavy flooding in June 2013 as the result of severe rains and melting glaciers in the Himalayas created heavy destruction in Uttarakhand, as the main river, the Ganges, swelled and washed away all the crop and agricultural belongings from the affected areas. All the summer crops of the affected area have been washed, and the sowing season for rice has been delayed as a result of the heavy accumulation of paddy fields with rainwater and by landslides. Around 2010 ha cropped area was affected, whereas 1209 ha of land was completely washed away (NIDM 2015).

In 2018, Kerala has suffered the massive and worst flood that occurred in the last century. The devastating flood caused extensive damage to agriculture in flood-affected districts of Kerala. All the major crops grown in the state were severely affected. Around 26,106 ha of rice-farming area was damaged. The flood had also delayed the farming activities in major rice production districts which is expected to bring down the annual paddy production in the state. The flooding has also affected cardamom and black pepper plantations, rubber, and tea with an estimated 500 acres of plantation land having been destroyed due to landslides in Nilambur, Malappuram, and Kalikavu districts (Holland 2018). Hence, these climate change events and extremities destabilize the agriculture production, thereby affecting food and nutritional security of the country.

3.5.3 Increase in Greenhouse Gas Emission

Increase in the emission of greenhouse gases, especially gases like methane, halocarbons, black carbon, etc., as a result of the uncontrolled man-made activity is a big threat not only to Indian agriculture but also at the global level. Greenhouse gases from human activities are the most significant driver of observed climate change since the mid-twentieth century. Worldwide, emissions of greenhouse gases from human activities increased by 35% from 1990 to 2010. Emissions of carbon dioxide, which account for about three-fourths of total emissions, are estimated to have increased by 42% over this period (USEPA 2017).

Methane emissions (mainly from natural gas production and livestock) are very harmful to plants because the gas increases surface ozone that causes harmful chlorosis or yellowing of the leaves (EOS 2019). The halocarbons (used in refrigeration and air-conditioning) and black carbon (from the burning of fossil and biomass fuels) all contribute to climate change to varying degrees, thereby affecting the agriculture adversely. The effect of carbon dioxide on agriculture is more complicated than other gases. Being a raw material for the process of photosynthesis, initially increase in the amount of carbon dioxide was beneficial in accelerating photosynthesis of the plants, which in turn initially increased yields as well. However, as carbon emissions continue to contribute to the greenhouse effect, the overall impact becomes negative.

Assessments by the Intergovernmental Panel on Climate Change (IPCC) suggest that the Earth's climate warmed 0.85 degrees centigrade (1.53 °F) between 1880 and 2012. Warming of the atmosphere and the ocean leads to changes in the global water cycle, melting the snow and ice resulting in global mean sea level rise and climate extremes (IPCC 2013) in turn leading to food and nutritional insecurity worldwide.

3.5.4 Urbanization

Urbanization in the developing country like India has increased exponentially during the last few decades due to express changes in economic activities, rapid industrialization in urban areas, and migration of rural population to urban centers for better employment and livelihood (Ellis and Roberts 2016). Rural migration toward the cities in many instances is triggered by challenges like reduction in land, water, and resource availability, low socioeconomic security, poor productivity and economic returns from agriculture, and the attraction of increased wage rate in cities, to cite a few. The land available for agriculture is declining very rapidly owing to urbanization along with infrastructure expansion and industrialization in rural areas (Sharma 2015). Concurrently, the percent of consumers (not producing food population) is increasing with urbanization, whereas the number of farming communities is declining rapidly (Satterthwaite et al. 2010). This, in turn, has widened the gap between demand and supply of food effectively pushing the food prices higher. High food prices have an adverse influence on the livelihoods of rural as well as urban poor and increasing hunger and deprivation thereby threatening food security.

3.5.5 Availability of Arable Land

The decline in agricultural land has been mainly due to the diversion of arable land for nonagricultural purposes such as urbanization, roads, industries, and housing. In 1961, the total agricultural land in the world was 35.64%, whereas it was 58.84% and 48.86% in India and the USA, respectively. An increase in the area available for agricultural activities was on a rise till 1991. However, later to 1991, a sudden decline in total agricultural land was observed on a global scale (World Bank 2018c, Fig. 3.3).

In India, the area under for nonagricultural purposes, primarily for housing, infrastructure, and industry, has increased from 21.3 million ha in 1991–1992 to 26.3 million ha in 2011–2012 which is almost 23% increase. This led to a total loss of 3.16 million ha of agricultural land. In addition, land fragmentation because of social reasons and declining landholdings along with low levels of technology have kept agricultural productivity at low levels.

3.5.6 Low Annual Crop Production Growth Rate

As discussed earlier, the population growth rate has shown an increasing trend over the last century. Consequently, there has been tremendous pressure on the farming

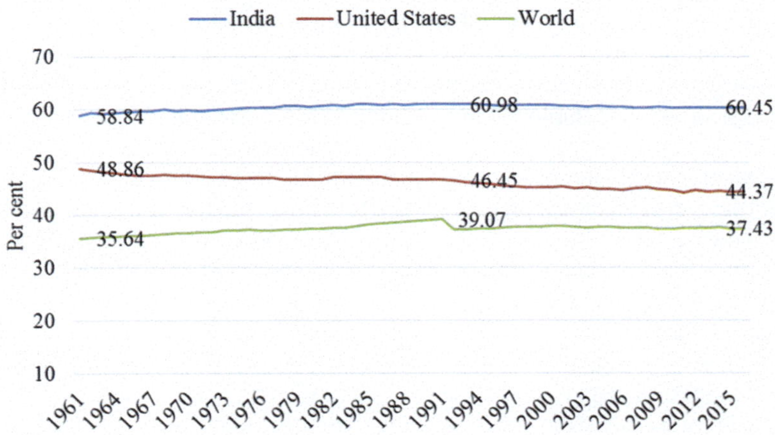

Fig. 3.3 Total agricultural land (% of total land area)

Table 3.1 Annual crop production growth rate

Particulars	1969–99	1979–99	1989–99	1999–2015	2015–30
	Percent				
A. All developing countries	3.1	3.1	3.2	1.7	1.4
Excl. China and India	2.7	2.6	2.5	2.0	1.7
Sub-Saharan Africa	2.3	3.3	3.3	2.6	2.5
Near East/North Africa	2.9	2.9	2.6	1.8	1.5
Latin America and the Caribbean	2.6	2.3	2.6	1.8	1.6
South Asia	2.8	3.0	2.4	2.1	1.5
East Asia	3.6	3.5	3.7	1.3	1.1
B. Industrial countries	1.4	1.1	1.6	0.9	0.9
C. World	2.1	2.0	2.1	1.5	1.3

sector to keep pace with the growth rate of the population to produce more and more food. However, the estimates indicate insufficient crop production growth rate. The crop production growth rate was considerably high between 1969 and 1999 but reduced noticeably by 1999–2015 and is expected to reduce further by 2030 (Table 3.1).

3.5.7 Labor Availability for Agricultural Activities

Agriculture has been the mainstay of livelihood and employment world over until 2000. Off late, however, the service sector has taken over this mantle. Although the employment growth in agriculture slowed down, the number of individuals engaged in this sector reached over a billion in 2009. In sub-Saharan Africa, agricultural employment accounted for half of all employment growth between 1999 and 2009, while in South Asia, nearly 33% of all employment growth since 1999 was in agriculture. In juxtaposition, agricultural employment fell in the developed economies, East Asia and Latin America and the Caribbean regions. At the global level, the percentage of women employed in agriculture (38%) was found to be higher than men (33%). As per the International Labour Organization (World Bank 2018d), the employment in agriculture (%) on a global scale fell to 28.305% in 2018.

Even in a developed country like the USA, this situation is nothing different. About a quarter of all farmers in the USA were in between the ages of 25 and 35 during 1965, and only about 10% of farmers were above 65. Around 70–80% population of the USA was in farming and agriculture allied activities. However, by the end of 2007, that ratio had flipped to adverse condition with more than 30% farmers above 65 and only around 5% were 25–35, and farming is no longer a key employer (Good Food World 2011).

Agriculture sector, being the backbone of the Indian economy, contributes around 17–18% to the country's GDP. Several new techniques of agriculture have also become prevalent. Despite this, the predicament of Indian farmers is getting worse with each passing year. This makes farmers, especially the small and marginal ones, to take the easier route of making money by either renting or selling their lands to corporates. The National Crime Records Bureau of India reported that a total of 296,438 Indian farmers had committed suicide since 1995. Of these, 60,750 farmer suicides were in the state of Maharashtra only. The National Bank for Agriculture and Rural Development All India Rural Financial Inclusion Survey (NAFIS) shows that average agriculture household income was very meager (Rs. 8931) per month in 2016–2017. It is not surprising that the farmer who feeds our entire nation is dying of hunger and malnutrition. According to the World Bank estimates, half of the Indian population would be urban by the year 2050. It is estimated that the percentage of agricultural workers in the total workforce would drop to 25.7% by 2050 from 58.2% in 2001. Thus, there is a need to enhance the level of farm mechanization in the country (ETMarkets 2018).

3.5.8 Postharvest Losses

Food losses can be quantitative as in terms of loss in weight or volume of edible grain or food available for human consumption or can be qualitative, such as deterioration of its nutrient value and undesirable changes in taste, color, texture, etc. of food (Buzby and Hyman 2012). Spillage of grain during storage and transportation, consumption by pets, and loss in moisture content due to overdrying lead to the quantitative loss (FAO 1980), whereas the qualitative loss occurs due to the incidence of insect pest, mites, rodents, and birds; improper handling; chemical deterioration of fat, carbohydrates, and protein; or contamination of fungal growth, pesticide residues, insect fragments, or excreta of rodents and birds and their dead bodies. This qualitative deterioration makes food unfit for human consumption and thus contributes to food loss.

Food waste is a subcategory of food losses (Buzby and Hyman 2012). Food waste is more of a social issue and occurs when the cooked food or edible food item goes unutilized as a result of human action or inaction (Bloom 2010; Buzby and Hyman 2012). In the UN's SAVE FOOD initiative, the FAO and UNEP agreed to the definition of food waste as any removal of food from the food supply chain which is or was at some point fit for human consumption, or which has spoiled or expired, mainly caused by economic behavior, poor stock management, or neglect. A 2013 report from the Institution of Mechanical Engineers (IME) estimated that 30–50% (or 1.2–2 billion tons) of all food produced remains uneaten (IME 2013). This is, in fact, a serious threat to food security.

3.5.9 Nutrition Security-Related Challenges

The green revolution and the subsequent revolutions like blue, golden, pink, red, round, silver, white, yellow, and evergreen revolution have successfully addressed the issue of food security in the twentieth century. Even the crop improvement for sustainable yield production is gaining momentum. Consequent to these mega-efforts, the percent of undernourishment in the global population has decreased from 19% in 1990 to 11% in 2016 (FAO 2017). However, despite this achievement, almost 2.5 billion people or one in three globally is at risk of nutritional deficiencies like vitamin A, iron, and/or zinc (Table 3.2). In fact, the big challenge in the twenty-first century, especially in developing countries, is not "food for all" but the access and availability to healthy, safe-to-eat nutritious food. This dearth has been recognized by the international nutrition community as most limiting in diets (Black et al. 2013) and also termed as "hidden hunger." Micronutrient deficiency (Table 3.2) particularly affects populations living in poverty. These people generally do not have the means to grow or purchase more expensive micronutrient-rich foods, thereby significantly contributing to the global disease burden of malnourished children.

Table 3.2 Total population at risk of major micronutrient deficiencies and top five staple crops, by region

Particulars	Asia	Africa	Latin America and the Caribbean	Total cases of deficiency/ inadequate intake
Total population at risk				
All	1,72,27,63,911	54,18,18,522	20,16,44,347	2,46,62,26,780
Iron	69,91,98,517	23,73,95,434	5,79,62,128	99,45,56,079
Zinc	90,13,36,413	23,68,01,679	13,55,67,293	1,27,37,05,384
Vitamin A	12,22,28,982	6,76,21,409	81,14,927	19,79,65,317
Total kilocalories per day (millions)				
Rice	31,46,030	2,01,275	1,41,990	34,89,295
Wheat	20,17,353	3,58,305	1,94,579	25,70,236
Maize	3,01,673	3,52,693	2,11,810	8,66,175
Potatoes	2,23,633	34,527	24,846	2,83,007
Cassava	71,263	1,40,542	31,554	2,43,359

Source: Saltzman et al. (2017)

3.5.10 Economic Slowdowns

In countries with a high dependence on agriculture, measured in terms of the number of people employed in the sector, the prevalence of undernourishment (PoU) was found to be 9.6% higher than in those countries where agriculture is not the primary occupation. For low-income countries, the increase was found to be equal to 13.6% points, whereas, in middle-income countries, the rise in PoU was less pronounced and occurred later between 2015 and 2016. This indicated that middle-income countries were able to absorb the impacts of increased exposure to climate extremes than the low-income groups. Factors like the economic slowdowns experienced in many Latin American countries, which led to reduce the fiscal environment to implement social programs and thus diminished the countries' capacity to cope with the aftermath of extreme climate events, may have also come into play in increasing food and nutritional insecurities in these countries (SOFI 2018).

3.5.11 Conflicts and Violence

Conflicts and violence in several parts of the world act as main drivers of hunger and food insecurity, suggesting that efforts to fight hunger must go hand in hand with those to sustain peace. The number of violent conflicts and the number of conflict-related deaths are estimated to have increased from 33 and 19,601, respectively, in 2006 to 49 and 102,000, respectively, in 2016 (FAO 2011; Allansson et al. 2017; UCDP 2018).

Countries currently considered to be at high risk of famine also experience significant violent conflict, totaling over 9000 conflict deaths during 2017 in South Sudan, Nigeria, Somalia, and Yemen (FAO 2011; UCDP 2018). A few of these conflict-affected countries also suffer from natural disasters such as prolonged droughts (Mali South Sudan or Syria), which further undermine food production, livelihoods, markets, and food consumption (FAO 2011; Sneyers 2017).

The war in Syria, for instance, has caused more than six million people to flee their homes to other locations in the country, where they face severe food insecurity (FAO 2011; Baliki et al. 2018), while another five million people have fled to neighboring countries and beyond (UNHCR 2018). These violent conflicts (and especially spikes of violence) cause forced displacement (FAO 2011; Bruck et al. 2018), further weakening food security in both countries of origin (where labor may be in short supply and rural markets collapse) and many host communities (which may face preexisting strong pressure on limited arable land) (FAO 2011; Bruck and d'Errico 2019).

3.6 Crop Improvement for Food and Nutritional Security

3.6.1 Conservation and Creation of Genetic Variability

Availability of abundant variation in a crop species is a prerequisite for successful plant breeding and crop improvement. A vast number of agricultural crops are known to have rich genetic variability known to exist in several agricultural crops and have been partly conserved as accessions in gene banks world over, while many genotypes are yet to be conserved. It is estimated that over 85–90% of the existing variability is still unexploited which lies in unused landraces and wild relatives. This variation exists in terms of wild relatives, landraces, mutant lines, modern hybrids, etc. Nevertheless, the majority of the commercial varieties and hybrids have been derived from a few parental lines in the collection. This has adversely affected the level of variability among the frequently and widely cultivated genotypes by narrowing down the genotypic diversity or genetic base. Owing to this, modern cultivars often become vulnerable to various biotic and abiotic stresses and most often reach a yield limit offering very little scope for further yield improvement.

It is a need of the hour to conserve and exploit the genetic traits of endangered and valuable species. Several successful attempts have been made in the last few decades to do so. In India, a seed vault has been established in the Himalayas with a naturally available cold condition. Settled at 17,500 feet high on a cliff top in the Himalayas, a carefully chosen site, Chang-La, has the subzero temperatures and low humidity necessary to store the valuable germplasm in terms of seed material. It is India's doomsday vault. Chang-La, opened in December 2017, now holds 5000 seeds from the Ministry of Defense, prioritized for qualities such as yield or resistance to temperature, pests, or humidity.

The Svalbard Global Seed Vault is a secure seed bank constructed in the remote Arctic Svalbard archipelago on the Norwegian island of Spitsbergen near Longyearbyen, about 1300 km from the North Pole (Daniel 2006). This is the natural site with the atmospheric storage temperature of −18 °C. The low temperature and limited access to oxygen ensure low metabolic activity and delay seed aging. Conservationist Cary Fowler, in association with the Consultative Group for International Agricultural Research (CGIAR), started the vault to preserve a wide variety of plant seeds collected worldwide by different gene banks (Siebert 2011). The seed vault is an attempt to ensure against the loss of seeds in other gene banks during large-scale regional or global crises.

Several attempts are being made by the researcher community around the world to broaden the existing genetic base by utilizing the available germplasm in different crops. Thanks to the modern molecular biology techniques like the recombinant DNA technology, the transfer of genes across the species and barriers has become plausible, thereby offering a wide scope for broadening the genetic base of crop species. For instance, Choudhary et al. (2013) reported an increase in genetic diversity of major Indian rice cultivars released from 1970 to 2010. The molecular geneticist at the International Rice Research Institute (IRRI) developed a large collection of rice varieties by crossing 46 recurrent parents with 500 donors which are reported to retain their stability and produce sustainable yields even in adverse growing condition (Jauhar and Santiaguel 2011). Several successful attempts have been made by researchers in crop species to exploit the existing and synthesized variability through crop breeding approaches, both conventional and modern, for the benefit of mankind.

3.6.2 Enhancement of Crop Productivity

Around 795 million people in the world do not get enough food to lead a healthy and active life. That's about one in nine people on earth. To satisfy the increasing worldwide demand for food, two broad options can be employed. The first approach is to increase the area under cultivation, and the second is to increase productivity from existing farmland.

One of the important success stories in productivity improvement is the development of New Rice for Africa (NERICA) varieties by Africa Rice Center, Cotonou, Benin, Africa. The NERICA lines were developed by crossing *Oryza glaberrima*, African rice, and *Oryza sativa*, Asian rice, by Dr. Monty Jones for which he received the world's food prize in 2004. These lines combine the hardiness and weed suppression of the African rice species with the productivity of the rice species of Asia. The most outstanding cultivar, NERICA 1, gave the highest yield (4.3 ton ha^{-1}) in 2005 and yielded more than 4.0 ton ha^{-1} in the other seasons (Kijima et al. 2006).

The green revolution in the late 1960s in India represents the most histrionic shift in agricultural production in human history. A tremendous increase in food production was evident. As a result of the crop improvement efforts, high-yielding non-lodging, semidwarf varieties responding positively to increased fertilizer and the irrigation use were developed in wheat and rice crops. Subsequently, cereal crop production tripled since the 1960s (Pingali 2012). Despite this, the rate of increase in food production was insufficient to keep pace with the rate of increase in the population. The concerted efforts taken under several 5-year plans implemented by the Indian government have bailed out the country from a ship to mouth existence during the pre-independence era and early 1960s to being a self-sufficient force to reckon with today.

3.6.3 Hybrid Technology

Hybrid technology is one of the successful breeding strategies to create variation as well as to exploit heterosis in order to increase production and productivity. New plant-type-based semidwarf varieties developed through hybridization during the 1970s led to the historic green revolution. Even with the intervention of genetic engineering tools to develop new varieties, the hybrid technology still holds the great potential to help meet the food demand of the increasing world's population. Thanks to Dr. Yuan Long Ping who is highly respected as the "Father of Hybrid Rice" for introducing hybrid technology in rice, China succeeded in increasing rice yields from 3.5 tons/ha (from 36.5 Mha in 1975) to 6.2 tons/ha (30.5 Mha in 2000) even while combating the threat posed by decreased land availability for cultivation (FAO 2004).

However, in spite of the great potential in hybrid technology, India has not yet achieved appreciable success, and reasons for this include high cost of hybrid seed, low quality, less attention toward the potential available in traditional cultivars, and insignificant yield edge over the local popular varieties besides lack of awareness too (Spielman et al. 2013). Despite these limitations, this technology has gained popularity in India which is evident from the release of approximately 100 rice hybrids which include first fine-grain aromatic hybrid, Pusa RH-10 from IARI, New Delhi (Siddiq et al. 2012).

3.6.4 Breeding Climate-Resilient Varieties

Climate change is a looming threat. Breeding strategies, both conventional and cutting-edge technologies, have been made use of in meeting the challenge of climate change with significant results.

3.6.4.1 Exploiting Variability through Conventional Breeding

Germplasm collections are a rich source of desirable traits that can help develop climate-smart varieties. Attempts to localize favorable genetic factors distributed among the natural population of the species and its relatives or by the creation of variability in the genome into a single genotype have been carried out successfully through conventional breeding in almost all the crop species and discussed in detail in the above section.

3.6.4.2 Utilizing and Exploiting the Functional Diversity Through Cutting-Edge Technologies

Climate change most likely modifies patterns of stresses affecting crop plants. This warrants a reassessment of the concept of plant ideotypes and the breeding objectives. The technological advancements in the field of genomics now offer the opportunity to discern the patterns of regulation by genes and assess their relevance in the trait expression such as water use efficiency, plant phenology, its response to CO_2, nitrogen use efficiency, etc. required to be focused during crop improvement program aiming to develop climate-smart varieties. Breeding climate-resilient varieties in staples, like rice, wheat, and maize, for mitigating the effect of climate change has employed both the traditional and modern biotechnological tools like genomics, proteomics, and phenomics.

Molecular tools have been employed to scan the diversity existing in the crop genome. Genetic engineering assists the breeders in their ability to select, transfer, and concentrate traits across the genetic barriers. A very classic example is the *cry* genes governing the production of crystalline (cry) proteins in *Bacillus thuringiensis* bacterium into cotton cultivars to impart plant resistance against cotton ball worm. Currently, the use of marker-assisted selection (MAS) and QTL mapping of climate-resilient traits, like drought, cold, heat, submergence, flooding, pest, and diseases, has become a day-to-day affair.

Modern biology being extremely data-intensive, technological advancement in the field of computational biology has been employed to rapidly gain in-depth information about the biological functions and their interactions within the plant system. The information thus gained usually with respect to functions of unprecedented complexity within the plant system is being translated into selection criteria (e.g., an intrinsic ability to use mineral resources, biochemical basis of combating stress, etc.) to develop climate-resilient crop varieties. After spending years in reading and deciphering plant genomes, researchers are now editing and rewriting them through site-directed genome editing tools to develop climate-resilient crop plants in crops like rice, tomato, wheat, and citrus.

3.6.5 Breeding for Safe-to-Eat Food

Pesticides are chemicals commonly used in agriculture for protection of crop plants from unwanted insects, weeds, animal pest, and disease-causing agents that might destroy crops or reduce the quality of produce. In all agrochemical used in agriculture, around 40% of the chemicals used in agriculture worldwide comprises pesticides and herbicides followed by insecticides (17%) and fungicides (10%). Though the use of these has noticeably increased the total crop production, there has been a huge negative consequence on the environment, soil, and human health. Application of these chemicals results in extremely harmful environmental pollution, water pollution, soil pollution, and reduced human health. It is estimated that more than 98% of sprayed plant-protective chemicals reaches a destination other than their targets (Tyler 2004).

Apart from conventional breeding techniques and recombinant DNA technology, marker-assisted breeding approaches like MAS, marker-assisted backcross breeding, gene pyramiding, gene stacking, etc. have been employed in almost all staples and vegetables to impart disease and pest resistance to cultivars, which reduces the usage of plant protection chemicals, thereby reducing environmental pollution and health hazards. Rapid advancement in genome editing has made the most important genetic tools in manipulating pathogen resistance in plants (Borrelli et al. 2018). Meganucleases, zinc-finger nucleases (ZFNs), transcription activator-like effector nucleases (TALENs), and clustered regularly interspaced short palindrome repeats (CRISPR)/CRISPR-associated protein 9 (Cas9) are a few of the recent tools employed for site-directed modification.

Off late, CRISPR/Cas9 has largely overtaken the other genome editing tools owing to it being more versatile and less expensive and having ease in designing and implementation and higher success rate in modifying the plant's response to viral, fungal, and bacterial pathogens. Imparting plant resistance against viral pathogens through this strategy mainly focuses on the integration of CRISPR-encoding sequences that target and interfere with the viral genome and the induction of a CRISPR-mediated targeted mutation in the host plant genome. In the case of fungal and bacterial disease resistance, CRISPR-/Cas9-targeted modification of susceptibility genes in crop species is usually employed.

3.6.6 Biofortification of Cereals and Legumes Through Breeding for Nutritional Security

As earlier discussed, a large number of people, especially people of the low-income group in developing countries, suffer from hunger and malnutrition. Biofortification is the idea of breeding crops to increase their nutritional value. Attempts at biofortification through breeding aim at increasing the micronutrient content of staple crops to improve the nutrition and thereby the health outcomes of populations whose diets comprise mainly staple crops. Biofortification of important food crop, especially cereals and legume, is seen as an upcoming strategy for dealing with deficiencies of micronutrients in low- and middle-income countries. The World Health Organization (WHO) estimated that biofortification could help to cure the two billion people suffering from iron deficiency-induced anemia (De Benoist et al. 2008).

In last two decades, research with the biofortification for nutrition has confirmed that this is an effective complement to these approaches in addressing micronutrient deficiency (Saltzman et al. 2013; Bouis and Saltzman 2017). Successful examples of biofortification are high-lysine maize, high-unsaturated fatty acid soybean, high-provitamin A and iron-rich cassava, and high-provitamin A golden rice.

In April 2002, the Consultative Group for International Agricultural Research (CGIAR) had initiated the "Biofortification Challenge Program (BCP)" to combat malnutrition. By mid-2003, they renamed the BCP as HarvestPlus. The main target of this program is to facilitate the research work for nutritional security. Under this program, CGIAR sponsored "country programs" in Brazil, China, and India. However, to reach one billion people by 2030, biofortification must move beyond a targeted development project like HarvestPlus. Policymakers must give higher priority to the role of agriculture to improve health. Public and private sector breeding partners must come together to mainstream the biofortified trait across their product lines. The consumers, in both rural and urban areas, must demand these nutritious foods. Only through a collaborative effort that reaches across the value chain will biofortification become business as usual and the vision of reaching one billion become a reality.

3.7 Marching Ahead in Time: Challenges in the Near Future

A "stomach full" of food is the right of every human being. After the green revolution, we have achieved near to sustainable production enough to feed all the hungry mouth of the world. Yet, almost 1 billion people remain starved. However, it is not only about producing more food, but it is critically essential to understand population dynamics and changes in food consumption to eradicate chronic hunger and malnutrition in the decades to come. The hunger challenge is multifaceted and concerted efforts to eradicate hunger needs to be undertaken. It warrants the participation of research agencies, policymakers, and farmers at national and international levels. If this can be achieved, then the world population can be fed impartially but also sustainably. The strategies enumerated are a comprehensive list of approaches that can be followed in order to achieve food and nutritional security in the short and long run (SADC 2014).

3.7.1 Improve Productivity of Diverse, Safe, and Nutritious Foods

The world is marching toward producing a surplus amount of food to feed the hungry mouths of a rapidly growing population. However, only producing more and more quantity of food is inadequate. We not only have to increase production but also have improved the productivity of diverse, safe, and nutritious food. To achieve the above goal, the following actions can be taken:

(i) Promote increased access to diverse and improved seeds without affecting traditional local varieties.
(ii) Promote eco-friendly and innovative production systems.
(iii) Resort to bioprotection or biological control.
(iv) Strengthen the management of information systems related to food and nutrition security.
(v) Promote good agricultural practices in production and extension services.
(vi) Facilitate documentation and sharing of best practices on incentives and empowerment of women and youth in food and nutrition security mentorship, skills development, and incubation.
(vii) Promote healthy lifestyle in schools, workplace, and communities.
(viii) Promote the inclusion of food- and nutrition-sensitive curricula at academic levels.

3.7.2 Improve Access to Land and Water for Agriculture

As discussed earlier, the area cultivable land and the natural resources of fresh water for the agriculture use are depleting. It is absolutely necessary to manage available land and water resources for sustainable production. Reconverting the polluted land into cultivable land is no doubt difficult; however, it is not an impossible task. The same applies to the recycling of water as well. To achieve the above, the following are priority actions:

(i) Promote access to land and water for vulnerable persons, in particular, women, youth, persons with disabilities, and other special categories of disadvantaged people.
(ii) Promote sharing of best practices on land reform emphasizing access to land for women, the youth, and other disadvantaged groups.
(iii) Increase land under irrigation.

3.7.3 Reduce Postharvest Losses

Prevention is not better than cure; prevention is the cure! Although with the intervention of new technologies, the world is now moving slowly toward producing a surplus amount of food, there is an urgent need to initiate measures to manage postharvest losses. Lack of proper management of produce postharvest includes that human consumption results in a noticeable amount of food being wasted. The Food and Agriculture Organization predicts that about 1.3 billion tons of food is globally wasted per year (Gustavasson et al. 2011). The high postharvest losses have a negative impact on food and nutrition security.

The involvement of public and private sector players onboard can be promoted to seek business opportunities in combating postharvest loss. Being an important part of the food chain, the simple postharvest technologies to reduce losses should be made available to the farmers with the subsidized cost. It is also essential to channelize the available fund for supporting the entire market system by reducing the constraints impeding the effective adoption by farmers. The microfinance aspect is still being neglected, and we need to encourage investment into postharvest management by providing access to credits.

To address the above, the following are the priority actions:

(i) Facilitate action-oriented research on food waste, and disseminate results to inform appropriate action.
(ii) Promote low-cost technologies on food processing, handling, preservation, and storage.
(iii) Encourage agro-processing and value addition of safe and diverse foods.
(iv) Promote the establishment of postharvest handling facilities particularly for horticultural crops.

With proper preharvest practices on the field to ensure safe transportation of food material, with proper storage facility, and with technological intervention, we hope all these efforts will help in combating postharvest loses.

3.7.4 Adaptation and Mitigation to Climate Change and Environment

The agriculture is highly vulnerable to climate change because of the heavy reliance on rain, temperature, and availability of water. There is a need to identify response strategies for climate change adaptation and mitigation. The priority actions are:

(i) Facilitate capacity building on adaptation and mitigation.
(ii) Facilitate and promote the dissemination of information and sharing of best practices on adaptation and mitigation.

3.7.5 Improve Access to Credit and Finance

Poor access to credit and finance remains one of the major hindrances to an agricultural revolution, speedy eradication of poverty, and food and nutrition security, especially in developing countries. It's very crucial to make rural poor and marginal farmers capable to earn their own meal by providing them with an initial credit. In order to promote access to credit and finance, the priority actions are:

(i) Explore and promote innovative approaches to agricultural financing and insurance schemes for farmers.
(ii) Explore and promote incentives including grants targeted to the youth and women in agribusiness.
(iii) Incentivize credit for the production of nutritious, diverse, and local foods.

3.7.6 Improve Access to Markets

Unavailability of proper market facility and the fluctuating mark*et al*ways remains a major hurdle in agriculture for the farming community. Poor access to markets always has a negative impact on food and nutrition security for the producers as well as consumers. This also makes it very difficult for marginal farmers to deliver their farm produce to the consumers. Strengthening market infrastructure, intelligence, and information systems are essential to the overall improvement of agri-based value chains and enhancing food and nutrition security. The priority actions are:

(i) Facilitate the removal of nontariff barriers, especially sanitary and phytosanitary measures.
(ii) Facilitate the development and/or upgrading of marketing infrastructure.
(iii) Facilitate the development of an agricultural market information system, including the use of information and communication technologies.

3.7.7 Ensure Stable and Sustainable Availability, Access, and Utilization of Food

There cannot be food and nutrition security if people do not have food available, accessed and utilized at all times when required. Hence, the issues of stability and sustainability are of vital importance for food and nutrition security. The priority actions are:

(i) Promote full ownership and commitment by the member states toward the attainment of food and nutrition security.
(ii) Promote sustainable mechanisms of funding for the implementation of the strategy at both the regional and national levels.

(iii) Promote the empowerment of youth and women in food and nutrition.
(iv) Develop, review, enact, and implement laws and policies that guarantee and protect food as a human right.

3.7.8 Spreading Social Awareness

In the developing countries, food is getting wasted because of postharvest losses; however, in developed countries, food gets wasted after it is cooked. Postharvest loss occurs because of lack of proper storage and market facility; however, the cooked food gets wasted because of a lack of social awareness. A key strategy for averting food waste in developed countries is to inhibit or use leftovers from foods prepared at home, purchased in supermarkets, or ordered in restaurants. The leftover food can be recycled, and there is a need for developing better recycling systems of food waste.

In the long run, it is more crucial to make people aware of the present scenario of food and nutritional security in order to achieve the goal of sustainability and to eradicate hunger and malnutrition. Food waste after cooking is a more social issue and can only be addressed by spreading awareness. Consumers can also considerably reduce the portion sizes of the meals they consume at home or outside of the home and can decide not to buy foods or food ingredients sold in excessive portion size. By limiting waste and overconsumption, consumers can reduce pressure on producers to supply them with unsustainable food.

3.7.9 Improved Technologies

Combating the problem of hunger and malnutrition requires more than existing technology, research method, and extensions. The new technological intervention has to be made in the areas with shortages of land or water or with particular problems of soil or climate. These are where most of the poor population stays, where such technology could play a key role in improving food security. Agricultural production could perhaps meet anticipated requirement over the period to 2030 even without major advances in modern tools of biotechnology. Nevertheless, the new techniques of molecular analysis could give a welcome boost to productivity, particularly in areas with special difficulties, thereby improving the incomes of the poor. What is needed for the twenty-first century is a second green revolution in agricultural technology. The requirement of an increase in productivity and production is still vital but must be combined with environmentally friendly ways or restoration and should be acclimatized with changing the climate. The new technologies must be affordable by the poor and undernourished. Biotechnology offers a great promise as a means of improving food and nutritional security while reducing pressures on the environment, provided the apparent environmental extortions from biotechnology itself are addressed.

Strategies to address food and nutritional securities require a focus on resilience. Targeted interventions aimed at anticipating, limiting, and adapting to the effects of climate variability and extremes are essential. It's not only about breeding climate-smart varieties but also about building the resilience of livelihoods, food systems, and nutrition to combat climatic shocks and stresses. This calls for integrated approaches in agriculture and allied sectors as well as cooperation cutting across sectors and a long-term vision to strengthen the resilience of livelihoods and food systems. Solutions require increased partnerships ensuring an error-free production/distribution/consumption system which is sustainable and environmentally, socially, and culturally sound to enable access to a diverse healthy diet by all.

3.8 Exploiting Crop Genetic Resources

It is perhaps unnecessary in this day and age to stress the need for the preservation of crop genetic resources as basic genetic material for plant breeding research. It is now generally agreed that a broad genetic base is essential, particularly as breeding objectives become more complex and more demanding. Breeders can no longer be content to use the basic stocks of cultivars which they inherited from their predecessors. Yield must be increased, adaptation to a wider range of environments must be sought, and more nutritious cultivars must be created with greater resistance to pests and diseases. It is clear that a much wider range of germplasm is needed by breeders today than in the past, including not only standard cultivars and breeding lines related to them but also wild species and the primitive landraces that still exist under cultivation in remote areas of mountains and forest where ancient crop plants were first domesticated.

Ensuring sustainable utilization of natural resources for agriculture in the developing world has focused attention on total resources required for crop production, including crop genetic resources (Cohen et al. 1991). Crop genetic resources are the key component of any agricultural production system. Without it, no natural, evolutionary adjustment of the system (agricultural or natural) to changing environmental and biotic condition would be possible. Agricultural science and forestry would not have the basic raw materials for their introduction, domestication, and improvement programs. Genetic resources include plants, animals, and microorganisms likely to contain genetic material of economic value. Genetic resources of crops, primarily cereals, have been the focus of both conservation and utilization efforts as the global food and nutritional security mainly depends on crop genetic resources for sustainable agriculture. Crop genetic resources essentially constitute the prime components of the food chain ever since the dawn of agriculture. According to Groombridge (1992) and Heywood (1995), the estimated existence of near about 400,000 and 300,000 plant species have been documented. As per Wilson (1992) report, 10%, i.e., 30,000, of these is edible plant species. Nature gave us 30,000 types of edible plants, but farmers today grow only about 170 plants, and out of these 170 edible plants species, only 30 feed the world, with the 3 major crops being maize, wheat, and rice (FAO 1996a). An assessment of the contribution of different plant sources toward the dietary energy supply at the global level shows predominance of only two crops, i.e., rice (26%) and wheat (23%). The search for new diversity is, therefore, important as we know that world population is increasing at an alarming rate. At present, world's population is approximately 6–7 billion (Jyoti et al. 2017), and it is projected to reach 9.8 billion in 2050 and 11.2 billion in 2100 (United Nations Department of Economic and Social Affairs Population Division 2017). To combat from future challenge of feeding 10 billion populations by 2050, there is a need to increase food crop production by 60% with present land and water resource challenges (FAO 2012), and crop genetic resources will play an important role in enhancing the crop production by planned and effective utilization of crop-specific genetic resources in different breeding programs for trait-specific improvement, wide hybridization, marker-assisted breeding, etc.

Crop genetic resources comprise genetic diversity present in landraces, cultivated varieties, obsolete varieties, wild relative and wild forms, genetic stocks, and elite breeding lines, and all of these act as a component of biological diversity which is the main source of basic needs of mankind. In contrast to biological diversity, Wilson and Frances (1988) coined the term biodiversity which refers to the variety of life forms. Biodiversity is divided into three categories as genetic diversity, species diversity, and ecosystems diversity. Crop genetic resources are an important component of agricultural biodiversity. Agricultural biodiversity is a subset of biodiversity which includes all crops and livestocks, their wild relatives, and all interacting species of pollinators, symbionts, pests, parasites, predators, and competitors (Qualset et al. 1995). Crop genetic resources (CGRs) are defined as all materials that are available for improvement of a cultivated plant species. Genetic resources are the rich source of genetic diversity and serve as essential raw material for improving crops and developing new value-added products. The genetic diversity of a given species extends beyond its taxonomic boundaries. The gene pool of a crop includes not only traditional local forms but also wild and weedy relatives. The entire gene pool forms the basic unit of conservation and must be the ultimate target of genetic resources collecting. There are well over a quarter of million plant species on this planet, and only a small proportion falls in the gene pools of current crops. Basing upon the extended gene pool concept, genetic resources are divided into primary, secondary, tertiary, and isolated gene pool (Harlan and de Wet 1971; Becker 1993). The primary gene pool consists of the crop species which can be crossed itself and with other species in an easy manner. The secondary gene pool is composed of related species that are more difficult to cross with primary gene pool members, and the crossing progenies are partially sterile. The tertiary gene pool consists of species which can be only be used by employing special techniques like embryo rescue or protoplast fusion. The fourth gene pool of isolated genes may derive from related or unrelated plant species, from animals or microorganisms. The worth of genetic resources in developing superior crop varieties is well recognized as utilization of *Norin 10* gene in wheat and *Dee Geo Woo Gen* in rice (sources of reduced plant height) has revolutionized the productivity of these crops worldwide.

3.9 International Efforts Toward Collections of Crop Genetic Resources

Collection of crop genetic resources is the preliminary step for the ex situ conservation and utilization of material for specific purpose. Collecting germplasm is not an easy task because it is not putting the few seeds in a bag at right place and right time. For this first of all, the target species must be identified and then which part of the plant we have to collect. Farmers, the earlier plant breeders, have always recognized the importance of exotic plants and novel crop and varieties. In the Andes region, farmers often gathered and exchange the planting material with each other. Also in Eastern Sierra Leone, to fulfill the consumption needs, women expect to take seeds and plants from each other's farm. In some parts of India, it is customary for a bride to bring a gift of rice seeds grown by her family to her husband's house. Thousands of years ago, Emperor of China Chen-Tsung introduced Champa varieties of rice from Vietnam to Yangtze Delta of China which is considered as the first large-scale germplasm introduction in the world. Frankincense trees which are well known for their perfumed gum property were collected from Deir al-Bahari temple by an expedition team sent by the Egypt Queen Hatshetput near about 3500 years ago from Punt region. The people of New World started a major germplasm exchange program with the Old World people 500 years ago which is still continuing in the nature. In the modern times, exploration of tropics by the emerging colonial powers continued, and the first botanic gardens outside Europe were created. These gardens served as introduction and acclimatization centers for a wide range of crops, fruit species, and ornamentals (Smith 1986; Heywood 1990). The Royal Botanic Garden, Kew, and other associated botanic gardens were responsible for the movement of vast amount of germplasm worldwide. Kew contains the most diverse collection of living plants of any botanic garden in the world. The collection contains plants from tropical, temperate, arid, and alpine climates and is grown out in the gardens and in controlled conditions within glasshouses and nurseries. Kew is well known for their *Rhododendron* and *Primula* species collections which are introduced by George Forrest from China and Bhutan during the eighteenth century. Simultaneously, in the USA also under the leadership of President Abraham Lincoln and his successors, the tradition of germplasm introduction continued by various outstanding germplasm collectors David Fairchild, Frank Meyers, Joseph F. Rock, and Wilson Popenoe (Berg et al. 1991). All-Union Institute of Plant Introduction was established in 1890 in St. Petersburg, Russia. It is amassed by extensive worldwide germplasm collections during 1920 by Vavilov and his colleagues, and later on, it is renamed as N. I. Vavilov All-Union Scientific Research Institute of Plant Industry (VIR). During the period of the 1920–1930s, Vavilov gathered some 50,000 samples of crop plants from over 50 countries. By the year 1970s, VIR acquired long-term seed storage capability. This is followed by the collection of potato accessions at the International Potato Center (CIP), Peru, which becomes one of the heirs of the Consultative Group for International Agricultural Research (CGIAR). It was in the 1960s when scientific community realized that developing countries must be integrated to a much greater extent into a global plant genetic resources system. A short chronology of events and actions associated with FAO/UN and Bioversity International is presented in Tables 3.3 and 3.4.

Table 3.3 Chronology of events and actions of FAO/UN with respect to plant genetic resources (PGRs) collection and conservation

Year	Events/actions
1957	Introduction of first international newsletter on crop genetic resources
1961	Plant production and protection division; FAO convened a technical meeting on plant exploration and introduction
1962	A proposal for a Crop Research and Introduction Center, Izmir (Turkey), was submitted to the United Nations Special Fund
1963	The 12th session of the FAO conference recommended the establishment of a panel of experts on plant exploration and introduction to advise FAO
1964	The Crop Research and Introduction Center, Izmir, becomes operative with UN Special Fund support
1965	A panel of experts on crop germplasm exploration and introduction was set up with Sir Otto Frankel as chair
1967	FAO and the International Biological Program (IBP) jointly sponsored a technical conference on exploration, utilization, and conservation of plant genetic resources
1968	A crop ecology and genetic resources unit and a panel on forestry were established in the plant production and protection division; FAO followed by an expert consultation on forest genetic resources
1971	CGIAR was established under joint sponsorship of the World Bank, FAO, and the UN Development Program (UNDP)
	A Technical Advisory Committee (TAC) was established to assist the CGIAR
1972	Under joint sponsorship of TAC, FAO, and CGIAR, a meeting was convened at Beltsville, Maryland, and a plan for a global network of genetic resources centers was drawn up
	The UN Stockholm Conference on the Human Environment called for action of genetic resources conservation and setting up of the United Nations Environmental Programme (UNEP)
1973	A second FAO/IBP technical conference was convened in Rome
	The CGIAR established a subcommittee on genetic resources
	Frankel's FAO/IBP survey of crop genetic resources in their centers of origin was published
1974	Portions of the global strategy devised at Beltsville began to be funded through bilateral agreements with donor governments; for example, Sweden agreed to support the Izmir
	Center for a time and the Federal Republic of Germany agreed to support genetic resources centers in Ethiopia and Costa Rica. Other similar agreements have been or are being arranged
	There were ten long-term seed storage facilities in the world
	The International Board for Plant Genetic Resources (IBPGR) was established by CGIAR with its headquartering at FAO, Rome
1979	FAO conference – the first political discussions began, which in only a few years led to the adoption of the nonbinding International Undertaking on Plant Genetic Resources for Food and Agriculture (international undertaking) and, much later, negotiations and approval of the legally binding ITPGRFA
1981	FAO held international technical conferences on plant genetic resources
1983	Conference requested the FAO council to develop the statutes of the intergovernmental commission on plant genetic resources as a global forum for plant genetic resources debate
1991	IBPGR became the International Plant Genetic Resources Institute (IPGRI)

(continued)

Table 3.3 (continued)

Year	Events/actions
1992	The Convention on Biological Diversity (CBD) was submitted for signature at the UN Conference on Environment and Development in June 1992
1994	IPGRI began independent operation as a CGIAR center and, at the request of CGIAR, took over the governance and administration of the International Network for the Improvement of Banana and Plantain (INIBAP)
1995	Commission on plant genetic resources was renamed as the Commission on Genetic Resources for Food and Agriculture
1996	Fourth international technical conference on plant genetic resources was held in Leipzig
	State of the world's plant genetic resources for food and agriculture was launched
	First global plan of action on plant genetic resources for food and agriculture was approved
2006	IPGRI and INIBAP became a single organization and subsequently changed their operating name to Bioversity International

Source: Harlan (1975)

Table 3.4 Highlights of IBPGR/IPGRI/Bioversity International since 1974

Year	Highlights
1974	Started a major effort to collect genetic resources as the International Board for Plant Genetic Resources (IBPGR)
1987	Initiated a research program on the conservation and study of plant diversity
1991	Became the International Plant Genetic Resources Institute (IPGRI)
1994	Merged with the International Network for the Improvement of Banana and Plantain (INIBAP)
2002	Founded the global crop diversity trust together with the UN Food and Agriculture Organization (FAO)
2004	Served as technical support in negotiations for the international treaty of plant genetic resources for food and agriculture
2005	Started work with partners and smallholder farmers in sub-Saharan Africa to conserve and use the African cherry tree in a sustainable way
2006	INIBAP and IPGRI begin working under the name Bioversity International
	Biodiversity for Food and Nutrition initiative launched at the Convention on Biological Diversity Conference
2009	Developed "Seeds for Needs" to help farmers adapt to climate change
2012	Became a co-organizer of the international collaboration "landscapes for people, food, and nature" led by eco-agriculture partners
	Launched a new research agenda and strategic priorities
	Contributed to an international effort on sequencing one of the founding genomes of banana
	Copublished "Sustainable Diets and Biodiversity: Directions and Solution for Policy, Research and Action" with FAO
2014	Refreshed strategy
2015	Launched three new research initiatives:
	1. Healthy diets from sustainable food systems
	2. Productive and resilient farms and forests
	3. Effective genetic resources conservation and use

(continued)

Table 3.4 (continued)

Year	Highlights
2016	Started to develop an Agrobiodiversity Index for use by countries and the private sector to measure and manage agro-biodiversity across three dimensions: diets, production, and genetic resources
2017	Launched the flagship publication – Mainstreaming Agro-biodiversity in Sustainable Food Systems: Foundations for an Agrobiodiversity Index. The book summarizes the most recent evidence on how to use agro-biodiversity to provide nutritious foods through harnessing natural processes

Source: https://www.bioversityinternational.org/about-us/who-we-are/history/

3.10 Types of Crop Genetic Resources

Genetic resources are the essential raw material for improving crops and developing new value-added products. The sum total of genes in a crop species is referred to as genetic resources or germplasm. The Convention on Biological Diversity (CBD) defines genetic resources as genetic material of actual or potential value. Germplasm encompasses the total "gene pool" of a species and forms the basic material as a source of variability in breeding for launching a crop improvement program. These plant genetic resources represent the full spectrum of their genetic potential with respect to economically important morpho-agronomic characters available in a crop species. FAO/IAEA (1977) have identified three kinds of genetic resources for exploitation in plant breeding, i.e., (1) the wild or weedy relatives of crop species together with alien species, (2) the landraces or primitive cultivars which were prevalent under primitive agriculture, and (3) the advanced cultivars of scientific agriculture produced during the last 100 years or so. Further, IPGRI (1993) broadly classified plant genetic resources as cultivated varieties (cultivars) in current use, newly developed varieties, obsolete cultivars, primitive cultivars (landraces), wild and weedy relatives of cultivated varieties, and special genetic stocks (elite lines, mutant lines, transgenic lines). Out of these plant genetic resources, the three (landraces, modern cultivars, wild and weedy relatives of cultivated plants) have been briefly described here.

3.10.1 Landraces

Landraces are the cultivars of prescientific or primitive agriculture that evolved over centuries or even millennia at farmers' field. They are the products of natural selection and maintain genetic heterogeneity in balance over time but without a systematic and sustained plant breeding efforts. They represent balanced populations variable in equilibrium with both environment and pathogens and genetically dynamic (Harlan 1965). They are the reservoirs of ancient diversity existing in the farmers' field especially in mountainous and tribal areas which makes them good sources of new quantitative variation for specific adaptations to biotic and abiotic stress conditions and breeding population for the development of improved locally adapted cultivars for the same or marginal areas. In addition to this, they act as donors for individual monogenic traits which probably contain several favorable linkage blocks. Landraces have broad genetic base which again provides them with wide adaptability as they are grown in extreme areas, i.e., semiarid to arid regions in Asia and Africa (Haussmann et al. 2004).

Landraces were first collected and studied by N. I. Vavilov in rice. They respond very well to selection but have low stable yield and less uniform. In Indian gene center, Himalayan foothills of northeast region include Assam, Meghalaya, and Arunachal Pradesh from where a large number of primitive races of rice and maize were collected. Similarly, local races of minor millets (*Paspalum*, *Setaria*, *Echinochloa*, *Panicum*) are available in plenty in tribal regions of Orissa, Madhya Pradesh, Bihar, and hilly regions of Uttarakhand. A systematic search for resistance to yellow mosaic virus of *Phaseolus* species has produced resistant stocks of mung bean (*Vigna radiata*) from Punjab (Gill et al. 1975) and of urdbean (*V. mungo*) from Gujrat. One cultivar of chickpea from Afghanistan (ICRISAT 1976) and 5 from India were resistant to wilt and 18 Indian cultivars to *Ascochyta*. Similarly, landraces offered sources of important diseases of wheat and rice. In rice, Intan, Mas, and Peta from Indonesia are the products of the cross of Indian variety Latisail and Chinese variety Cina. Peta is parent of IR8. In wheat, Kharchia Local/Kharchia 65 has been used for salinity tolerance, Hindi 62 for heat tolerance, and NP4 for grain quality in several countries (Gautam et al. 2004). Landraces have been collected in maize, sorghum, pearl millet, and many other crops especially in South Asia.

3.10.2 Modern Cultivars

The currently cultivated yielding varieties are referred to as modern cultivars. Modern cultivars are also known as improved cultivars or advanced cultivars. These varieties have high yield potential and are uniform as compared to obsolete varieties and landraces. Modern cultivars constitute a major part of working collections and are extensively used as parents in the breeding programs for further genetic improvement in various characters. Hence, these cultivars are in great demand. These varieties are the outcome of scientific plant breeding and have been developed for modern intensive agriculture. However, modern cultivars have narrow genetic base and low adaptability as compared to landraces. Hard red winter wheat varieties cultivated in the USA are developed from just two lines imported from Poland and Russia (Harlan 1987). Similarly, the soybean cultivars in North America traced back to China strains. These are the modern crop varieties developed from deliberate selection for genetic homogeneity, attractiveness, and high productivity under scientific crop improvement programs. The basic materials for advance varieties were provided by landraces, primary varieties (derived straightway from primitive races), supplemented or strengthened particularly for disease resistance by alien (wild species) (Sharma 1994). Currently, over 200 wheat varieties have been released after the green revolution in India, and out of these, near about 28% are direct release of foreign introduction or selection in the introduction; 48% of the cultivars involve exotic germplasm as parent (Sharma and Jagshoran 2000). Similarly, in soybean, exotic gene pool collections have been used for developing varieties, and contribution of indigenous collections toward varietal development is low (Gautam et al. 2004). Utilization of landraces in the evolution of modern cultivars has been increased substantially over the period as evident from CIMMYT wheat and IRRI rice breeding programs where over 60 and 46 landraces were used for the development of wheat and rice varieties worldwide, respectively (Witcombe 1999).

3.10.3 Wild Forms and Weedy Relatives

As Hawkes (1977) questioned "why use wild species when so much diversity exists in the crops themselves?", the answer to this depends on the nature of crop. For crops like faba beans, cassava, barley, and maize, most of the useful allelic variation exists in the crop itself. Thus, their wild forms have little interest. Wild species are the reservoir of the many useful genes/alleles as they have evolved under natural selection pressure under extreme climatic conditions (Jyoti et al. 2017). The wild species when brought under domestication based on their property for immediate adaptation and attractiveness are called crops. According to Mangelsdorf (1966), at least 3000 plant species have so far been domesticated, 150 of them entering the world commerce (Verma and Kumar 1974). The wild relatives of crop plants are of considerable importance in crop improvement programs as a source of disease and pest resistance, abiotic stress tolerance, etc., providing basic information on species relationship and giving clues to crop evolution patterns (Singh and Srivastava 2013). The concentration of these wild taxa lies mainly in the centers of diversity of crop plants. The diversity of the wild relatives has enabled them to survive longer than the oldest cultivated variety. The estimated strength of wild relatives of crop plants and related taxa occurring in India is about 320 species (Table 3.5) (Arora and Nayar 1984). The wild edible plant gene pool includes the naturally occurring wild types, weedy types, and protected, semi-protected, and domesticated diversity (Hoyt 1988). A large number of varieties of cultivated plants used by mankind today have been derived from the wild relatives through selection/hybridization for bringing out desired improvement. Utilization of wild species for crop improvement has enjoyed great success for a few crops while being disappointing for numerous others. The contribution of wild relatives toward economic returns is well exemplified by crop plants such as wheat, rice, and maize (Table 3.6). Working with wild species and relatives is very difficult

Table 3.5 Wild species of agri-horticultural importance

Crop group	Species number
Cereals and millets	51
Legumes	31
Fruits	109
Vegetables	54
Oilseeds	12
Fiber plants	24
Spices and condiments	27
Others	26

Source: Arora and Pandey (1996)

Table 3.6 Contribution of different wild relatives of wheat, rice, and maize for desirable genes/traits

Crop	Genes/traits
Wheat wild species	
Aegilops comosa	Yr8/Sr34
A. intermedium	Wsm1, Lr38, BYDR
A. longgissima	Pm13
A. speltoides	Lr28, Lr35, Lr36, Pm12, Sr32, Lr35/Sr9, Gb5
A. squarrosa	Lr21, Lr22a, Lr32, a gene on 1DS
A. ventricosa	Pch1, Sr38/Lr37/Yr17
A. umbellulata	Lr9
Agropyron elongatum	Lr19/Sr25, Lr24, Lr29, Sr43, Lr19, Sr24/Lr24, Sr26, WSMR (wheat streak mosaic virus resistant), Cmc2 (wheat curl mite resistant)
Haynaldia villosa	Pm21
Secale cereale	Lr25, Lr26, Yr9, Pm7, Pm8, Pm17, storage glutennins, Lr26/Sr31/Yr9/Pm8, Gb2/Sr/Pm17, RWA (Russian wheat aphid resistance), Lr25/Pm7, Pm8/Sr31, Lr26/Yr9, Gb2/Pm17, Gb6, Lr45, Sr27, Pm20, H21, H25 11
Triticum compactum	Pm15
T. dicoccoides	Zur15, Pm16, glutenins, and many other genes
T. dicoccum	Pm4a, Pm5
T. distichum	Lr19/Sr25
T. machasubletschumicum	Pm3b
T. spelta	Yr5
T. speltaduhamelianum	Pm10, Pm11
T. sphaerococcum	Pm3b
T. timopheevii	Sr36/Pm6, Sr37, Lr18, Sr40
Rice wild species	
Brown plant hopper resistance	
Oryza nivara (AA)	
O. punctata (BB/BBCC)	Bph10 and Bph18(t) (*O. australiensis*)
O. longistaminata (AA)	bph11(t), bph12(t), Bph13(t), Bph14, and Bph15 (*O. officinalis*)
O. barthii (AA)	Bph12 (*O. latifolia*)
O. rufipogon (AA)	Bph16(t), Bph20(t), Bph21(t), and Bph23(t) (*O. minuta*)
O. officinalis (CC)	Bph22(t) (*O. glaberrima*)
O. australiensis (EE)	Bph24(t), bph29, and Bph30 (*O. rufipogon*)
O. minuta (BBCC)	
O. latifolia (CCDD)	
O. glaberrima (AA)	
Blast resistance	
O. minuta (BBCC)	~100 resistance (R) genes and 350+ QTL; three major R gene clusters (*Piz*, *Pik*, and *Pita*) were subjected to extensive characterization
O. autraliensis (EE)	
O. rufipogon (AA)	
O. rhizomatis (CC)	

(continued)

Table 3.6 (continued)

Crop	Genes/traits
Bacterial blight resistance	
O. longistaminata (AA)	~41 resistance genes have been reported
O. rufipogon (AA)	Xa21 (*O. longistaminata*); Xa23 (*O. rufipogon*)
O. minuta (BBCC)	Xa27 (*O. minuta*); Xa29(t) (*O. officinalis*)
O. officinalis (CC)	Xa30(t), Xa38 (*O. nivara*); Xa34(t) (*O. brachyantha*)
O. nivara (AA)	
O. brachyantha (FF)	
Grassy stunt virus resistance	
O. nivara (AA)	*Gs (O. nivara)*
Tungro bacilliform virus tolerance	
O. longistaminata (AA) *O. rufipogon (AA)*	Ongoing efforts include gene/QTL identification and using *O. longistaminata* and *O. rufipogon* as donors in developing tolerant lines
Maize wild species	
Fall armyworm tolerance	
Z. mays parviglumis	Leaf toughness and leaf trichome
Z. diploperennis	Chemical composition of leaves
Z. mays subsp. parviglumis	The higher expression of wip1, RP1, and chitinase genes
Teosinte	Emission of herbivore-induced volatiles such as indole and a large number of mono- and sesquiterpenes resulted from FAW leaf herbivory attracting larval parasitoids, *Cotesia marginiventris*, and *Meteorus laphygmae*
Disease resistance *Zea diploperennis*	Resistance to maize chlorotic dwarf virus resistance, maize streak virus resistance, maize stripe virus resistance, maize rayado fino virus resistance, southern corn leaf blight resistance, corn leaf spot resistance, and maize bushy stunt mycoplasma resistance

Source: Gautam et al. (2004) and Mammadov et al. (2018)

as (i) wild species are often more difficult to cross and the hybrids when formed may be completely or partially sterile; (ii) there may be difficulties with varying ploidy levels and consequent sterility; (iii) it requires intensive efforts and extensive knowledge of taxonomy, reproductive biology, cytogenetics, genetics, and, in many cases, cell culture techniques; and (iv) all wild species possess a whole syndrome of undesirable characters (low yield, poor flavor, and unfavorable agronomic features). Further, Hawkes in 1977 described five levels of difficulty associated with transfer of wild germplasm into cultivated species as (1) wild and cultivated forms of overlapping distributions, often exchanging genes through introgressions; (2) wild and cultivated species in separate gene pools, but fertility of hybrids still high; (3) wild and cultivated species of different chromosome number, often with different genome formula, but hybrid fertility still high or can be restored by amphidiploidy; (4) difficulties requiring special techniques in the transfer of characters from wild to cultivated species; and (5) very wide hybridization.

3.11 Genetic Diversity and Center of Crop Origin

During pre-Mendelian era, genetic variability present in the form of visual differences with respect to agriculturally important traits with respect to morphological, developmental, and stress traits has been utilized by plant breeders (Aniol 2001). According to the Alfonse de Candolle (1890) hypothesis, crop plants were domesticated in the centers of respective species diversity. After the re-discovery of Mendel's law which laid to the foundation and establishment of a new science called genetics by Bateson in 1906. Modern plant breeding is essentially an applied genetics. Genetics has given to breeding a better knowledge of the processes involved in the mechanism of development of variability and utilization of variability in crop improvement programs (Aniol 2001). Genetic variation provides the foundation for any breeding program, and natural genetic diversity represented historically the major source of variability for crop domestication, improvement, and adaptation (Tavakol et al. 2017). Genetic diversity is referred to as any variation in the nucleotides, genes, chromosomes, or genomes of a species at a level of individual, population, species, or region for a given time (Fu 2015). Crop genetic diversity (CGD) is an important aspect of plant genetic resources (PGRs). Unlike PGRs, CGD is the result of conscious selection from wide range of natural genetic diversity. After Mendel's period, Candolle's concept regarding centers of diversity was refined with Vavilov's (1926) concept of centers of crop origin (center of diversity) of cultivated species and Harlan and de Wett's (1971) gene pool concept which distributed genetic diversity over primary, secondary, and tertiary gene pools basing upon their crossing compatibility.

A study of the origin of agriculture/cradles of agriculture and the spread of agriculture provides clues to the geographical distribution of centers of plant domestication. de Candolle (1890) was perhaps the first who located the areas of maximum diversity of crops where the initial plant domestication might have taken place. Nikolai Ivanovich Vavilov (1887–1943), an outstanding Russian geneticist, plant breeder, and crop geographer, was the first to realize the essential need for a really broad genetic base for the crop plant improvement. He believed that time is the only factor that influenced the dispersal of a species and initiated the collection of crop genetic resources through plant exploration approach to study plant diversity and breeding (Tavakol et al. 2017). Based upon de Condolle's three primary plant domestication areas (China, Southwest Asia/Egypt, and Tropical Africa) and Charles Robert Darwin's theory of natural selection, Vavilov proposed the concept of center of crop and named eight centers where agriculture developed independently. Center of origin of crop plants was characterized by the accumulation of dominant genes in the center and the recessive genes in the periphery. Center of origin is defined as a geographic area where a group of plants were first domesticated and identified (Hummer and Hancock 2015), whereas center of diversity is defined as a geographic area where the genetic diversity of a particular crop is especially high (Simmonds 1970; Dvorak et al. 2011). The amount and quality of genetic variation in crop plants were Vavilov's basic percept for his center of origin concept. He began with five in 1926, recombining some and adding others from time to time,

Table 3.7 Vavilovian centers of origin of crop plants

Centers of origin	Regions crop-wise
East Asian (Chinese)	Western China, Korea, Japan, and Taiwan – native regions of soybeans, millet, and many vegetables
Hindustan	Tropical India, Indochina, southern China, and the islands of southeastern Asia – rice, sugarcane, and tropical fruit originated in this area
Southwest Asia	Syria, Palestine, trans-Jordania, Afghanistan, northwestern India, Transcaucasia, Iran, and Turkmenistan. The native plants of this area are wheat, rye, as well as species of temperate fruit trees
Mediterranean	Countries bordering on the Mediterranean Sea. While considering this area less important than the other eastern centers, Vavilov proposed it as native to some wheat, barleys, and forage plants
Abyssinian (Ethiopian)	This was a refuge area for crops from other regions, especially wheats and barleys, spices, and local grains
The south Mexican	Southern North America, Mexico, and the west Indian Islands. Important plants such as maize, cotton, beans, sweet potato, arrowroot, melons, pumpkins, cocoa, papaya, avocados, and subtropical fruits originated in this area
The central American	High mountainous regions of Peru, Bolivia, Ecuador, Columbia, parts of Chile and Brazil, and whole Paraguay. Potatoes, maize, lima bean, peanut, pineapple, pumpkin, Egyptian cotton, quinine tree, cassava, and rubber originated in this center
Andean regions	The Andean mountain range, Bolivia, Peru, and Ecuador. This area produced many tuber-bearing crops such as potatoes, quinine tree, and the coca bush

Source: Tavakol et al. (2017)

until he reached eight, with three subcenters (Vavilov 1935, 1951). Later on, Darlington (1973) has increased the number of centers to 16, while Vavilov's colleague Zhukovsky (1965) proposed a series of 12 megagene centers which almost covered the whole world.

Vavilov conducted many scientific expeditions all around the world, with the aims of identifying those centers, through the study and classification of crop plants and their wild ancestors (Hummer and Hancock 2015). Finally, Vavilov proposed that domestication of a wide range of crops can be traced back to eight centers of origin (Table 3.7) (Vavilov 1935; Vavilov and Dorofeev 1992).

3.12 Status of World Crop Genetic Resources: An Overview

3.12.1 Consultative Group for International Agricultural Research (CGIAR) Gene Banks

Plant and animal genetic resources are the foundation of sustainable agriculture and global food security. Genetic diversity enables plants to adapt to new diseases and pests as well as to threat from climate change, drought, soil erosion, and more. Our gene banks have thousands of crop varieties available for research. Today, we are losing genetic resources at unprecedented rates while the world's capacity to maintain food security decreases. Agricultural sustainability is dependent on a strong

Fig. 3.4 CGIAR Gene Banks. (Source: https://www.genebanks.org/genebanks/)

conservation capacity (http://www.iita.org/research/genetic-resources/). CGIAR is a global research partnership for a food-secure future. CGIAR science is dedicated to reducing poverty, enhancing food and nutrition security, and improving natural resources and ecosystem services. CGIAR have 15 independent research centers (Fig. 3.4). These 15 research centers are, viz., the African Rice Center, Bioversity International, Center for International Forestry Research (CIFOR), International Center for Agricultural Research in the Dry Areas (ICARDA), International Center for Tropical Agriculture (CIAT), International Crops Research Institute for the Semi-Arid Tropics (ICRISAT), International Food Policy Research Institute (IFPRI), International Institute of Tropical Agriculture (IITA), International Livestock Research Institute (ILRI), International Maize and Wheat Improvement Center (CIMMYT), International Potato Center (CIP), International Rice Research Institute (IRRI), International Water Management Institute (IWMI), World Agroforestry (ICRAF), and WorldFish across the globe. Out of these 15 research centers, 11 centers have gene bank facility for the conservation of crop genetic resources. The 11 CGIAR gene banks (AfricaRice, Bioversity, CIAT, CIMMYT, CIP, ICRAF, ICARDA, ICRISAT, IITA, ILRI, and IRRI) conserve near about 768,576 accessions of cereals, grain legumes, forages, tree species, root and tuber crops, and banana (Table 3.8). Many of these accessions are crop wild relatives.

3.12.2 Svalbard Global Seed Vault/Doomsday Vault

The idea of having a global security storage facility in Svalbard to house duplicates of seed conserved in gene banks all over the world began being discussed in the 1980s. In 1984, the Nordic Gene Bank (now NordGen) had established a backup

Table 3.8 Number of accessions per crop as per CGIAR gene banks

S. No.	Crop	Gene bank	Number of accessions
1.	Andean roots and tuber crop	CIP	2526
2.	Banana	Bioversity and IITA	1865
3.	Barley	ICARDA	32,560
4.	Beans	CIAT	37,987
5.	Cassava	CIAT and IITA	10,041
6.	Chickpea	ICRISAT and ICARDA	36,344
7.	Cowpea	IITA	15,115
8.	Faba bean	ICARDA	10,033
9.	Forages	CIAT, ICARDA, and ILRI	70,514
10.	Grass pea	ICARDA	4289
11.	Groundnut	ICRISAT	15,622
12.	Lentil	ICARDA	14,512
13.	Maize	CIMMYT and IITA	29,900
14.	Mung bean	AVRDC	6742
15.	Pea	ICARDA	6121
16.	Pearl millet	ICRISAT	23,092
17.	Pigeon pea	ICRISAT	13,788
18.	Potato	CIP	6809
19.	Rice	IRRI and AfricaRice	149,217
20.	Small millet	ICRISAT	11,466
21.	Sorghum	ICRISAT	39,948
22.	Tree: fruit and multipurpose	ICRAF	9194
23.	Wheat	CIMMYT and ICARDA	197,945
24.	Yam	IITA	5839

Source: https://www.genebanks.org/resources/crops/

seed storage facility in an abandoned coal mine outside Longyearbyen, and the idea of establishing a worldwide backup storage gradually evolved. During the year 2001, several negotiations that led to the International Treaty on Plant Genetic Resources for Food and Agriculture (ITPGRFA) were finalized, and the treaty opened to signatures by national governments. The treaty calls for establishing a multilateral system for plant genetic resources that includes global rules for access to and benefit sharing of those resources. Keeping this in mind, Bioversity International (then IPGRI) approached and encouraged the government of Norway to consider the establishment of a global facility at Svalbard. Based on the results of a 2004 feasibility study and the endorsement and welcoming of the initiative by the FAO Commission on Genetic Resources for Food and Agriculture, in October 2004, the Norwegian government committed to fund and establish the Svalbard Global Seed Vault. The seed vault became operational on 26 February 2008 (https://www.seedvault.no/history/). It is located deep inside a mountain on a remote island in the Svalbard archipelago, halfway between mainland Norway and near the North Pole. Worldwide, more than 1700 gene banks hold collections of food crops for safekeeping as these collections are prone to natural as well as man-made disasters and loss of a crop variety is as irreversible. In order to safeguard, a long-term seed storage facility was built at Svalbard. The purpose of the vault is to store duplicates (backups) of seed samples from world's crop collection. Currently, the vault holds 983,524 samples originating from almost every country in the world. These samples consist of more than 6000 crop species comprising 1,075,954 seeds deposited by 76 depositors. The seed collections range from unique varieties of major African and Asian food staples such as maize, rice, wheat, cowpea, and sorghum to European and South American varieties of eggplant, lettuce, barley, and potato. In fact, the vault already holds the most diverse collection of food crop seeds in the world (https://www.croptrust.org/our-work/svalbard-global-seed-vault/). ICRISAT, Hyderabad, maintained a collection of 126,000 germplasm accessions assembled from 144 countries of the 6 mandate crops, i.e., sorghum, pearl millet, chickpea, pigeon pea, groundnut, and finger millet, and 5 small millets – foxtail millet, little millet, kodo millet, proso millet, and barnyard millet. At the moment, 111,173 accessions out of these are duplicated in the seed vault, and in the future, it will continue to ship the rest of the germplasm accession seeds to Svalbard until copies of the entire collection are deposited there (https://www.seedvault.no/news/close-to-90-of-the-icrisat-collection-duplicated/). India has one of the largest plant germplasm collections in the world, hosted in the National Gene Bank at ICAR-National Bureau of Plant Genetic Resources (NBPGR), New Delhi. The National Gene Bank holds about 440,991 seed accessions, in addition to around 1868 accessions in vitro and 13,579 accessions in cryopreservation. NBPGR signed the depositor agreement in 2014 and has so far made two small deposits of seeds to the seed vault, in 2014 and 2017, of pigeon pea, sorghum, and rice (https://www.seedvault.no/news/eager-to-deposit-more-seeds-in-svalbard/).

3.12.3 Indian Gene Bank Scenario: NBPGR Gene Bank

With an area of 3.287 million km^2, India is seventh largest country of the world with spectacular climatic and geographical diversity. It is one of the seventeen mega-diverse countries of the world. Four global biodiversity hot spots of the total 35 in the world have been identified in India, and these are the Western Ghats, the Himalayas, the Indo-Burma region, and the Sundalands. Moreover, India is 1 of the 8 Vavilovian centers of origin and 1 of the 12 primary centers of diversity of crop plants. It also ranks among the top ten nations of the world on the species richness index. Therefore, from the plant genetic resources point of view, India is a very important country, and this is the reason that the country is having one of the most robust plant genetic resources conservation and utilization programs in the world. The National Bureau of Plant Genetic Resources, New Delhi, under the aegis of the Indian Council of Agriculture Research is the nodal organization for all PGR-related activities in India. The Indian Plant Genetic Resources System comprises the National Active Germplasm Sites (NAGS) comprising ICAR institutes, the Project Directorates, the National Research Centers, the All India Coordinated Research Projects, the State Agricultural Universities, and the regional stations of the NBPGR. The NBPGR has ten regional stations located across the country representing different agroclimatic zones, and these are Srinagar, Bhowali, Shimla, Shillong, Ranchi, Cuttack, Hyderabad, Thrissur, Akola, and Jodhpur. The regional stations are mandated to collect, characterize, and conserve the locally important species of cultivated plants and their wild relatives. The germplasm collected and characterized by the regional stations of the NBPGR is conserved in the field gene banks and medium-term storage of the stations and in the long-term storage module, cryobank, and tissue culture repository of the National Gene Bank at the NBPGR headquarters at New Delhi. The Indian National Gene Bank located at the National Bureau of Plant Genetic Resources, New Delhi, is the second largest gene bank of the world with 440,991 accessions (Table 3.9) belonging to 1947 species of crop plants and their wild relatives. The bureau has also established a cryopreservation facility for difficult-to-conserve species, and 13,579 accessions have already been cryopreserved (Table 3.10). Besides, permafrost facility has also been developed in collaboration with the Defence Research and Development Organisation at Chang-La, Leh, for conservation of the safety duplicates of the accessions conserved at the National Gene Bank for retrieval in case of any eventuality.

3.13 Narrowing Down of Crop Diversity and Its Impact on Genetic Gain

3.13.1 Genetic Bottlenecks

Plant breeding has a strong influence on food production and will play a vital role in the world food security in the future. It is the science, art, technology, and business of improving crop plants for the betterment of humankind. Genetic variation is the building block of any crop improvement programs. Genetic variation can be created by domestication, germplasm collection, plant introduction, hybridization, mutation, polyploidy, somaclonal variation, and genetic engineering approaches. These different plant breeding methods showed different impacts on plant breeding diversity as plant introduction increased genetic diversity, selection enhanced genetic differentiation at the expense of genetic diversity, and intraspecific hybridization reduced genetic diversity. Plant breeding collects, induces, and rearranges genetic diversity upon which selection is acting and results in the development of uniform cultivars by the plant breeders across the world. It is the beauty of nature which maintains a perfect balance of diversity among the crop plants; it is the plant breeding which decides increase/decrease in the genetic diversity through various breeding procedures. As we know, plant breeding intends to produce uniformity in the cultivars which becomes a limitation of plant breeding as crop uniformity is vulnerable to biotic and abiotic stress which will lead us toward the disease disasters such as the world-famous Irish Potato Famine in the 1840s, US Victoria oat blight and corn blights epidemics in 1946 and the 1970s, and East Africa Ug99 wheat stem rust a new virulent race in 1999, respectively. Such disease disasters present very tough and complicating situation and forced farmers and breeders to compromise between crop yield and crop failure. This seems to be challenging for researchers to

Table 3.9 Status of base collection at National Gene Bank of NBPGR, New Delhi

S. No	Crop/crop group	Number of accessions conserved	S. No	Crop/crop group	Number of accessions conserved
A.	**Cereals**	**1,64,401**	**H**	**Oilseeds**	**59,170**
1	Paddy	1,09,327	1	Groundnut	13,838
2	Wheat	33,547	2	Oilseed brassica	11,404
3	Maize	11,237	3	Safflower	7367
4	Other	10,290	4	Sesame	10,306
B.	**Millets**	**59,270**	**5**	Soybean	4222
1	Sorghum	26,091	6	Sunflower	1410
2	Pearl millet	8235	7	Others	10,623
3	Minor millet	24,944	I	**Fiber**	**15,704**
C.	**Forages**	**7088**	**1**	Cotton	10,045
1	Oats	1384	2	Jute	3303
2	Clover	598	3	Mesta	2017
3	Teff	298	4	Others	339
4	Marvel grass	334	**J**	**Vegetables**	**26,327**
5	Others	4474	1	Tomato	2595
D	**Pseudo cereals**	**7590**	**2**	Brinjal	4461
1	Amaranth	6156	3	Chilli	4972
2	Buckwheat	1054	4	Okra	3714
3	Others	380	5	Onion	1132
E	**Legumes**	**66,634**	**6**	Others	9453
1	Chickpea	14,724	**K**	**Fruits and nuts**	**276**
2	Pigeon pea	11,617	1	Buchanania	97
3	Mung bean	4224	2	Others	179
4	Pea	4466	**L**	**Medicinal and aromatic plants and narcotics**	**8068**
5	Cowpea	3842	1	Opium poppy	466
6	French bean	3985	2	Ocimum	608
7	Cluster bean	4315	3	Tobacco	2271
8	Horse gram	3066	4	Others	4723
9	Rice bean	2171	**M**	**Spices and condiments**	**3152**
10	Others	14,224	1	Coriander	1078
F	**Ornamental**	**659**	2	Fenugreek	1310
1	Marigold	366	3	Others	764
2	Others	293	**N**	**Duplicate safety samples (lentil, pigeon pea)**	**10,235**
G	**Agroforestry**	**1646**	**O**	**Trial material (wheat, barley)**	**10,771**
1	Sesbania	647			
2	Others	999			
	Total	**4,40,991**[a]			

Source: http://www.nbpgr.ernet.in/Research_Projects/Base_Collection_in_NGB.aspx

[a]The figure includes 5034 released varieties and 4316 genetic stocks; regenerated accession not included. No. of crop species conserved: 1762

Table 3.10 Status of tissue culture and cryopreservation at NBPGR, New Delhi

Crop/crop group	Present status	Crop/crop group	Present status
In vitro bank		Cryobank	
Tropical fruits	430	Recalcitrant	–
Temperate and minor tropical fruits	350	Intermediate	6782
Tuber crops	518	Orthodox	3902
Bulbous crops	171	Dormant bud (mulberry)	389
Medicinal and aromatic plants	172	Pollen	572
Spices and industrial crops	227	DNA	1934
Total	1868	Total	13,579

Source: http://www.nbpgr.ernet.in/Research_Projects/Base_Collection_in_NGB.aspx

develop effective breeding strategies for the understanding of crop genetic diversity (CGD) for achieving sustainable agriculture goal. There is no doubt that domestication and crop evolution, which took place in short window which clearly indicates that majority of crop plants have experienced several genetic bottlenecks in the history of a crop, led to the present diversity level. Such domestication bottleneck, population bottleneck, dispersal bottleneck which was facilitated by founder effects and genetic drift, and modernization bottleneck which includes modern plant breeding methodologies and policies, respectively, are responsible for narrowing down of genetic diversity among the highly productive crop plants at wild ancestor, landrace, and cultivar levels.

3.13.2 Domestication Bottleneck

Since the dawn of agriculture near about 10,000 years ago, transition from gathering wild plants to cultivation under human management resulted in the domestication of wild plants into highly productive crop plants required for the supply of food, fiber, fuel, and other uses (Tanksley and McCouch 1997; Vaughan et al. 2007; Hallauer 2011). As earlier stated, domestication is an important historical event which is experienced by majority of todays' food crop and also shaped the genetics, breeding, and diversity of crop plants (Louwaars 2018). Hence, it is assumed that during the process of domestication, earlier farmers select useful variation from the wild types that were most productive and provide need-based products for the development of improved varieties with trait of interest. The domesticated species resemble to their wild ancestors because of domestication gene selection. Domestication bottleneck severely reduces the diversity in the improved varieties as the first generation of plant breeders, i.e., farmers selected only few useful genes and left other genes in the wild progenitor that seems to be not useful at the time but has become useful in the present generation of plant engineers (Vaughan et al. 2007). Several morphological, physiological, biochemical, and genetical changes occurred in domesticates with respect to his progenitor during domestication. Besides phenotypic changes, genomes of the domesticates also undergo changes which lead to moderate-level reduction in the genetic diversity in domesticates relative to their wild ancestors due to sampling process as very few wild plants were selected for domestication (Flint-Garcia 2013).

3.13.3 Population Bottlenecks

In addition to domestication bottleneck, a genetic bottleneck also occurs when the genetic diversity of a population reduced due to decrease in its population size named as population bottleneck. Population bottlenecks occur when a population's size is reduced for at least one generation. Because genetic drift acts more quickly to reduce genetic variation in small populations, undergoing a bottleneck can reduce a population's genetic variation by a lot, even if the bottleneck doesn't last for very many generations. Coupling of population bottleneck with domestication bottleneck severely reduced diversity among domesticates. The severity of reduction in the diversity associated with domestication varied much in differently crop species. Grass family members had less severe bottlenecks because of their large effective population size (Buckler et al. 2001). With the progression of time, genetic diversity likely to have lost due to random genetic drift in small population size as a result of new crop introductions or crop failure. A founder effect may work when cultivation of crop plants moved away from their center of diversity, and it occurs due to reduced genetic diversity and nonrandom sampling of genes in the original population. There is no second opinion on this that polyploidization and mutation increase the genetic variation, and it is evident from the results of Lelley et al. (2000) which clearly mentioned that *Aegilops tauschii,* one of the parents of hexaploid bread wheat, has more genetic variation compared to *A. tauschii*-derived D-genome. This is because polyploidization or particular mutations are associated with domestication process in case of small effective population size. Outbreeding crop plants more prone to inbreeding and in small outcross population results in the deleterious effects and loss of fitness. Population bottlenecks severely degrade heterozygosity and genetic diversity (Pimm et al. 1989). Therefore, it is well suggested to researchers to maintain an effective population size for planned utilization of genetic diversity in crop improvement programs. From this, a clear picture can be drawn that there are still chances of diversity enhancement through introgression of desired alleles from the wild population.

3.13.4 Dispersal Bottleneck

Regions outside the center of crop domestication where several crop species spread by the agency of man, wind, water, or natural means or introduced by humans where a considerable amount of crop diversity exists are defined as secondary center of crop diversity. This process established a secondary bottleneck, dispersal bottleneck, which caused a severe reduction in the original crop genetic diversity away from the primary center of crop diversity. Such crop migration has been reported in soybean and mung bean which cause narrow genetic base. Mung bean dissemination from its center of origin in Western Asia to Southeast Asia led to a reduction in seed protein diversity (Tomooka et al. 1992). Similarly, soybean originated in China, and its diffusion to other parts of the world subjected to dispersal bottleneck which narrows the genetic base of soybean outside Asia (Shoemaker 1986). Several breeding barriers such as crop breeding system subsequently restricted the gene flow between domesticates and wild ancestors (Cooper et al. 2001). If the gene flow is either very less or absent with wild relatives than in that case, diversity must have increased after initial bottlenecks. But when gene flow is very high, it gradually increases genetic diversity for those genes which remain unselected as a result of both domestication and dispersal bottlenecks. Sometimes, what happens is genetic diversity increased in landraces after initial bottlenecks and reached as near as of wild relatives as evident for barley landraces in Eastern Mediterranean (Jana and Pietrzak 1988) and in Mexican *Capsicum annuum* (Hernandez-Verdugo et al. 2001). Such bottlenecks make introgression somewhat difficult between domesticates and wild relatives, but landraces can be utilized for development of improved cultivars.

3.13.5 Modernization Bottleneck

The rediscovery of Mendel's laws of genetics in 1900 and Darwin's theory of natural selection both together led the evolution of plant breeding methodologies for the genetic improvement of crop plants. So many new cultivars have been developed through different breeding methods ranging from domestication, introduction, and phenotypic selection followed by hybridization to genomic approaches such as marker-assisted selection, genomic selection, etc. for trait of interest (Allard 1999). Plant breeding always aimed toward yield enhancement, adaptation, genetic uniformity, agronomic stability, biotic and abiotic stress resistance, and quality enhancement of end product. Manipulation of crop genetic systems to develop superior high yielding varieties (HYVs) utilizing genetic diversity which can be created/introduced by crossing morphologically or physiologically or geographically or genetically divergent selected plants with desired traits is the central theme of plant breeding (Breseghello 2013). With the advancement and refinement in the breeding methodologies, it becomes easy and possible to develop genetically uniform varieties which replaced ancient landraces from cultivation at farmers' field and initiate the decay of genetic variability. Modernization bottleneck comprised two important aspects: replacement of landraces with modern cultivars and trends in crop diversity in newly released cultivars (van de Wouw et al. 2010a). During green revolution, the improved and uniform modern varieties replaced the most of the cultivated landraces across the world as evident from the first semidwarf rice variety IR8 which is cultivated on large area in Asia and Latin America (Evenson and Gollin 2003). Similarly, semidwarf wheat varieties replaced a wide range of landraces in many countries. This takes breeder's attention toward the importance of acquisition, characterization, evaluation, and utilization of genetic diversity found in the crop genetic resources of food crop plants. Initially, at the time of introduction of modern cultivars, diversity increased, and this transition of replacement of landraces cultivars with modern cultivars occurred at a slow pace. It is a gradual process, and duration of this change varies across the crops and regions as still it is ongoing in developing countries. Also, it is important to know about the level of diversity existing in the landraces and cultivars during the transition phase of displacement of landraces with modern cultivars (van de Wouw et al. 2010a).

With technological advancement over the last three decades (1985–1995, 1995–2005, 2005–2105), literature is full of too many crop genetic diversity assessment studies. Rodgers et al. (1983), Ortiz et al. (2003), Cox et al. (1985), and van Beuningen and Busch (1997), basing upon phenotypic and pedigree data, revealed that satisfactory progress has been achieved in yield enhancement at the cost of reduction in the genetic diversity of crop plant gene pools (Cox et al. 1986; Smith et al. 2004). Some studies revealed increase in diversity (Fu et al. 2007; White et al. 2008), while some studies found decrease in diversity over the time (Fu et al. 2003; Reif et al. 2005; Hao et al. 2006; Malysheva-Otto et al. 2007). Besides this, few workers (Roussel et al. 2004; Qi et al. 2006) observed increase in diversity after decrease in diversity level. From meta-analysis reviews of 23 articles by Fu (2006) of DNA markers RFLP, RAPDs, and SSRs published during 2000–2005 in 8 reputed journals associated with plant breeding, it is revealed that allelic reduction in CGD at individual level accompanying with genetic improvement was minor over. By reviewing 110 published research articles, van de Wouw et al. (2010a) reached the conclusion that genetic erosion occurs at three levels, i.e., crop, variety, and allele. In an another meta-analysis of 44 published papers by van de Wouw et al. (2010b), it is observed that there is no gradual narrowing of genetic base in the varieties released by breeders. A significant diversity reduction of 6% occurred before the 1960s, and after onward, increased in diversity was found through plant breeding methodologies. During the same year, Rauf et al. (2010) reviewed 230 publications and found that losses of genetic diversity were found in released cultivars followed by wild ancestors and ancient landraces. In general meta-analysis of diversity papers helpful in studying the diversity trends in crop plants, and from this, it is found that overall no substantial diversity reduction occurred (van de Wouw et al. 2010b).

3.14 Genetic Erosion and Decay of Diversity: The Global Scenario

Genetic erosion is defined as the loss of genetic diversity in general. In particular, it is the process of replacement of ancient indigenous landraces and varieties by modern high-yielding varieties, which is equated to loss of genes (Qualset et al. 1997). It results in the reduction in the genetic diversity among crop plants at species and allele level, reduction in the number of plants of a species or in geographic range of species at species level (Friis-Hansen 1999), and reduction in the richness and evenness of alleles at allelic level (van de Wouw et al. 2010a). Genetic erosion occurs as a result of several genetic bottlenecks experienced by crop plants at three levels – crop, variety, and allele. Baur (1914) already warned the world about the consequences of loss of genetic diversity. Harlan (1970) spoke that loss of varietal diversity is not affordable for the humankind. At the same time, Frankel and Bennett (1970) also stated that the genetic variation treasure present in the crop diversity center may become invisible. According to an estimate of the committee on the threatened plants of IUCN, about 10% of the flowering plants are under threat of extinction (Swaminathan 1978). During the 1990s, loss of diversity has drawn the attention of all the researchers across the globe, and genetic erosion was noted at the top of Agenda 21 during United Nations Conference on Environment and Development (UNCED), Earth Summit conference in 1992. The FAO report on the State of World's Plant Genetic Resources, released at Leipzig in 1996, listed the main causes of plant genetic erosion in 154 countries. These mainly included over-replacement of local varieties (80 countries), deforestation (over 60 countries), overexploitation (nearly 55 countries), and population pressure and urbanization (45 countries). Other causes noted in the order of importance were environmental effects, overgrazing, legislation and policy, agricultural changes, diseases and pests, and shifting cultivation.

3.14.1 Examples of Genetic Erosion in Crops, Landraces, and Wild Relatives (Source: FAO 1996b)

1. A survey of far household in the Republic of Korea showed that in 14 crops cultivated in home gardens, an average of only 26% of the landraces cultivated there in 1985 were still present in 1993.
2. In China, in 1949, nearly 10,000 wheat varieties were in used in production. By the 1970s, only about 1000 varieties remained in use.
3. In Malaysia, the Philippines, and Thailand, it is reported that local fruit varieties are gradually being replaced with better-quality varieties, such as guava in Malaysia and rambutan in the Philippines.
4. In Ethiopia, traditional barley and durum wheat varieties are suffering serious genetic erosion due to displacement by introduced modern varieties. In highlands of West Shewa, of 14 barley landraces described by farmers, only 4 were still found in cultivation (Eticha et al. 2010). About 65% of barley landraces were lost in North Shewa during 1994–2010 due to introduction of improved varieties, recurrent drought, changed land use pattern, etc. Above 70% of tetraploid wheat varieties were lost because of expansion of improved wheat varieties and tef, difficulties in seed supply, reduction in farm size, etc. (Teklu and Hammer 2006; Tsegaye and Berg 2007).
5. A total of 19,000 rice varieties were cultivated in Chhattisgarh state in India. In the 1960s, almost all the local varieties were replaced by HYVs of rice (Chaudhuri 2005).
6. In Italy, several field and garden crop loss corresponds to 72.8% (Hammer et al. 1996).
7. In the Federal Republic of Yugoslavia (Serbia and Montenegro), it was estimated that the area sown with old varieties of wheat is now less than 0.5%.

8. The large-scale genetic erosion of local varieties of native Andean crops, such as *Ullucus tuberosus*, *Oxalis tuberosa*, *Tropaeolum*, *Polymnia sonchifolia*, *Mirabilis expansa*, and *Pachyrhizus tuberosus*, is reported in the Country Report of Ecuador. Argentina reported the genetic erosion of Amaranthus and quinoa.
9. The Country Report of Uruguay stated that many landraces of vegetables and wheat have now been replaced by modern varieties, and in Costa Rica, genetic erosion is reported within the native gene pools of cultivated maize and *Phaseolus vulgaris*, due to landrace replacement by modern varieties.
10. Chile reports genetic erosion of local potato varieties as well as other crops such as oats, barley, lentils, watermelon, tomato, and wheat.
11. Only 20% of the local maize varieties cultivated in 1930 are now known in Mexico.
12. Indigenous potato varieties are lost due to pests and diseases and introduction of improved varieties in Chile, Peru, and Bolivia (Ochoa 1975).
13. Traditionally cultivated vegetable diversity decreased in Varamin plains in Iran (Davari et al. 2013).
14. Einkorn wheat is reported to be extinct and emmer wheat subject to strong genetic erosion (Laghetti et al. 2009).
15. Mali reported that 60% of local varieties of sorghum were lost in one region over the previous 20 years due to expansion of cotton production, introduction of maize cultivation, and saturation of the available cropping area.
16. Average rates of landrace losses were estimated to be 32% for cassava in 21 villages of Republic of Congo (Kombo et al. 2012).
17. Erosion in a number of wild fruit trees species was observed on the Italian island Sardinia due to degradation of natural environment (Chessa and Nieddu 2005).
18. African wild rice (*Oryza longistaminata*, *O. barthii*, *O. punctata*, *O. eichingeri*, and *O. brachyantha*) populations were threatened by extensive land use and overgrazing (Kiambi et al. 2005).

3.15 Conserving Natural Genetic Variation: An Overview

According to the Convention on Biological Diversity (2007), international conservation policy recognizes biodiversity at three levels, ecosystem, species, and genetic, and that management should aim to retain all three. Yet current approaches to biodiversity conservation are largely based on geographic areas, ecosystems, ecological communities, and species, with less attention on genetic diversity and the species-population continuum. The species metric of biodiversity is focused with some consideration of intraspecific units within intensively managed threatened species. Hence, there is a natural desire to categorize species, and other taxa (i.e., subspecies), and sometimes conservation units within species. As a fundamental unit of conservation, species are often used to quantify biodiversity value through lists of species within protected areas, identification of threatened species within jurisdictions, and as a basis for biosecurity procedures.

Like many fields of biological science, systematics, the naming of taxa, and understanding their relationships and formation are being significantly advanced by genomics and statistical models. Whether through whole-genome sequencing (Nater et al. 2017) or genome sampling (SNPs, target capture sequencing) (Jones and Good 2016; Leache and Oaks 2017), we now have unprecedented resolution of patterns of genetic diversity (Funk et al. 2012). Coupled with powerful statistical models for inferring histories of genomic and species divergence, these data are providing new insights into the evolutionary processes that generate species and genetic variation.

Insights to many studies are revealing highly divergent genetic populations within named species, so-called cryptic species (Bickford et al. 2007; Jorger and Schrödl 2013; Struck et al. 2017). On the other hand, genome-scale analyses are also revealing that (as long been recognized for plants) genetic exchange (introgression) among animal species is more common than previously thought, both during and after speciation, and can even drive new adaptive radiations (Mallet 2007; Rieseberg and Willis 2007; Arnold and Kunte 2017). From the very beginning of agriculture, natural genetic variability has been exploited within crop species to meet subsistence food requirement, and now it is being focused to surplus food for growing populations. In the middle of the 1960s, developing countries like India experienced the green revolution by meeting food demand with help of high-yielding and fertilizer-responsive dwarf hybrids/varieties especially in wheat and rice. Overall, these insights further emphasize that speciation is protracted (Roux et al. 2016) and reveal the opposing forces operating through the phylogeography-phylogenetics continuum (Edwards et al. 2016).

Diversity in plant genetic resources (PGR) provides opportunity for plant breeders to develop new and improved cultivars with desirable characteristics, which include both farmer-preferred traits (yield potential, large seed, etc.) and breeder-preferred traits (pest and disease resistance, photosensitivity, etc.). Situation again worsened with genetic erosion (loss of genetic diversity and loss of adaptive genes (landraces)). Today with an advancement of agricultural and allied science and technology, we still ask ourselves whether we can feed the world in 2050; this question was recently sensitized at the world food prize event in 2014 and remains that unanswered in every one hands since global population will exceed 9 billion in 2050.

Furthermore, high-throughput sequencing has also been used to obtain measurements on aspects of the genome other than nucleotide sequence variation. GBS is a versatile and inexpensive method for natural and experimental populations of crops and their wild relatives. Archaeogenetics – the extraction, sequencing, and analysis of ancient DNA fragments – has transformed our concepts of the history of human and animal species. Until recently, plant genetics has not had access to this window into the past, mainly owing to a paucity of archaeobotanical remains containing sufficient amounts of well-preserved DNA. Recent studies in maize and barley retrieved

DNA sequences from samples preserved under arid conditions and analyzed them together with sequences from extant individuals. da Fonseca et al. (2015) thus reconstructed the past demography of maize in the southwest USA.

3.16 Significance of Plant Genetic Conservation

After the historical events, the importance of PGR had only got popular recognition when the spread of green revolution across cultivated crops threatened the conservation of landraces. Green revolution technologies introduced improved crop varieties that have higher yields, and it was hoped that they would increase farmers' income. Consequently, the Consultative Group for International Agricultural Research (CIGAR) initiated gene banks and research centers of domestication for conserving PGR in most of the stable food crops around the world. The Food and Agriculture Organization (FAO) supported the International Treaty on Plant Genetic Resources (ITPGR) and UN supported the Convention on Biological Diversity (CBD) which is the international agreements that recognize the important role of genetic diversity conservation. Such treaty still plays in current and future food production as one of the major supremo.

Genetic diversity is the key pillar of biodiversity and diversity within species, between species, and of ecosystems (CBD, Article 2), which was defined at the Rio de Janeiro Earth Summit. Obstacle is that modern high-yielding crop varieties, especially, have been developed primarily under well-endowed production conditions. Such varieties are often not suitable for low-income farmers in marginal production environments as they are facing highly variable stress conditions. Landraces or traditional varieties have been found to have higher stability (adaptation over time) in low-input agriculture under marginal environments; thus, their cultivation may contribute farm-level resilience in the face of food production shocks.

The aim of conservation genetics is to maintain genetic diversity at many levels and to provide tools for population monitoring and assessment that can be used for conservation planning. Conservation efforts and related research are rarely directed toward individuals, but genetic variation is always measured in individuals. It is possible to identify the genetic variation from phenotypic variation by either quantitative traits (traits that vary continuously and are governed by many genes, e.g., plant height) or discrete traits that fall into discrete categories and are governed by one or few major genes (e.g., white, pink, or red petal color in certain flowers) which are referred to as qualitative traits. Genetic variation can also be identified by examining variation at the level of enzymes using the process of protein electrophoresis. Further, genetic variations can also be examined by the order of nucleotides in the DNA sequence.

3.17 Genetic Variation Assessment

Various techniques such as (i) morphological; (ii) biochemical characterization/evaluation (allozyme), in the pre-genomic era; and (iii) DNA (or molecular) marker analysis especially single nucleotide polymorphism (SNPs) are utilized in post-genomic era that can exhibit similar modes of inheritance, as we observe for any other traits, that is, dominant/recessive or codominant. If the genetic pattern of homozygotes can be distinguished from that of heterozygotes, then a marker is said to be codominant. Generally codominant markers are more informative than the dominant markers.

Morphological markers are based on visually accessible traits such as flower color, seed shape, growth habits, and pigmentation. No expensive technology is required, but big land area is required for carrying out field experiments making it possibly more expensive. These marker traits are often susceptible to phenotypic plasticity. These still have advantage and are mandatory for distinguishing the adult plants from their genetic contamination in the field, for example, spiny seeds, bristled panicle, and flower/leaf color variants.

The second type of genetic marker is called biochemical markers, allelic variants of enzymes called isozymes (codominant) detected by electrophoresis and specific staining. Simple inheritance studies and functional genetics could be studied. Only small amounts of plant material have been utilized for its detection. However, with the limited number of enzyme markers available, these enzymes do not fulfill complex structural problems; thus, genetic diversity is limited to explore.

The third and most widely used genetic marker type is molecular markers, comprising a large variety of DNA molecular markers, which can be employed for analysis of genetic and molecular variation. These markers can detect the variation due to deletion, duplication, inversion, and/or insertion in the chromosomes. These markers themselves do not affect the phenotype of the traits of interest because they are located only near or linked to genes controlling the traits. These exist in both dominant and codominant patterns. Different markers have different genetic qualities. A molecular marker can be defined as a genomic locus, detected through probe or specific starter (primer) which, in virtue of its presence, distinguishes unequivocally the chromosomic trait which it represents as well as the flanking regions at the 3′ and 5′ extremity. Molecular markers may or may not correlate with phenotypic expression of a genomic trait. They offer numerous advantages over conventional, as they are stable and detectable in all tissues regardless of growth, differentiation, development, or defense status of the cell. Also, they are not confounded by environmental, pleiotropic, and epistatic effects. Now germplasm characterization based on molecular markers has gained importance due to the speedy and quality of data generated.

Molecular markers may be broadly divided into three classes based on the method of their detection: hybridization-based, polymerase chain reaction (PCR)-based, and DNA sequence-based. Restriction fragment length polymorphisms (RFLPs) are hybridization-based markers developed first in human-based genetic study during the 1980s (Botstein et al. 1980; Martinville et al. 1982), and later they were used in plant research. RFLP is based on the variation(s) in the length of DNA fragments produced by a digestion of genomic DNAs, and hybridization to specific markers of two or more individuals of a species is compared. RFLPs have been used extensively to compare genomes in the major cereal families such as rye, wheat, maize, sorghum, barley, and rice. RFLPs, being codominantly robust, help detect unlimited number of loci. Presently, polymerase chain reaction (PCR)-based marker systems are more rapid and require less plant material for DNA extraction. Rapid amplified polymorphic DNAs (RAPDs) were the first of PCR-based markers and are produced by PCR machines using genomic DNA and arbitrary (random) primers which act as both forward and backward primers in creation of multiple copies of DNA strands with advantage of having multiple loci from a single primer with only small amounts of DNA. Amplified fragment length polymorphisms (AFLPs) combine both PCR and RFLP (Mohan et al. 1997). AFLP is generated by digestion of PCR amplified fragments using specific restriction enzymes that cut DNA at or near specific recognition site in nucleotide sequence. Being highly reproducible, these markers enable rapid generation and high frequency for identifying polymorphisms and for assessing linkages among individuals from a segregating population. Another class of molecular markers depends on the availability of short oligonucleotide repeat sequences in the genome of plants such as SSR, STS, SCAR, EST-SSR, and SNP.

Microsatellites are also known as simple sequence repeats (SSRs), short tandem repeats (STRs), or simple sequence length polymorphisms (SSLPs) which are short tandem repeats, their length being 1–10 bp. Some of the literatures define microsatellites as 2–8 bp, 1–6 bp, or even 1–5 bp repeats. SSRs are highly variable and evenly distributed throughout the genome and common in eukaryotes, their number of repeated units varying widely among crop species. The repeated sequence is often simple, consisting of two, three, or four nucleotides (di-, tri-, and tetranucleotide repeats, respectively). These polymorphisms are identified by constructing PCR primers for the DNA flanking the microsatellite region. The flanking regions tend to be conserved within the species, although sometimes they may also be conserved in higher taxonomic levels. The loci identified are usually multi-allelic and codominant. Bands can be scored either in a codominant or as present or absent.

The microsatellite-derived primers can often be used with many varieties and even other species because the flanking DNA is more likely to be conserved. These are evenly distributed universally in the genome, easily automated, and highly polymorphic and have good analytic resolution with high reproducibility making them a preferred choice of markers (Matsuoka et al. (2002) for individual genotyping, germplasm evaluation, genetic diversity studies, genome mapping, and phylogenetic and evolutionary studies.

An alternative source of SSRs development is development of expressed sequence tag (EST)-based SSRs using EST databases. With the availability of large numbers of ESTs and other DNA sequence data, development of EST-based SSR markers through data mining has become fast, efficient, and relatively inexpensive compared with the development of genomic SSRs. However, the development of EST-SSRs is limited to species for which this type of database exists as well as being reported to have lower rate of polymorphism compared to the SSR markers derived from genomic libraries.

Single nucleotide polymorphisms (SNPs) are DNA sequence variations that occur when a single nucleotide (A, T, C, or G) in the genome sequence is changed, that is, single nucleotide variations in genome sequence of individuals of a population. These polymorphisms are single-base substitutions between sequences. SNPs occur more frequently than any other type of markers and are very near to or even within the gene of interest. SNPs can be identified by using either microarrays or DHPLC (denaturing high-performance liquid chromatography) machines. These are used for rapid identification of crop cultivars and construction of ultrahigh-density genetic maps. They provide valuable markers for the study of agronomic or adaptive traits in plant species, using strategies based on genetic mapping or association genetics studies.

A DArT marker is a segment of genomic DNA, the presence of which is polymorphic in a defined genomic representation. A DArT was developed to provide a practical and cost-effective whole-genome fingerprinting tool providing high-throughput and low-cost data production. It is independent from DNA sequence; that is, the discovery of polymorphic DArT markers and their scoring in subsequent analysis does not require any DNA sequence data. DArT markers can be used as any other genetic marker. With DArT, comprehensive genome profiles are becoming affordable regardless of the molecular information available for the crop. DArT genome profiles are very useful for characterization of germplasm collections, QTL mapping, reliable and precise phenotyping, and so forth. DArT markers are primarily dominant (present or absent) or different in intensity, which limits their value in some application.

Diverse genetic resources are priceless assets for humankind that cannot be lost. These must be increasingly required to assess the feeding burgeoning world's population in the near future by the 2050s. This genetic variability is much important to breeders to develop new hybrids and varieties which can be achieved through use of molecular and phenotypic characterization. Low cost and affordable high throughput analysis or automation are important factors to be considered when choosing a technology. With NGS, time and costs are reduced for sequencing the whole genomes. Many other softwares and packages are nowadays available to assess molecular and phenotypic diversity parameters that increased the efficiency of germplasm curators, thereby helping plant breeders to pace up crop improvement.

3.18 Broadening of Genetic Base of Crops

Plant breeding is a man-directed crop evolution which comprises creation of genetic variation (natural and induced) and exploitation of genetic variation through selection, evaluation, and characterization of genetic variation which leads to the development of a superior genotype that will be released as cultivar after multiplication for the benefit of mankind. Plant Breeding is practiced worldwide by plant breeders for feeding the global population fighting against hidden hunger. The pace of climate change is very high in the present time, and designing climate-smart crop requires planned and effective utilization of genetic resources for introgression of useful traits from the wild species under varied diverse regions. For the success of any crop improvement programs, genetic variability is the utmost importance. The genetic diversity of the crop plants acts as the foundation for the sustainable development of new varieties for present and future challenges. Ample amount of genetic diversity existed in all the crops which are stored in their wild relatives or primitive landraces. Base collection of most of the crops has a huge germplasm collection, and accessing genetic diversity for desired traits in such collections becomes very difficult. Quantification of genetic diversity and access to desirable variation in the usable form for the development of cultivars suitable to climate vagaries, curators, and breeders must be empowered with basic knowledge of the techniques indispensable for identification of desirable genotypes with trait of interest in large germplasm collections. There is a great concern that genetic diversity is limited both within production systems and in breeding programs and that there is a need for concerted efforts to broaden the genetic base of the crop plants. Genetic base of a crop includes all the genetic diversity that is readily available to breeders and growers for adaptation to any particular environment of interest. The needs and priorities for base broadening vary from crop to crop and target production area. There are many criteria and indicators to determine or predict the likely need for base broadening for a given crop in a particular situation which are the following: (1) when a yield plateau or low rate of progress (genetic gain) is apparent in breeding programs, (2) when breeders are increasingly forced to source qualitative genes from the secondary and tertiary gene pools (often using costly high-technology approaches), (3) when there are recurring instances of crop failure or crop vulnerability to particular biotic or abiotic stresses, and (4) when the needs of farmers for a sufficiently diverse range of planting material are not being met. Genetic base of a crop includes all the genetic diversity that is readily available to breeders and growers for adaptation to any particular environment of interest. For widening the genetic base of crops, pre-breeding and core and mini-core collections are the most suitable approaches which make breeder's work simple.

3.18.1 Conventional Approaches

Conventional methods include hybridization of cultivated crop varieties with wild species. The major concern in this approach is to selectively transfer agronomically important genes from wild species while avoiding linkage drag. To achieve the precise transfer of genes from wild species, a combination of conventional plant breeding methods along with molecular approaches, tissue culture, and genetic engineering techniques has become important. Several mapping populations have been generated from interspecific crosses between the following distant crossing programs. Conventional mapping populations including F_2, backcross, recombinant inbred lines (RILs), and doubled haploids (DH) have been used for mapping both major genes and quantitative trait loci. Advanced mapping populations like backcross inbred lines (BILs), near-isogenic lines (NILs), and chromosome segment substitution lines (CSSLs), multiparent advanced generation intercross (MAGIC) population, and association mapping panels are powerful tools for identifying the naturally occurring, favorable alleles in unadapted germplasm (Ali et al. 2010; Jacquemin et al. 2013). The mapping population like MAGIC is developed through crossing multiple parents that has a broad genetic base. This strategy has been proposed to interrogate multiple alleles and to provide increased recombination (Cavanagh et al. 2008). The principal target of developing MAGIC populations is to promote intercrossing and shuffling of the genome.

3.18.2 Base Broadening Using Genetic Engineering Approaches

Genetic engineering has become a versatile platform for cultivar improvement as well as for discerning the functions of genes in the plant system. The combination of genetic engineering and conventional breeding programs allows useful traits to be introduced into commercial crops within a short time span unlike in conventional breeding approach. Genetic transformation is an advanced biotechnical approach employed to enable directed desirable gene transfer from one organism to the other overcoming the barrier of crossability existing between genomes and the subsequent stable integration and expression of the incorporated foreign gene. Thus, genetic transformation facilitates the introduction of desirable genes without cotransfer of any undesirable genes from donor species which normally occurs when conventional breeding methods.

3.18.3 Widening Crop Diversity: The Concept of Pre-breeding

The crop wild relatives (CWR) contain several valuable traits related to quality, biotic and abiotic stress tolerance, and yield ceiling traits; pre-breeding term was first coined by Rick (1984) and defined as the process of transferring of useful genes from exotic or wild (unadapted sources) types into agronomical acceptable background/ breeding material (FAO 1996b). Genetic enhancement is the alternate term to pre-breeding and often used recently. It aims to transfer these traits to an intermediate set of the breeding population that breeders can further use in producing new varieties. It is the basic step in the "linking genetic variability to utilization" use of diversity arising from wild relatives and other unimproved materials. Pre-breeding offers a unique opportunity to exploit this genetic diversity by means of transferring genes from the unadapted germplasm that cannot be used directly in breeding populations to cultivated ones. It aims to enhance genetic variability in the germplasm, and the improved germplasm can be readily used in a regular breeding program for cultivar development (Lokanathan et al. 2003). The success of any pre-breeding program depends mainly upon three factors: (1) identification of promising donor with the trait of interest, (2) type of germplasm (cultivated/cross-compatible wild type/cross-incompatible wild type), and (3) agronomic performance of the donors. New and diverse sources of variation for agronomic and nutrition-related traits and resistant sources for biotic and abiotic stresses are now available in both cultivated and wild-type germplasm and can be utilized to develop new pre-breeding populations having greater variability for various traits and better adaptation to changing the climate.

The approach of pre-breeding, particularly wide hybridization exploiting wild relatives, has established credentials with successful stories of overcoming challenges in many crops (Table 3.11). Near about 50% of genes conferring resistance

Table 3.11 Pre-breeding for quantitative and qualitative traits in different crop plants

Crop	Remarks	References
Domestic bean	Wild relatives are a potential source of novel alleles that can be exploited for the improvement of yield and other quantitative traits	Acosta-Gallegos et al. (2007)
Soybean	Useful traits have been identified and introgressed in cultivated species from wild species through interspecific hybridization	Sebolt et al. (2000)
Rice	Continuous efforts are being made to transfer the desired traits into the cultivated varieties from the rice accessions which are stored in gene banks due to narrow genetic base	Plunkett et al. (1987)
Maize	Value of exotic resource has yet to be explored in polymorphic genome which is resulted due to gene flow between cultivated and wild species	Cantrell et al. (1996); Luciano and Peterinain (2000); Nass and Paterniani (2000)
Tomato	Different genes for disease resistance have been incorporated from various wild resources in commercial hybrids through recurrent backcross, and each resides on a small independent chromosome segment from one of the diverse donor species. An important gene was introduced from the wild tomato species (*Lycopersicon pennellii* B.), which resulted into raised level of provitamin A in the fruit by more than 15-fold	Ronen et al. (2000)
Cotton	Pre-breeding has been attempted in cotton (since 1970) in different countries, and several varieties were developed using wild species for resistance to biotic stress (sucking pest), abiotic stress (drought tolerance), and quality improvement (fiber quality) (Texas, USA)	Lokanathan et al. (2003)
Groundnut, pigeon pea, chickpea, sorghum, and pearl millet	Improved the existing cultivars using wild species for resistance to biotic stress and abiotic stress and quality improvement	ICRISAT (2004)
Spring barley	Pre-breeding has been attempted for improvement of biotic/abiotic stresses and agronomic and nutrition-related traits	Vellve (1992)
Wheat	Successful introduction of useful genes from wild species into cultivated species for specific traits	Seetharam (2007); Dalrymple (1986); Valkoun (2001)
Sugarcane, sunflower, cotton	Successful introduction of useful genes from wild species into cultivated species for specific traits	Seetharam (2007)
Mulberry	Crosses between cultivated (*M. indica*) and different wild species (*M. cathayana*, *M. pendulata*, and *M. serrata*) to improve the quality	Tikader and Thangavelu (2002); Tikader and Dandin (2001); Tikader and Ananda Rao (2003); Tikader and Kamble (2007)

Source: Jain and Omprakash (2019)

to potato late blight disease are reported in *Solanum demissum* (Ross 1979). Similarly, tomato genes for the improvement of most of the traits are concentrated in nine *Solanum* species of tomato (Rick 1979). Wild species (*Oryza rufipogon, O. longistaminata*, and *O. glaberrima*) of rice provide resistance against several biotic and abiotic stresses (Brar 2005). In several other crops, such as wheat, maize, cotton, sugarcane, groundnut, potato, tomato, etc., wild relatives have been used for incorporating diverse traits, such as disease and insect resistance (wheat, rice, potato, tomato), yield (oat, tomato), quality (fruit size and total soluble sugars in grasses, pigeon pea, tomato), earliness and adaptation (rye, potato, tomato, grape, strawberry), modes of reproduction (cytoplasmic male sterility in several crops), and miscellaneous traits (hard-seediness, color, leaf texture, delayed ripening). In addition to the above, in the past, a few promising wild accessions as detailed below have been utilized by researchers for the improvement of crops of commercial importance (Sharma et al. 2013).

3.19 Shifting Through Natural Variations: The Concept of Core and Mini-Core Collections

3.19.1 Core Collection

Improvement of yield potential or nutritional qualities and resistance to biotic and abiotic stresses are the focused objectives in almost all crop improvement programs. Plant breeders have successfully achieved the incremental increase in yield potential of most crops, resulting in large food production in the past few decades. However, yields have become stagnant in several crops, and no further significant progress has been witnessed in the twenty-first century. One reason for this is that breeders generally restrain themselves to their working collection, which is most highly adapted materials, and rarely use more diverse germplasm sources or wild relatives. In India's chickpea program, the 184 breeding lines evaluated in 2001 involved only 13 germplasm lines (mostly for stress resistance) in their pedigrees. This represents only a small fraction of available germplasm diversity in the crop as ICRISAT's gene bank repository alone contains over 17,000 chickpea accessions. Studies by Frey (1981) indicated that the introduction of new alien germplasm in advanced breeding programs often increases yield potential. However, it is almost impossible to predict which germplasm accessions(s) will be suitable for use in breeding programs because of high genotype × environment interactions.

The large numbers of accessions accumulated in the gene banks are often poorly described. Their use for the breeding purpose could be greatly increased if more information on the amount and kind of variation in these collections is available. However, in most cases, such information is not available. This poses difficulty in effective utilization of germplasm for plant breeders and other research workers. In addition, the sheer size of many collections has frequently been cited as a barrier to increased utilization of collections (Holden 1984).

The extent of variation in the germplasm collections and their accessibility to biologists and breeders are essential factors affecting their utilization in the crop

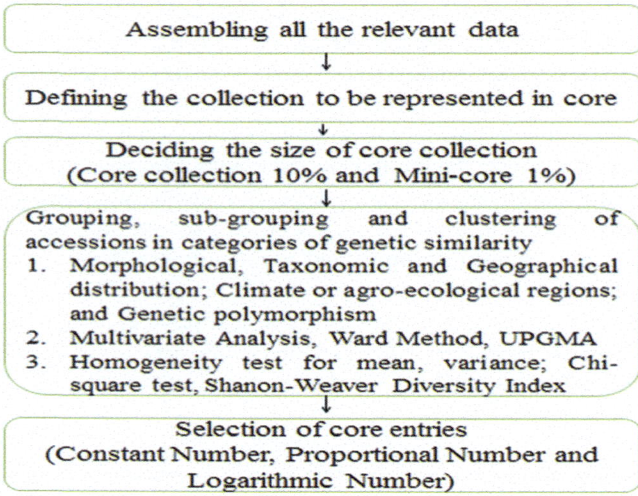

Fig. 3.5 Procedure of core and mini-core collection development in crop plants

improvement program. Frankel and Brown (1984) proposed the concept of "core collection" (CC) as a collection that encompasses a representative sample of the entire collection with minimum repetitiveness and maximum genetic diversity of a crop species and its relatives. With the core collection, it is convenient to study and utilize germplasm resources. Establishment of CC (Fig. 3.5) has proven to be a favored approach to facilitate efficient exploration of novel variation from genetic resources (Brown 1989; Ellis et al. 1998; Holbrook et al. 2000; Malvar et al. 2004). The core subset can be evaluated extensively and the information derived used to guide more efficient use of the entire collection (Brown 1989).

The first core collection established was that of perennial *Glycine* by Brown et al. (1987). The core consisted of 111 accessions developed from a collection of 1400 accession of twelve different species of *Glycine* at Canberra, Australia. The grouping was made initially at the species level, and then ecogeographic factors were used to select the entries for the core. Since then, core sets have been derived in several crop species such as rice (Liang et al. 2004; McClung et al. 2004; Pkania et al. 2006; Yan et al. 2007; Jia et al. 2009; Weiguo et al. 2010; Zhang et al. 2011; Ling et al. 2011; Rao et al. 2012; Saini 2012), wheat (Moody et al. 1988; Mackay 1990; Zeuli and Qualset 1993; Balfourier et al. 2007), maize (Crossa et al. 1993; Radovic and Jelovac 1994; Taba et al. 1998; Tabare et al. 1999; Yao et al. 2008), barley (Weltzein 1989; Bothmer et al. 1990; Knupffer and van Hintum 1995; Igartua et al. 1998), sorghum (Rao and Rao 1995; Grenier et al. 2001), pearl millet (Bhattacharjee 2000), finger millet (Gowda et al. 2007; Haradari 2009; Upadhyaya et al. 2010), proso millet (Upadhyaya et al. 2011), little millet (Gowda et al. 2008), barnyard millet (Gowda et al. 2009), annual *Medicago* species (Diwan et al. 1994, 1995; Basigulp et al. 1995; Skinner et al. 1999), *Pisum* (Mathews and Ambrose 1994), common bean (*Phaseolus vulgaris* L.) (Galwey 1995; Tohme et al. 1995), cassava (Corderio et al. 1995), *Coffea liberica* (Hamon et al. 1995), pecan (*Carya illnoinensis* (Grauke and Thompson 1995), sweet potato (Huaman et al. 1999), chickpea (Upadhyaya et al. 2002a), pigeon pea (Reddy et al. 2005), perennial rye-grass (*Lolium perenne*) (Balfourier and Charmet 1994; Balfourier et al. 1999), okra (Hamon and Van Sloten 1989; Hamon and Noirot 1990; Mahajan et al. 1996), groundnut (Holbrook et al. 1993; Holbrook 1999; Upadhyaya et al. 2002b, 2003; Swamy et al. 2006; Jiang et al. 2008; Kottapalli et al. 2010), soybean (Cho et al. 2008; Oliveira et al. 2010), and chillies (Mongkolporn et al. 2015) for research globally.

3.19.2 Mini-Core Collection

In several crops having thousands of accessions in the ex situ collections in the gene banks, even a core collection could be unmanageably large and unwieldy. The International Rice Research Institute (IRRI) rice collection contains over 80,000 accessions. Hence, even a core subset would be expensive and time-consuming to evaluate. The challenge is to further reduce the size of the core subset without losing the spectrum of diversity. A strategy for sampling the entire and core collections was developed to build up a mini-core subset involving two stages (Upadhyaya and Ortiz 2001). First, a representative core collection (10%) is developed from the entire collection (Fig. 3.5), using the information on the origin, geographical distribution, and characterization and evaluation data. The core collection is then evaluated for various morphological, agronomic, and quality traits selected and, ultimately, a subset of 10% accessions from the core subset (i.e., 1% of the entire collection) that captures most of the useful variation in the crop. At both stages, standard clustering procedures are used to separate groups of similar accessions, and various statistical tests are used to evaluate representatives of the core and mini-core collections. Further, the mini-core collection can be used as a diversity panel for assessing population structure, diversity, allelic richness, and association genetics for identifying the promising accessions for economically important traits.

The basic requirement for the success of plant breeding programs is the availability of variation in plant species. Pre-breeding is often considered to be an activity at the interface between germplasm conservation and utilization. Knowledge of nature and extent of genetic variation and diversity available in the germplasm helps the breeder to plan sound breeding programs. However, identification of potential parental material from the diverse collections is often tedious, owing to the huge size of germplasm collections. This could be overcome if a subset of genotypes or the core set that has funneled out or retained the maximum diversity in the original base population can be drawn. The core set and mini-cores will thereby help the breeder to pinpoint and effectively select the appropriate genotype based on the objective of the crop improvement program.

3.20 Exploiting Natural Variation Through Molecular Approaches

3.20.1 Unlocking Favorable Natural Genetic Variation Through Sequencing and Resequencing of Crop Genomes

With the aid of technology revolution, from Sanger's DNA sequencing with chain terminating inhibitors in 1977 to next-generation sequencing (NGS) in the present period, crop plant genome sequencing becomes very easy and routine. The first full DNA genome to be sequenced was that of bacteriophage in 1977. The most awaited human genome sequence becomes available 15 years of inception of human genome project. The first plant genome sequence was *Arabidopsis thaliana* that provided and improved the genetic landscape for studying all plants. After that, so many crop plant genomes have been sequenced including cereals, millets, legumes, oilseeds, vegetables, fruits, and tree plants. But selection of crops for sequencing has been mainly based on the cost efficiency and avoidance of genome complexity. Genome complexity is mainly contributed by genome size which is further driven by polyploidy, repetitive sequences, transposable elements, noncoding RNAs, heterozygosity, etc. It enables the researchers to select small genome crop plants. *Arabidopsis thaliana* (120 Mb split in 5 chromosomes) and *Oryza sativa* (420 Mb split in 12 chromosomes) have small genome size, whereas *Triticum aestivum* (16 Gbp split in 21 chromosomes) and pines (22-2 Gbp split in 12 chromosomes) have large and complex genome. Only rice and Arabidopsis have been sequenced through Sanger's sequencing using BAC-by-BAC strategy. But NGS technology has changed the impact of sequencing on our knowledge of crop genomes and gene regulation, and it allowed rapid and efficient development of genomic resources in orphan/underutilized crop plant species. These techniques are today acquiring a great potential in genome sequencing, variant discovery, exome sequencing, and metagenomic and epigenomic research while showing prospects for their utilization in plant breeding. Besides this, genome sequence helps in finding genes much more easily and quickly, development of sequence-based markers (SNPs), evolutionary study (polyploidization events), comparative genomics, and thorough understanding of biological processes. Now it is possible to obtain new and beneficial information about gene regulation on the cellular as well as whole plant level through RNA sequencing and subsequent expression analyses of genes participating in biotic and abiotic stress tolerance. Sequencing of genome provides information regarding polyploidy events which occur during the course of evolution of crop plants that molds and shapes the plant genome. Genome sequence information is very useful as it tells us about the insight of crop evolution and their domestication events. For example, soybean is the result of three polyploidy events which are the consequences of having a high-quality genome sequence (Schmutz et al. 2010). Genotyping by sequencing (GBS), whole-genome sequencing (WGS), and resequencing of crop genome can lead to the development of sequence-based DNA marker systems suited to study relationships among breeding materials and creation of detailed genetic mapping of targeted genes (Jana and David 2016). GBS is a simple highly multiplexed system for generating large numbers of SNP for use in genetic analyses and genotyping. It is an ideal platform for studies ranging from single gene markers to whole-genome profiling. GBS allowed plant breeders to implement genome-wide association studies (GWAS) genomic diversity, genetic linkage analysis, molecular marker discovery, and whole-genome prediction (WGP) under a large scale of plant breeding programs. Availability of high-quality whole-genome sequence for staple foods provides access to thousands of many useful genes that constitute the genetic architecture of the organism. Nowadays, for most of the crop plants, reference genome is available which permits rapid identification of candidate genes and SNPs with computational biology tools utilizing sequence comparisons of cultivars with reference genome sequence (Edwards and Batley 2010). Resequencing of crop genomes is becoming more practical, enabling thorough analysis and cataloging of genetic variation. The 3000 rice genome project involving resequencing at an average sequencing depth of 14X, average genome coverages (94%), and mapping rates (92.5%) could discover 18.9 million SNPs in rice using Illumina-based NGS, which serves for large-scale discovery of novel alleles for important rice phenotypes. It also serves to understand the genomic diversity within *Oryza sativa* (Kang et al. 2014).

3.20.2 Accessing Natural Genetic Variation Through Pangenome

Capturing of entire genetic variation which includes structural variation (SV), copy number variants (CNV), and presence/absence variants (PAV) is not possible through sequencing of a single species/cultivar genome as the genes which may be present in the one genome; they might not be present in the other genome of the same species. So in order to targ*et all* the genetic variation existing in the particular crop species, it requires the construction of pangenome. Tettelin et al. (2005) proposed the concept of pangenome while working on *Streptococcus agalactiae* for the production of first ever pangenome. In the beginning, this approach was used for microorganisms. A pangenome refers to the full complement of genes of a biological clade, such as species, which can be partitioned into a set of core genes and dispensable genes (Tettelin et al. 2005). Pangenome covers all the useful genes which are divided into two groups: (1) core genes which are present in all of the individuals and (2) dispensable genes which are present in some individuals but not in all. Several strategies are available for the development of pangenome, but comparative de novo approach (Li et al. 2014; Schatz et al. 2014; Gordon et al. 2017; Zhao et al. 2018), iterative assembly approach (Golicz et al. 2016b; Montenegro et al. 2017; Hurgobin et al. 2018), and the map-to-pan approach (Wang et al. 2018) are found to be the most promised approaches among all. Recently, it has been utilized in the higher organisms including maize (Hirsch et al. 2014), soybean (Li et al. 2014), *Brassica rapa* (Lin et al. 2014), wheat (Montenegro et al. 2017), and rice (Schatz et al. 2014). At the time of pangenome construction, it is found that addition of each genome identifies new gene sequenced which leads to increase in the pangenome size and simultaneously decrease the core genes percentage (Vernikos et al. 2015; Golicz et al. 2016a). The same is observed for the crop species which indicates beyond a finite genomes addition of genome would not affect the pangenome expansion. In plants, pangenome is very much useful for speedy exploitation of wild species, and their relatives in crop breeding as CWR contain very useful genes related to plant architecture, root architecture, quality and nutritional enhancement, biotic and abiotic stress tolerance, and of course grain yield/grain size in multiple crop species such as rice (Xiao et al. 1998; Li et al. 2002; Ram et al. 2007; Thalapati et al. 2012), wheat (Huang et al. 2003), sorghum (Tao et al. 2017, 2018), and soybean (Concibido et al. 2003). Resequencing has not been tapped the full potential of diversity present in wild species and their relatives. Therefore, construction of pangenome for crop species emerged as most efficient and effective approach as it covers full landscape of natural genetic diversity sitting in the wild relatives, facilitates genetic dissection of economically important agronomic traits, and also helps in genomics-assisted breeding strategy development for crop improvement.

3.21 Elucidating the Genetic Structure of Crop Plants Through Molecular Fingerprinting

To distinguish the genotypes/cultivars among the crop plants, utilization of molecular markers technique is the most promising approach. Markers are the visible impression that can distinguish one individual from another individual. DNA marker is a DNA sequence having a known chromosomal location with identifiable feature/phenotype and whose inheritance pattern can be tracked. During the pre-molecular marker era, morphological markers were the used for differentiation among the various genotypes of same or different species. Later on, isozymes concept was introduced. Isozymes are the enzymes which catalyze the same chemical reaction but having different amino acid and travel on electrophoretic gel basing upon charge. Morphological markers are stage specific, are small in numbers, and have pleiotropic effects, whereas isozymes are also stage specific. Both morphological and isozyme markers have limitations and were replaced by DNA markers. DNA markers have three categories mainly as (1) southern hybridization-based (RFLP), (2) PCR-based (RAPD, AFLP, SSR, ISSR, CAPS, SCAR, EST, DArT), and (3) sequence-based (SNPs, GBS). In the year 1980, Botestein proposed the first DNA-based marker named as restriction fragment length polymorphism (RFLP) which utilizes the different pairs of restriction endonuclease enzymes and southern blotting techniques. This marker requires large quantity of high-quality DNA and involves radioactive substances. Such type of marker provides polymorphism basing upon the length variations of different DNA fragments. Hence, their large-scale use in practical plant breeding may be restricted. After the invention of polymerase chain reaction (PCR) by Saiki et al. (1987), Williams et al. (1990) developed a new PCR-based molecular marker called as random amplified polymorphic DNA (RAPD) in which a random 10-base (decamer) single primer acts as both forward and reverse primer. No sequence information was required for primer designing in RAPD, but it suffers from the limitation of low level of reproducibility, and it is a dominant marker which results in polymorphism basing upon presence and absence of alleles. A few years later, Vos et al. (1995) from Keygene company invented a new molecular marker AFLP which combines both hybridization and PCR methodology. It gives more number of amplified bands and polymorphism results from variation in DNA sequence. An ideal marker should have high level of polymorphism, can distinguish heterozygote from homozygote, occurs in genome frequently and uniformly, should be selectively neutral, and should have high rate of reproducibility. The simple sequence repeat (SSR) or microsatellites are 1–5 base pair long repeats which were first used by Litt and Luty (1989) in humans, and the first SSR-based linkage map was developed in rice during 1995. SSRs are abundantly distributed throughout the genome, provide high level of polymorphism, and are codominant in nature. Following this, several molecular markers (ISSR, CAPS, SCAR, STS, DArt, TRAP, SRAP, RGAP, etc.) have been developed by different researchers time to time. With the technological revolution in the twenty-first century, so many advancements have taken place in the genetic engineering and molecular biology, and a new branch emerged with the name of genomics which revolutionized the marker technology. With this, plant breeding entered in the genomics-assisted breeding era which utilized the recent trends of molecular biology for the crop improvement. In these recent trends, sequencing platforms enable the development of sequence-based molecular marker such as single nucleotide polymorphism (SNPs). SNPs are biallelic in nature. In plants, one SNP is typically per approximately 100–500 bp of the DNA. SNPs are well-established marker system across the major crop plants especially those whose genomes have been fully sequenced. Molecular markers have been used in most of the crop plants including cereals, oilseeds, legumes, vegetables, fruits, and tree plants for the varietal/genotype identification. A wide array of molecular marker are applied for genotype differentiation in wild species, seed propagated, vegetatively propagated crop plants, their genotyping, assessment of genetic diversity, population structure, and genetic relatedness by researchers across the globe. Besides these applications, molecular markers are well deployed for genetic mapping through linkage as well as association mapping approach, genome-wide selection, and marker-assisted breeding programs for varietal development.

Since the molecular markers have been employed for distinguishing different genotypes of commercial varieties or natural population of an important crop species is the prerequisite of plant breeding. Basing upon reproductive system, plants exhibits inbreeding and outbreeding which have profound influence on the amount and portioning of genetic variability between and within various entries such as cultivars and population. The reproductive aspect will have impact on molecular fingerprinting of different genotypes or cultivars for the elucidation of genetic structure of the plant populations. Life history traits such as mode of propagation and seed dispersal influenced the wild and cultivated plant relationships. Self-pollinating (inbreeders) species produce the identical or similar plants, i.e., homozygotes, whereas cross-pollinating (outbreeders) species exhibit different genotypes, i.e., heterozygote. Clonal propagation or apomixis, a third mode of propagation, results in identical progenies just like inbreeders. Molecular fingerprinting is very useful for drawing clear picture between the reproductive systems which produce seeds by both sexual and asexual (apomixis) mode and also for the determination of factors involved in shaping the populations. Triploid plants of *Taraxacum* population reproduce apomictic clones which were differentiated through SSR and AFLP analysis (Majeský et al. 2012). Similarly, in the apomictic *Ranunculus carpaticola*, AFLP analysis has been done and more amount of genetic variation was detected (Paun et al. 2006; Paun and Hörandl 2006). There will be existence of some amount of genetic variation within the lines/genotypes in case of cross-pollinating crops propagated through seed which becomes a problem for cultivar identification through molecular markers. In the later phases of seed production, the situation becomes more complicated due to foreign pollen which results in the modification in the allele as well as genotypic frequency and introduces new genetic variation. To resolve this problem, a large number of molecular markers should be required for assessing the changes occurring in the gene and genotypic frequency in the populations and also for discrimination among the cultivars. In case of self-pollinating crops, it is quite easy to identify the genotype/lines among the populations because of the homogeneous nature of populations and the entire genetically identical individual. The primitive cultivars of inbreeding crops still contain intra-variability variation. Highly polymorphic markers are required for differentiated cultivars. Inbreeding crops have identical DNA fingerprinting. But, sexual propagation-derived progenies of inbreeding are expected to have nonuniform DNA fingerprints. The same could be expected to have in vegetatively propagated crops which have identical DNA fingerprints. Irrespective of propagation and breeding methods, the value of accessions in plant genetic material collections benefits tremendously from DNA marker-aided identification. Thus, molecular markers are very much useful for identification of genetic nature (homozygote/heterozygote/homogeneous/heterogeneous) of the different cultivars/genotypes derived from inbreeding and outbreeding species.

3.22 Discovery of Genes Underlying Quantitative Traits

3.22.1 Association Mapping

Association mapping (AM) is a forward genetics approach which utilizes linkage disequilibrium for the determination of significant marker-trait association. Association mapping is widely used in humans and other model organisms, whereas linkage disequilibrium (LD) tells about the nonrandom association of alleles but does not necessarily correlate/associate with a particular phenotype. AM is actually a major application of LD which exploit the historical recombination and mutational events for the discovery of genes/QTLs which remains unidentified through linkage mapping. Association mapping overcomes the limitations of linkage mapping as no mapping population is required for AM which is a time-consuming and skilled work. AM utilizes unstructured populations, i.e., germplasm collections which are the sources of huge genetic diversity and useful genes/QTLs linked with yield, quality traits, and biotic and abiotic stress tolerance. The development of high-throughput, dense genotyping platforms enables the molecular biologists to assess thousands of markers simultaneously; almost all major crops can now be subjected to AM. With a large diversity panel and dense genome-wide marker coverage, association mapping can potentially map causative loci to individual nucleotide changes. AM studies have been successfully conducted in almost all the crop plants. Zhao et al. (2011) carried out AM in 413 rice accessions and identified more than 44,000 SNPs. Similarly, in maize, 56,110 SNPs have been analyzed for chilling tolerance in 375 inbred lines (Strigens et al. 2013). Choice of population and appropriate marker density are the key determining factors of the success of an association mapping study. One of the sources of false positive in AM is the population structure, which is a division of the population into distinct subgroups related by kinship factor. Several bioinformatics tools and software are available which have been used to determine the population structure. It offers greater precision in QTL location than family-based linkage analysis and should therefore lead to more efficient marker-assisted selection, facilitate gene discovery, and help to meet the challenge of connecting sequence diversity with heritable phenotypic differences. Association mapping has emerged as a tool to resolve complex trait variation down to the sequence level by exploiting historical and evolutionary recombination events in the population. As a new alternative to traditional linkage analysis, association mapping offers three advantages of increased mapping resolution, reduced research time, and greater allele number.

3.22.2 Genome-Wide Prediction (GWP)

Genome-wide prediction (GWP) or genome-wide selection (GWS) or genomic selection (GS) is also a variant of marker-assisted selection (MAS), which is based on the simultaneous selection for a large number of markers covering the entire genome in such a way so that all genes/QTLs are expected to be in linkage disequilibrium with a minimum of one marker locus in the entire target population. It minimizes the efforts required for the identification of individual marker-QTL association as it covers the whole genome (Dhillon and Chhuneja 2014). It is widely practiced in animal breeding, but nowadays, it becomes popular in plant breeding programs. It is an exciting new tool that promises to accelerate genetic gains by aiding rapid line development with economical empirical evaluation and heterotic pool information, predicting high-yielding newer combinations, and improving grain quality and stress tolerance with precision across crops. It becomes feasible in plants due to the discovery and development of large numbers of single nucleotide polymorphism (SNP) markers by genome sequencing. GWP has two major components which include training and breeding populations. Training population is a group of related individuals that are both phenotyped and genotypes. Genetic markers are used across the whole genome to predict complex traits with accuracy sufficient to allow selection on that prediction alone. Selection of desirable individuals is based on genomic estimated breeding value (GEBV) which is predicted breeding value calculated using statistical simulations models. Breeding population, on the other hand, comprises descendants of a training population. GEBVs are used for the selection of desirable individuals in the breeding phase. GS strategy allows for fixing the large known large effect alleles while simultaneously selecting for multiple unknown loci of cumulative effects for a range of traits. When implemented along with a number of other tools in the breeding pipeline such as Rapid Generation Advancement (RGA), Accelerated Trait Introgression (ATI), and precision phenotyping for biotic and abiotic stress tolerance, GS promises to amplify the genetic gain manifold (Ramesh Babu 2019). A number of studies for dissection of complex traits, viz., grain yield, quality, biotic and abiotic stresses, etc., through genomic selection have been carried out for several crops (de los Campos et al. 2009; Crossa et al. 2010, 2013; Jannink et al. 2010; Burgueño et al. 2012; Poland et al. 2012; González-Camacho et al. 2012; Heslot et al. 2013; Rutkoski et al. 2014; Spindel et al. 2015; Grenier et al. 2015; Lado et al. 2016; Pierre et al. 2016).

3.22.3 Next-Generation Mutagenesis

Recombination and independent assortment of favorable alleles is indispensable to produce new and unique individuals from which varieties with desired trait of interest can be picked or produced. But the limitation is that most crop species, which have been selectively bred for centuries, have large portions of their genome essentially fixed. The number of traits available for reshuffling is reduced and is only capable of creating combinations of traits that already exist. Now it is very important to breed new plant varieties that can maintain production under the varied agroclimatic conditions of diverse regions. The drawbacks of conventional plant breeding can only be avoided through the act of mutation in the genome, which can ultimately lead to new species or allelic series of useful traits. Mutagenesis can be described as the process of inducing any heritable change in the genetic material which is subsequently transmitted to daughter cells where it gives rise to a mutant cell or individual (Rieger et al. 1976). Mutagenesis, the act of inducing mutations within the genome of an organism, has been used in plant breeding since Muller's discovery of the mutagenic effects of X-rays on *Drosophila* flies (Muller 1927). In the current epoch, mutation breeding and mutant discovery have been accepted as viable options to achieve desired crop traits with several advantages such as (1) it is possible to achieve instant progress in elite material; (2) single trait improvement can be made to an established variety; (3) limited breeding efforts are required; and (4) single gene mutant with no negative pleiotropic effects is possible. Induced mutagenesis, a crucial step in crop improvement program, is now successful in application due to the advancement and incorporation of large-scale selection techniques, molecular biology tools, and techniques in modern crop breeding performance and can be called "next-generation mutagenesis." Recent developments in mutation scanning permit the identification of novel alleles of target genes within both germplasm collections and mutagenized populations, allowing significant progress in functional genomics within model species and in the assessment of candidate genes for crop improvement. The common strategies used to create a mutant population of self-fertilizing crops (e.g., barley, wheat, and rice) as a resource for both forward (mutant phenotype leads to gene sequence and function) and reverse (mutant sequence leads to possible phenotype and function) genetics approaches are MutMap and TILLING. MutMap and TILLING are based on whole-genome sequencing and enzymatic mismatch cleavage, respectively.

3.22.3.1 TILLING (Targeting Induced Local Lesion in Genomes)

TILLING (Targeting Induced Local Lesion in Genomes) is a novel reverse genetics approach that combines the advantages of point mutations provided by chemical mutagenesis with the advantages of PCR-based mutational screening (McCallum et al. 2000). It allows the identification of single-base-pair (bp) allelic variation in a target gene in a high-throughput manner. DNA is first extracted from test samples and pooled (typically eightfold). Screening for mutations begins with PCR amplification of a target fragment using gene-specific infrared dye-labeled primers. After PCR amplification, samples are denatured and annealed to form heteroduplexes between mutant and wild-type DNA strands. Samples are then incubated with a single-strand-specific nuclease to digest mismatched base pairs. Each sample is then loaded onto a denaturing polyacrylamide slab gel and using LI-COR DNA analyzer system DNA visualize fluorescently labeled. The exact nucleotide change is then determined using standard DNA sequencing methods. Higher throughput and economy of scale can be achieved by using multiple thermal cyclers and analyzers. Mapping the location of causal mutations using genetic crosses has traditionally been a complex, multistep procedure, but next-generation sequencing now allows the rapid identification of causal mutations at single nucleotide resolution even in complex genetic backgrounds (Schneeberger 2014).

3.22.3.2 MutMap

MutMap, a forward genetics approach based on whole-genome resequencing, was developed to rapidly identify a loss-of-function mutation (Abe et al. 2012). MutMap approach could markedly accelerate crop breeding and genetics, as mutant plants and associated molecular markers can be made available to plant breeders. MutMap includes use a mutagen (e.g., EMS) to mutagenize a rice cultivar (X) that has a reference genome sequence. Mutagenized plants of this first mutant generation (M_1) are self-pollinated and brought to the second (M_2) or more advanced generations to make the mutated gene homozygous. Through observation of phenotypes in the M_2 lines or later generations, identify recessive mutants with altered agronomically important traits such as plant height, tiller number, and grain number per spike. Once the mutant is identified, it is crossed with the wild-type plant of cultivar X, the same cultivar used for mutagenesis. The resulting first filial generation (F_1) plant is self-pollinated, and the second-generation (F_2) progeny (>100) is grown in the field for scoring the phenotype. Because this F_2 progeny is derived from a cross between the mutant and its parental wild-type plant, the number of segregating loci responsible for the phenotypic change is minimal, in most cases one, and thus segregation of phenotypes can be unequivocally observed even if the phenotypic difference is small. All the nucleotide changes incorporated into the mutant by mutagenesis are detected as single nucleotide polymorphisms (SNPs) and insertion-deletions (Indels) between mutant and wild type. Among the F_2 progeny, the majority of SNPs will segregate in a 1:1 mutant/wild-type ratio. However, the SNP responsible for the change of phenotype is homozygous in the progeny showing the mutant phenotype. Collect DNA samples from recessive mutant F_2 progeny, and bulk sequence them with substantial genomic coverage (>10x coverage); we expect to have 50% mutant and 50% wild-type sequence reads for SNPs that are unlinked to the SNP responsible for the mutant phenotype. However, the causal SNP and closely linked SNPs should show 100% mutant and 0% wild-type reads. SNPs loosely linked to the causal mutation should have >50% mutant and < 50% wild-type reads.

The application of mutation techniques has generated a vast amount of genetic variability and is playing a significant role in plant breeding and genetics and advanced genomics studies. The widespread use of mutation techniques in plant breeding programs throughout the world has generated thousands of novel crop varieties in hundreds of crop species and billions of dollars in additional revenue. Thus, next-generation mutagenesis is expected to play a crucial role in the generation of "climate-resilient crops" to address the uncertainties of global climate variability and change and the challenges of global plant-product insecurity.

3.22.4 Gene Cloning Approaches

Gene cloning is done to conserve the natural diversity in crop plants. By using DNA molecular marker, not only can we protect the genetic integrity of germplasm resources, maintain minimum breeding population and seed amount, screen important germplasm, and preserve large amount of germplasm resources, but also we can study the genetic diversity and evolutional relation of germplasm resources. The information about their DNA level diversity and their origin and evolution relationship will greatly help us to make better use of the excellent germplasm resources of crop and provide an important scientific basis for the protection of these germplasm resources.

Detection for the positive markers or positive locus related to heterosis in any period or organ, and then prediction of the heterosis based on the heterozygosity of these loci, is easy through the use of DNA markers. This technique overcomes the deviation of genetic distance-based heterosis prediction. Furthermore, the genetic maps constructed by DNA molecular marker technique give us chance to analyze the linkage relation of the genetic loci at the molecular level directly. The genetic loci we can get from the genetic map are far more than that we can get from the traditional genetic markers methods constructed based on morphology, physiology, or biochemistry. In addition, the construction cycle is shorter, and the density of markers is higher compared with the traditional ways. Most of the field and economic crops have finished the construction of high-density and even saturated genetic linkage maps, and this will be of great help to understand the information of genomic composition and structure and will finally benefit to the breeding practice. On the basis of precise location of the target gene by using DNA molecular markers, breeders can take advantage of the molecular markers which closely linked to or co-separated with the target genes, by which they can identify the existence of the target gene in breeding offspring, and determine whether they obtain the desired individuals.

3.22.4.1 Map-Based Cloning (MBC)

Map-based gene cloning technology is suitable for gene cloning, which is developed on the basis of DNA molecular markers and genetic linkage map. Based upon the accurate location of target gene and closely linked DNA marker as probe to screen genomic library, we can clone the target gene sequence. Map-based cloning techniques have great advantages for novel gene cloning and have been successfully used in the separation and cloning of excellent agronomic-, growth-, development-, and resistance-related genes in rice and maize. For example, Tamura et al. (2014) and Gao and Lin (2013) successfully cloned salt- and insect-resistant genes in rice by map-based cloning technique, and Lu et al. (2012) cloned the key enzymes genes of terpenoid metabolic pathway in maize.

Furthermore, the expression of gene is always earlier than the appearance of phenotype, so monitor and diagnose the growth and development status of crops at gene level, making it possible to predict the unfavorable change or harm before it happens, and take remedial measures in advance. The development of modern molecular biology technique provides extremely rich and even redundant gene information to the researchers. Even though the gene expression profile has been widely used to address the relationship between ecologically influenced or disease phenotypes and the cellular expression patterns, the information is mainly limited to laboratory experience; it is necessary to apply the crop growth and development-related gene information to agricultural production, so the techniques or methods that are able to convert these gene information to convenient and easy access information are urgent.

3.22.4.2 Gateway Cloning (GC)

The Gateway Cloning Technology is based on the site-specific recombination system used by phage l to integrate its DNA in the *E. coli* chromosome. Both organisms have specific recombination sites called *att*P in phage l site and *att*B in *E. coli*. The integration process (lysogeny) is catalyzed by 2 enzymes: the phage l encoded protein Int (Integrase) and the *E. coli* protein IHF (Integration Host Factor). Upon integration, the recombination between *att*B (25 nt) and *att*P (243 nt) sites generates *att*L (100 nt) and *att*R (168 nt) sites that flank the integrated phage l DNA. The process is reversible, and the excision is again catalyzed Int and IHF in combination with the phage l protein Xis. The *att*L and *att*R sites surrounding the inserted phage DNA recombine site specifically during the excision event to reform the *att*P site in phage l and the *att*B site in the *E. coli* chromosome. The Gateway Reactions are in vitro versions of the integration and excision reactions. To make the reactions directional, two slightly different and specific sites were developed, *att*1 and *att*2 for each recombination site. These sites react very specifically with each other. For instance, in the BP *reaction*, att*B1* only reacts with *att*P1 resulting in *att*L1 and *att*R1, and *att*B2 only with *att*P2 giving *att*L2 and *att*R2. The reverse reaction LR reaction shows the same specificity. One of the main advantages of the GC is that once you have made an *Entry Clone*, the gene of interest can be easily subcloned into a wide variety of *Destination Vectors* using the LR reaction.

3.22.4.3 T-DNA Insertion

With genome-wide mutational changes and the collection of these mutants, the combination of classical forward genetics is beginning to revolutionize the way in which gene functions are studied in plants. High-throughput screening using these mutant populations should provide a means to analyze plant gene functions of the phenome on a genomic scale. Biological agents that mediate the transfer of DNA or RNA molecules into plant cells can also be used as mutagens. Whether the DNA or RNA is used to disrupt the sequence of a gene or to alter its activity indirectly, one can differentiate between insertional and gene-replacement mutagenesis versus overexpression and gene-silencing methods. Individual components of plant transformation are often mutagens. T-DNA insertion (using *Agrobacterium*) and tissue culture have both been used as such, either in plant breeding or to identify (by disruption) functional gene sequences in model organisms (Alonso et al. 2003; Jain 2001; Krysan et al. 1999). Less well known is that pathogen infection and antibiotics may also cause mutations in plant genomes (Bardini et al. 2003; Lucht et al. 2002; Madlung and Comai 2004). The plant genome differs both inter- and intraspecifically despite the highly conserved roles in plant kingdoms for specific elements, for example, element storage requires transport across the plasma membrane and commonly deposition within the central vacuole. Many studies have been conducted to study the use of natural variation in *Arabidopsis thaliana* for identifying genes involved in elemental accumulation. In some situations when highly complex traits may go uncharacterized in case of small sample size with limited natural genetic diversity. Thus, the advent of techniques which integrate wider sets of accessions and RILs can be employed such as multiparent advanced generation intercross (MAGIC) lines (Kover et al. 2009) and Arabidopsis multiparent RIL (AMPRIL) (Huang et al. 2011). *Agrobacterium*-mediated transformation has been used in research for over 15 years and has frequently been applied to create commercial transgenic cultivars. A few studies have examined unselected T-DNA insertions for chromosomal rearrangements and deletion of host DNA (Gheysen et al. 1987; Kim et al. 2003; Kumar and Fladung 2002). Insertion of superfluous DNA is also a consistent feature of *Agrobacterium*-generated insertion sites (Chen et al. 2003). This superfluous DNA may consist of extra whole or partial copies of the transgene, vector backbone DNA, or filler DNA. Filler DNA is DNA newly created at DNA-DNA junctions. It usually has some homology to the T-DNA or the transgene, alternatively it may resemble nearby chromosomal DNA, or it may be of unknown origin.

Only a handful of studies have provided detailed data on the chromosomal mutations resulting from particle bombardment insertion (Windels et al. 2001; Shimizu et al. 2001; Makarevitch et al. 2003; Hernández et al. 2003). None of these have been large scale or systematic, and all chose to examine relatively "simple" insertions identified by southern blotting as containing only a single copy of the transgene (Forsbach et al. 2003). Although relatively simple insertion events from particle bombardment are rare, they are important because they are more likely to be relevant to events presented for regulatory approval. Analysis of insertion-site mutations requires DNA sequence analysis of large stretches of flanking DNA and a careful comparison with the original target site (Ichikawa et al. 2003). Without this, deletions or rearrangements will probably not be detected. The sequence of a functional transgene insertion site resulting from particle bombardment has therefore never been definitively compared to its undisrupted site of insertion, either in the scientific literature or in applications submitted to US regulators. In even more sophisticated approaches, transposons can be introduced into the plant genome using T-DNA-mediated transformation. Once inserted, the transposon can hop from one chromosomal location to another, as long as an active transpose is present, with the potential of creating mutations at both the landing and excision sites. Although most transposons tend to hop to linked sites, a strategy has been devised to select for transpositions that land at unlinked loci. Consequently, the minimal extent of mutation possible at a functional particle bombardment insertion site is unknown. Due to the small number of events analyzed (even partially), any conclusions regarding particle bombardment insertion events can only be provisional. However, it appears that transgene integration resulting from particle bombardment is usually or always accompanied by substantial disruption of plant DNA and insertion of superfluous DNA.

3.22.4.3.1 Molecular Evidences of T-DNA Insertions

Mutations at insertion sites have the potential to result in inadvertent loss, acquisition, or mis-expression of important traits, in part because transgenes insert into or near functional gene sequences. In the plant species most studied (*A. thaliana* and rice), approximately 27–63% of T-DNA insertions disrupt known gene sequences (Ryu et al. 2004). Many researches illustrate few good genes identifiably exploiting natural variation using T-DNA insertion approach like ICARUS-1, gene for cell proliferation and growths at higher temperatures, and yellow seedling 1 gene with photosynthetic acclimation (Zhu et al. 2015). Large-scale studies of insertion patterns of transgenes delivered by particle bombardment have never been conducted in any species. Deletions or rearrangements associated with transgene insertion further increase the likelihood of alterations to the plant phenotype. Among many examples, the 78 Kbp deletion recorded in *A. thaliana* resulted in loss of 13 genes and disruption of two others (Kaya et al. 2000). The integration sites within plant genomes are largely randomly distributed (Kim et al. 2007), although preference for certain genomic regions is still debatable. Also, preferences are for transcription initiation sites and polyadenylation sites as well as regions outside of centromeric, at least after selecting insertion events using a marker gene that needs to be expressed (Szabados et al. 2002a, b; Li et al. 2006). In contrast, it has been reported that the integration sites are distributed within the genome in a completely random manner under nonselective conditions (Kim et al. 2007).

Gene disruption and deletion are not the only mechanisms by which transgene insertion may affect the phenotype of a transgenic plant. When transgene insertion is associated with rearrangements or insertion of superfluous DNA, then juxtaposition of promoter sequences and coding fragments may lead to sense or antisense transcripts which, similar to siRNAs and miRNAs, can interfere with the expression of genes containing homologous or similar sequences (Bartel and Bartel 2003). A naturally occurring instance of this phenomenon has been reported in the non-transgenic rice low-glutelin content mutation (Kusaba et al. 2003). Here, a deletion resulted in transcription into a neighboring member of the glutelin gene family and was thought to have caused gene silencing of the entire glutelin gene family. Studies of transcription patterns at insertion sites are rare. Very few studies described the transcription patterns at the transgene integration sites. Little focus on flanking DNA aberrant transcription. Bacterial chromosomal DNA, plasmid sequences (bacterial origins of replication in particular), or antibiotic resistance genes accidentally inserting adjacent to the transgene may significantly enhance the probability of horizontal gene transfer. Availability of sequence homology is considered one of the major obstacles to horizontal gene transfer from plants to bacteria.

Both insertion-site and genome-wide mutations may result in transgenic plants with unexpected traits. Despite the supposed precision of genetic engineering, it is common knowledge that large numbers of individual transgenic plants must be produced in order to obtain one or a few plants that express the desired trait in an otherwise normal plant. Even after selection, there are many reports of apparently normal transgenic plants exhibiting aberrant behavioral or biochemical characteristics upon further analysis. These examples show that unexpected transformation-induced phenotypes can affect any aspect of plant phenotype, including those of value or concern to humans. Although in many cases *Agrobacterium* transformation resulted in the expected outcome where the gene of interest is present and functional, it is common knowledge (yet usually unpublished) that the majority of transformation events are unsuccessful (Gelvin 2017). For example, transgenes can be only partially present or present but differentially or not expressed (Gelvin 2003; Peach and Velten 1991). The majority of studied cases enforce the conclusion that a transgene's destiny is determined by alterations to the genome structure at the site of insertion or the structure of the insertion itself, whereby both can induce epigenetic features with detrimental effects on the transgene function. In order to understand these structural effects better, these need to be resolved. However, all attempts were limited due to the short-read length of sequencing technologies (Szabados et al. 2002a, b) and the barely proven repeativity of concatenated identical T-DNA (transgene) and vector backbone insertions (Kim et al. 2003). Recent advances in the DNA sequencing space enabled the here presented detailed study of transgene insertions in the model plant *Arabidopsis thaliana*. We identified and analyzed perturbations to the genome. Furthermore, the incidence of unintended phenotypes in transgenic plants seems to be high, indicating that plant transformation is currently not predictable. Thus, mutational consequences are considered to be the unpredictable phenotypes.

3.22.5 Genetic Modifications Approaches: Transgenesis and Cisgenesis

Transgenic plants carry additional, stably integrated and expressed foreign gene(s) usually from trans- species and are commonly called as GMOs/Biotech crop. The whole process involving isolation, introduction, integration, and expression of foreign gene(s) in the host is called genetic transformation or transgenesis. Transgenesis has emerged as an additional tool to carry out single gene breeding or transgenic breeding of crops. Unlike conventional breeding, only the cloned gene of agronomic importance is being introduced without cotransfer of other undesirable genes from the donor. The first transgenic plant was developed in tobacco in 1984, and *Flavr Savr* tomato becomes the first marketed transgenic plant in 1996 in the USA. After that, rapid and remarkable achievements have been made in the production, characterization, and field evaluation of transgenic plants in several field, fruit, and forest plant species across the world. Using different gene transfer methods and strategies, transgenic varieties/hybrids carrying useful agronomic traits have been developed in several crops. *Bt* cotton is a successful example of transgenic technology which carried *Cry* gene from a soil bacterium *Bacillus thuringiensis* that provides resistance against lepidopteran insect *Helicoverpa armigera*. During 2002, *Bt* cotton was released in India, and after staying 15 years in the field, the resistance of *Bt* cotton was broken down in 2017. Presently, 30 approved biotech crops are grown commercially in 44 countries. In 21 years (1996–2016), an accumulated 2.15 billion hectares of biotech crops have been grown commercially, comprising 1.04 billion hectares of biotech soybean, 0.64 billion hectares of biotech maize, 0.34 billion hectares of biotech cotton, and 0.13 billion hectares of biotech canola. Biotech products derived from 2.15 billion hectares significantly contribute food, feed, fiber, and fuel to the current 7.6 billion people (ISAAA 2017). Virus-resistant transgenics have been developed in tomato, melon, rice, papaya, potato, and sugar beet. A variety of yellow squash called Freedom II has been released in the USA. Likewise, transgenic papaya resistant to papaya ring spot virus has been released for commercial cultivation in the USA. Golden rice is an excellent example where three genes, namely, *phy (plant phytoene synthase)* and *lycopene beta-cyclase* from daffodil (*Narcissus pseudonarcissus*) and bacterial *phytoene desaturase (Crt1)* from *Erwinia uredovora*, were incorporated into Japonica rice variety to using *Agrobacterium*-mediated genetic transformation system which results into golden rice (Ye et al. 2000).

The concept of cisgenics was first introduced by the Henk J. Schouten and Evert Jacobsen in 1999 from Wageningen University, the Netherlands. Cisgenesis is the genetic modification for transfer of beneficial genes from crossable species to the recipient genotypes. Traditional plant breeding introduces beneficial genes/alleles from crossable species through sexual hybridization, and transgenic breeding, on the other hand, transfers genes from another species or even from microorganism to the plants through genetic transformation. Cisgenesis is the combination of these two techniques where genes are called cisgenes and are isolated from crossable species and transferred to the crop plants through genetic transformation. Cisgenesis has great potential to overcome a major bottleneck in traditional breeding. During introgression breeding, linkage drag can slow down the breeding process tremendously. To reduce linkage drag, plant breeders usually need successive generations of recurrent backcrossing with the cultivated plant. A crop plant genetically modified with one or more genes isolated from a crossable donor plant will not have any accompanying linkage drag, and isolation of genes from wild crossable species and developing cisgenic plants takes only a few years. This can enhance the breeding speed, particularly if several genes from different relatives must be combined into an elite variety, for example, to obtain durable multigenic resistance. Cisgenesis is a particularly efficient method for cross-fertilizing heterozygous plants that propagate vegetatively, such as potato, apple, and banana. It can directly improve an existing variety without disturbing the genetic makeup of the plant. Resistance to late blight in potato (2008), red fruit flesh in apple (2009), growth and architecture in poplar tree (2010), resistance to apple scab (2011), and improved phytase activity in barley (2012) are the some success stories in cisgenesis.

3.22.6 Genome-Wide Editing (GWE): CRISPR/Cas9 System

Genetic variation is a key source of successful crop improvement. Induction of genetic variation in the crop gene pool is the foremost requirement for development of superior cultivars. Plant breeders have developed new plant hybrids and varieties through traditional plant breeding methods for thousands of year. Improvements in breeding methods and supporting technologies have been key drivers of this success to date. Since the last three decades, progress has been made in transgenic technology for crop improvement. Transgenic technology results in the random insertion of a foreign gene from one organism to another organism. Techniques have been developed for precise genome modification, and these techniques do not involve transfer of genes from one to another. Genome editing involves site-specific nucleases which precisely cleave the target gene, and break can be repaired through nonhomologous end joining (NHEJ) or homology-directed recombination (HDB). Meganucleases, zinc-finger nucleases (ZFNs), and transcription activator-like effector nucleases (TALEN) are the first-generation genome editing technologies which involve tedious procedures to achieve target specificity and are labor intensive and time-consuming. Clustered regularly interspaced palindromic repeats (CRISPR/Cas9) referred to as second-generation genome editing technology enable a more targeted way to discover and develop valuable traits with crop's own genome. The general methodology for implementing targeted mutagenesis using CRISPR/Cas9 technology starts with the selection of a specific target site having a short PAM sequence at 3′ end. The guide RNA (gRNA) are designed by using different online tools, and Cas9 is guided by the gRNA at two adjacent positions at the target site producing a single-stranded break on each of the two DNA strands (Arora and Narula 2017). CRISPR-cas9 has potential applications in different fields. In 2013, CRISPR was demonstrated on rice, wheat, and maize, whereas, in 2014, the technique was applied to tomato, soybean, and citrus. It was adopted in cotton and potato during 2015 followed by watermelon, grapes, and alfalfa in 2016. CRIPSR/Cas was also applied to cassava, ipomoea, and legumes during 2017. It is also applied to carrot, cacao, salvia, and lettuce during 2018, and many more crops are yet to come with CRISPR/Cas9 in the future (Jaganathan et al. 2018). Success with this technology builds on growing knowledge of the crop's own genome and ability to resolve key phenotypic responses down to a validated candidate gene, and development of a suite of technologies allows us to create the specified variation within the crop. The possible challenges for the successful utilization and deployment of genome-wide editing technology could be (1) development of key competencies and capabilities for CRISPR deployment, (2) defining specific trait opportunities that will have a positive impact on farmers, and (3) defining key hurdles for building regulatory and public acceptance of CRISPR technology (Greene 2019).

3.23 Harnessing Diversity: A Huge Investment

3.23.1 Ecosystem to Crop Genes: Bridging Landraces and Molecular Modifications for Economical Important Traits

There are many crop genes which involved in transcriptional response to several stresses such as drought, high salinity, or low temperature, and the expression level of these genes increased or decreased under the stress condition. Through the expression information of these resistance-related genes, researchers will be able to know whether the crop encounters unfavorable growth conditions and takes timely remedial measures. Before the designing field experiments, molecular designed breeding would integrate the information related to breeding program on the computer and simulate the implementation plan, thus considering more factors, more comprehensively, and putting forward the best strategies of selection of parental and progeny, to enhance predictability in the breeding process, thereby substantially improving breeding efficiency. This is on the basis of the combination of the DNA technology and cultivar improvement. However, with stable and diverse biosphere, we need to adapt science to an open research environment where data and genetic resources are effectively shared giving clear demonstration of environments.

The application of omics-scale technologies including genomics, transcriptomics, proteomics, and metabolomics, have provided alternative opportunities for gene discovery, global analysis of regulatory genes, expressed proteins, or metabolite candidates underlying important traits in CWRs. These omics approaches also are particularly suitable for dissection of the variation in complex traits such as drought tolerance and pest resistance. By characterizing CWRs under diverse treatments using omics strategies, a number of stress-resistant genes have been identified in various wild relatives of crops. For example, the dehydrin genes in both wild barley (*H. spontaneum*) and wild tomato species (*S. chilense* and *S. peruvianum*), as well as ABA/water stress/ripening-induced (*Asr*) gene family members (*Asr2* and

Asr4) from wild *Solanum* species, are known to be involved in drought tolerance (Fischer et al. 2013; Suprunova et al. 2007).

CWRs do require a higher density of genomic markers for metabolomics association studies because they typically have much lower levels of linkage disequilibrium than are found in domesticated crops. Fortunately, many markers now are publically available or can be genotyped at a reduced cost. Several laboratories recently have successfully developed high-density SNP markers for wild soybean (Song et al. 2015; Zhou et al. 2015), wild tomatoes (Aflitos et al. 2014), and wild rice (Xu et al. 2012), all of which can be or have been used in metabolomics mapping in CWRs.

Genetic modification (GM) technology has been considered a revolutionary solution to transfer target genes to crop cultivars to obtain desired traits. Commercial GM crops typically produce their target product and yield as expected, and they have the advantage of not suffering from the introduction of other linked genes (linkage drag). Genetic engineering techniques are particularly useful when the desired trait is not present in the germplasm of the crop or when the trait is very difficult to improve by conventional breeding methods. A well-known example of the use of this technology was in the production of transgenic *Bacillus thuringiensis* (Bt) crops (Tabashnik 2010).

3.24 Ensuring Genetic Gain Through Molecular Breeding Approaches

Assessment of the genetic gain acquired through various molecular breeding approaches utilizing natural genetic variation in crop plants faces two major limitations: first, to remove deleterious alleles of gene and accumulation of novel genes and second, to reconstruct/modify genetic network and molecular and biochemical pathways through breeding by design approach (Xu et al. 2017). Plant breeding utilizes the natural and induced genetic variation for selection of superior genotypes for crop improvement. The genotypes are selected on the basis of phenotype of plant. The phenotype is not only the resultant of genotype and environment interaction; it also involves the crop management practice and socioeconomic factors (GEMS) which provide information and tools for precision decision support (CIAT and IFPRI 2018). From this, the inference can be drawn as phenotype is not a good indicator of genotype. DNA-based molecular markers are not influenced by environment, can be detected at any stage of plant development, and are long-sought objective of plant breeders for making effective selection. Marker-assisted selection involves selection for the desirable gene/QTLs linked with desired trait. Molecular breeding involves utilization of genotyping data generated by molecular markers for enhancing the effectiveness of various breeding activities such as germplasm characterization, parental selection basing upon diversity assessment, gene introgression, gene and trait stacking, marker-assisted selection, and genetic purity testing. Molecular breeding approaches such as marker-assisted backcrossing (MABC), marker-assisted recurrent selection (MARS), marker-assisted gene pyramiding (MAGP), and the most recent one genomic selection (GS) have been practiced in several crops. With the development of wide array of molecular markers and genetic maps, marker-assisted breeding (MAB) can be used for qualitative and quantitative traits. In breeding programs using MAB, the rates of genetic gain have been two times higher as compared to the genetic gain by phenotypic selection. MABC has become a standard application in the major crop breeding programs due to two major advantages. Firstly, the marker-based selection for the target gene (foreground selection) combined with marker-based monitoring of the recurrent parent genome recovery (background selection) in the backcross generations shortens the time required for introgression of target gene in an elite background by 2–3 generations. Secondly, a good performing elite variety represents a very valuable fixed combination of alleles, and keeping this combination intact, when introgressing one or two desirable traits, is very remunerative for many crop improvement programs. Another advantage offered by MABC is the control on the linkage drag by selecting for the recombinants or double recombinants in the chromosomal region around the gene of interest. In India, rice varieties like Pusa 1460, RP Bio-226 (bacterial blight resistant), Punjab Basmati 3 (dwarf and bacterial blight resistant), and Swarna Sub1A and IR64 Sub1A (submergence tolerant); wheat varieties Unnat PBW 343 and Unnat PBW550; and maize hybrid QPM 9 (high-quality protein maize) developed through MAS are under cultivation. Disease resistance genes were the first to be targeted for marker-assisted gene pyramiding for providing durable resistance to the crop cultivars. Pyramiding favorable alleles to multiple traits, quantitative and qualitative, will increase the opportunity for the plant breeders to create novel genotypes by monitoring the segregation of traits during breeding process and focusing the selection on novel combinations of favorable alleles through markers. Dissections of the traits into trait components which can be mapped separately have the potential to yield better success. Using markers to select/pyramid multiple genes/QTLs is more complex. Recurrent selection is an effective strategy for the improvement of polygenic traits, and MARS is a recurrent selection scheme using molecular markers for the identification and selection of multiple genomic regions involved in the expression of complex traits to assemble the best-performing genotype within a single or across related populations. It can be particularly helpful in integrating multiple favorable genes/QTLs from different sources through recurrent selection based on a multiple parental populations (Dhillon and Chhuneja 2014).

3.25 Future of Plant Natural Diversity

3.25.1 From Plant to Crop: Three Bs (Biotechnology/Biodiversity/Biomimicry)

Biodiversity is the core strengths and resource of developing countries. In a country like India, conservation and sustainable use of the country's biodiversity is central to all developmental planning because its mainstay is agriculture and animal husbandry being subsidiary to it. Agro-ecologists argue that the reason to protect species and ecosystems is that we humans are but one species in a wholly interdependent milieu of species and everything depends on everything else. This argument is the most scientific despite our skeptical attitude toward science is among the most forceful since it ties our own survival as a species to the survival of other organisms claiming numerous effects on human as well as animal life. The real value of biodiversity is in the information that is encoded in genes and molecules. However, the extent of losses in biodiversity due to the introduction of GM crops depends on the degree in which local GM crops are adopted rather than on a single generic GM.

Geneticists argue that wide variability in genotypes and phenotypes within a species increases its chances for adaptation to changing environments and thus increases the likelihood of long-term survival of at least some members of a species. But even at this level, we cannot think that human biodiversity is always good. Therefore, reduction of diversity in our own environment may not be good if maximum genetic diversity at other loci is indeed a species survival value. We do not develop formal measures of biodiversity here, but we assume that an increase in the acreage that uses generic GM biotechnology, and especially replacement of traditional varieties with the generic GM crops, is undesirable from the crop-biodiversity perspective. Not only can biodiversity be preserved through biotechnology, these methods may help to restore previously lost crop diversity. Biotechnology already provides alternative sets of tools to address problems that were treated in the past through the use of chemicals or classical breeding. A sustainable strategy to provide food security for a growing population must promote biodiversity conservation and avoid further habitat loss of natural ecosystems. Biomimicry, a term coined by Janine Benyus in 1997, from mother nature to creating technologies with strategies must also seek to reduce unsustainable technologies such as the overuse of chemical fertilizers and pesticides, unsustainable irrigation procedures, and soil preparation methods that promote soil erosion; increase nutritional composition; reduce postharvest storage losses; and increase production from the present 2 billion metric tons per year to 4 billion. The strategy must also consider ethics, biosafety, and intellectual property rights (IPR) in the use of new biotechnologies. Biodiversity-rich countries can take advantage of their biological/genetic resources from wild land diversity, locally adapted varieties and races, and wild relatives of crops to increase yields. Genetic modification replaces selective breeding as a technology used to improve seeds and hybrids. While selective breeding generated green revolution seed varieties by introducing genetic materials that were a distinct departure from traditional varieties, biotechnology slightly alters existing seed varieties, modifies a few genes (sometimes only one), and leaves the others intact. Once a new genetic modification has been discovered, it can be inserted in all the traditional crop varieties by backcrossing. This genetic modification of all the existing seed varieties allows crop biodiversity to be maintained with only a slight change in the original genetic structure of the altered seeds. With advancement in biotechnology, we expect more genetic interventions. The challenge now is to bridge the fields of biotech and conservation in a way that is thoughtful and keeps pace with innovation. Increased public-private partnership will be essential to developing new biotech innovations. Hence, it will allow building proper use of biotech tools and allow the public acceptance for the same. The Biotechnology Industry Organization (BIO) is a helpful guideline to meet the common goal of conserving and sustaining biological diversity in all levels. Together with the traditional techniques, biotechnology leads us to more impact in plant genetic resources and biodiversity in general and in return meets the needs of the massively growing population and sustains life under rapidly changing climate.

3.26 Wildly Optimistic: Conservation for Sustainable Future

China is one of the world's richest countries in terms of plant diversity and also has a high level of endemism. However, due to rapid industrialization, urbanization, and explosive economic growth, the diversity is continuously threatened. Twenty percent of China's total higher plants are threatened with extinction (Huang 2011, Huang et al. 2013). As a result, effective protection of plant diversity is a major problem. Compared with other groups of organisms, notably birds and mammals, the conservation of plants is poorly funded and the great bulk of the literature on conservation biology and practice refers to animal examples and is not necessarily applicable to plants. This situation has been exacerbated, especially in developed countries, by the decline of botany in universities as an academic discipline and the widespread closure of departments of botany or their loss of identity when they are included within schools of biology, ostensibly due to lack of student demand. Curiously, zoology has not suffered the same fate.

Plant conservation is largely dependent in most countries on the creation of a system of protected areas. This is complemented by both in situ and ex situ actions at the species and population level, notably species recovery actions, reintroductions and conservation translocations, and the creation of gene banks for storing germplasm such as seed, pollen, cell, and tissue cultures. Also, much effort is now being placed on ecological restoration. Species recovery programs are complemented by plant reintroductions involving the deliberate movement of individuals of a species to parts of its natural range from which it has been lost, with the aim of establishing a new population. They are difficult and complex operations, and reported success rates are low, due to range of factors such as poor planning and execution, overoptimistic expectations of what is possible, and lack of suitable habitat.

In most countries, the management of national parks and protected areas is the responsibility of different ministries or agencies from those charged with ex situ and in situ conservation and species recovery. The need for close integration of area-based and species-based approaches in conservation planning is, however, essential. Conservation of wild plants species ex situ is now increasingly being recognized as an important method of conservation, complementing in situ approaches, although for many years actively discouraged largely on the grounds that it might encourage governments to rely on it as a more economical option than maintaining plants in the wild. In the case of plants of agricultural importance, ex situ conservation of material in gene banks, especially seed samples of landraces and cultivars, has long been the main conservation approach, and most of the technology and protocols for seed sampling, storage, germination, and regeneration were developed for the agricultural sector, largely under the aegis of FAO and IBPGR (today Bioversity International). Of course, botanic gardens are characterized by their holdings of (ex situ) living plant collections although most of these neither were nor established with conservation in mind.

In China, a program for the conservation of plant species with extremely small populations (PSESP) – defined as those having a narrow geographical distribution as a result negative external factors over a long period and whose numbers are below the minimum required to prevent extinction (State Forestry Administration of China 2012) – first promulgated in Yunnan Province (Yang 2017), is now becoming more widely adopted, and several national- and regional-level conservation strategies and actions for conserving them are being implemented over the coming years. In 1992, the United Nations Convention on Biological Diversity (CBD) embraced the imminent sixth mass extinction. Signatory countries assumed the responsibility for complete inventory of the biodiversity within their borders including domesticated species of plants and animals that sustain agricultural needs (UNEP 2011). Signatory states made efforts to comply by conducting thorough inventories of biological diversity including cultigens, their wild relatives, and systematic collection of cultivars throughout their sovereign lands (e.g., CONABIO in Mexico) (Brush 2007).

Throughout the world, biodiversity is under serious threat from factors such as intensive agriculture and silviculture, increased habitat fragmentation, and exposure to pollution and mass tourism (Wilson 1992; Hodkinson and Parnell 2007). Biotechnology represents a tool for enhancing genetic diversity in crop species through the introduction of novel genes. This does not aim at the single transgene inserted, but is based on the fact that beneficial characters can now be inserted in a variety of crops that have been neglected because of the limitations of traditional breeding methods, which failed to enhance the traits (Slabbert 2004; Gressel 2004).

References

Abbo S et al (2011) Experimental growing of wild pea in Israel and its bearing on near eastern plant domestication. Annals Bot 107:1399–1404

Abbo S, Yadun SL, Gopher A (2012) Plant domestication and crop evolution in the near east: on events and processes. Crit Rev Plant Sci 31(3):241–257

Abbo S, Pinhasi van-Oss R, Gopher A, Saranga Y, Ofner I, Peleg Z (2014) Plant domestication versus crop evolution: a conceptual framework for cereals and grain legumes. Trends Plant Sci 19(6):351–360

Abe A, Kosugi S, Yoshida K, Natsume S, Takagi H, Kanzaki H, Matsumura H, Yoshida K, Mitsuoka C, Tamiru M, Innan H, Cano L, Kamoun S, Terauchi R (2012) Genome sequencing reveals agronomically important loci in rice using MutMap. Nat Biotech 30:174–178

Acosta-Gallegos J, Kelly JD, Gepts P (2007) Prebreeding in common bean and use of genetic diversity from wild germplasm. Crop Sci 47:44–59

Aflitos S, Schijlen E, de Jong H, de Ridder D, Smit S, Finkers R, Peters S (2014) Exploring genetic variation in the tomato (*Solanum* section *Lycopersicon*) clade by whole-genome sequencing. Plant J 80:136–148

Ali ML, Sanchez PL, Yu S, Lorieux M, Eizenga GC (2010) Chromosome segment substitution lines: a powerful tool for the introgression of valuable genes from *Oryza* wild species into cultivated rice (*O. sativa* L.). Rice 3:218–234

Allansson M, Melander E, Themner T (2017) Organized violence, 1989-2017. J Peace Res 55(4):535–547

Allard RW (1999) Principles of plant breeding. Wiley, New York

Allard RW (2013) Evolution during domestication. In: Principles of plant breeding (Reprint 2nd Edition). Wiley, New York, pp 24–35

Alonso JM, Stepanova AN, Leisse TJ (2003) Genome-wide insertional mutagenesis of *Arabidopsis thaliana*. Science 301(5633):653–657

Aniol A (2001) Genetic variation, development and availability of useful germplasm for plant breeding. Plant Breed Seed Sci 45:33–43

Arnold ML, Kunte K (2017) Adaptive genetic exchange: a tangled history of admixture and evolutionary innovation. Trends Ecol Evol 32:601–611. https://doi.org/10.1016/j.tree.2017.05.007

Arora L, Narula A (2017) Gene editing and crop improvement using CRISPR-Cas9 system. Front Plant Sci 8:1–21

Arora RK, Nayar ER (1984) Wild relatives of crop plants in India. NBPGR Sci. Monog. No. 7, NBPGR, New Delhi

Arora RK, Pandey A (1996) Wild edible plants of India: diversity, conservation and use. National Bureau of Plant Genetic Resources, New Delhi

Badr A et al (2000) On the origin and domestication history of barley. Mol Biol Evol 17:499–510

Balfourier F, Charmet G (1994) Etude methodologique de la conservation de resources genetiques de ray-grass anglais (*Graminee fourragere*) par multiplication en pools de populations naturelles. Genet Selection Evol 30:215–235

Balfourier F, Prosperi JM, Charmet G, Goulard M, Monestiez P (1999) Using spatial patterns of diversity to develop core collections. In: Johnson RC, Hodgkin T (eds) Core collections for today and tomorrow. International Plant Genetic Resources Institute, Rome, pp 37–49

Balfourier F, Roussel V, Strelchenko P, Exbrayat-Vinson F, Sourdille P, Boutet G, Koeng J, Ravel C, Mitrofanova O, Beckert M, Charmet G (2007) A worldwide bread wheat core collection arrayed in a 384-well plate. Theor Appl Genet 114:1265–1275

Baliki G, Bruck T, Stojetz W (2018) Monitoring and impact analysis of the BMZ and EU-funded FAO resilience programme in Syria. ISDC, Berlin

Ballini E, Berruyer R, Morel JB, Lebrun MH, Notteghem JL, Tharreau D (2007) Modern elite rice varieties of the "Green revolution" have retained a large introgression from wild rice around the *Pi33* rice blast resistance locus. New Phytol 175:340–350

Barchi L, Lefebvre V, Sage-Palloix AM, Lanteri S, Palloix A (2009) QTL analysis of plant development and fruit traits in pepper and performance of selective phenotyping. Theor Appl Genet 118:1157–1171. https://doi.org/10.1007/s00122-009-0970-0

Bardini M, Labra M, Winfield M, Sala F (2003) Antibiotic-induced DNA methylation changes in calluses of *Arabidopsis thaliana*. Plant Cell Tissue Organ Cult 72(2):157–162

Bartel B, Bartel DP (2003) MicroRNAs: at the root of plant development? Plant Physiol 132(2):709–717

Basigulp DH, Barnes DK, Stuckev RE (1995) Development of a core collection for perennial Medicago plant introductions. Crop Sci 35:1163–1168

Baur E (1914) Die Bedeutung der primitiven Kulturrassen und der wilden Verwandten unserer Kulturpflanzen fu¨r die Pflanzenzu¨chtung. Jahrbuch der Deutschen Landwirtschafts Gesellschaft 29:104–110

Becker H (1993) Pflanzenzu¨ chtung. Verlag Eugen Ulmer, Stuttgart

Berg T, Bjornstad A, Fowler C, Skroppa T (1991) Technology options and the gene struggle: a report to the Norwegian Research Council for Science and Humanities (NAVF). NORAGRIC/ Agricultural University of Norway, pp 1–146

Bhattacharjee R (2000) Studies on the establishment of a core collection of pearl millet (*Pennisetum glaucum*). Ph. D. Thesis, CCS Haryana Agricultural University, Hisar – 125 004, India, p 162

Bickford D, Lohman DJ, Sodhi NS, Ng PK, Meier R, Winker K et al (2007) Cryptic species as a window on diversity and conservation. Trends Ecol Evol 22:148–155
Black RE, Victora CG, Walker SP, Bhutta ZA, Christian P, de Onis M, Ezzati M, Grantham-McGregor S, Katz J, Martorell R, Uauy R (2013) Maternal and child undernutrition and overweight in low-income and middle-income countries. Lancet 832(9890):427–451
Blanca J, Montero-Pau J, Sauvage C, Bauchet G, Illa E, Díez MJ et al (2015) Genomic variation in tomato, from wild ancestors to contemporary breeding accessions. BMC Genomics 16:257. https://doi.org/10.1186/s12864-015-1444-1
Bloom J (2010) American wasteland. Da Capo Press, Cambridge, MA, p 360
Borrelli VMG, Brambilla V, Rogowsky P, Marocco A, Lanubile A (2018) The enhancement of plant disease resistance using CRISPR / Cas9 technology. Front Plant Sci 9:1245
Bothmer VR, Fishbeek G, van Hintum T, Hodgkin T, Knupffer H (1990) The barley core collection. Final report of the BCC working group, Weihenstaphan, p 14
Botstein D, White RL, Skolnick M, Davis RW (1980) Construction of a genetic linkage map in man using restriction fragment length polymorphisms. Amer J Human Genet 32:314–331
Bouis HE, Saltzman A (2017) Improving nutrition through biofortification: a review of evidence from HarvestPlus, 2003 through 2016. Glob Food Sec 12:49–58
Brar DS (2005) Broadening the gene pool of rice through introgression from wild species. In: Toriyama K, Hong KL, Hardy B (eds) Rice is life: scientific perspective for 21st century. IRRI, Muñoz, pp 157–160
Breseghello F (2013) Traditional and modern plant breeding methods with examples in rice (*Oryza sativa* L.). J Agric Food Chem 61:8277–8286
Brown AHD (1989) Core collections: a practical approach to genetic resources management. Genome 31:818–824
Brown AHD, Grace JP, Speer SS (1987) Designation of a core collection of perennial glycine. Soybean Genet Newsl 14:59–70
Brown TA, Jones MK, Powell W, Allaby RG (2008) The complex origins of domesticated crops in the Fertile Crescent. Trends Plant Sci 24(2):103–109
Bruck T, d'Errico M (2019) Reprint of: food security and violent conflict: introduction to the special issue. World Develop 119:145–149
Bruck T, Días Botía OM, Ferguson NTN, Ouédraogo J, Ziegelhöfer Z (2018) Assets for alimentation? The nutritional impact of assets-based programming in Niger. Innocenti Working Papers 5:1–56
Bruke JM et al (2007) Crop evolution: from genetics to genomics. Curr Opin Genet Dev 17:525–532
Brush SB (2007) Farmers' rights and protection of traditional agricultural knowledge. World Develop 35:1499–1514
Buckler ES, Thornsberry JM, Kresovich S (2001) Molecular diversity, structure and domestication of grasses. Genet Res 77:213–218
Burger Y et al (2002) A single recessive gene for sucrose accumulation in *cucumis melo* fruit. J Am Soc Hortic Sci 127:938–943
Burgueño J, de los Campos G, Weigel K, Crossa J (2012) Genomic prediction of breeding values when modeling genotype – environment interaction using pedigree and dense molecular markers. Crop Sci 52:707–719
Buzby JC, Hyman J (2012) Total and per capita value of food loss in the United States. Food Policy 37:561–570
Cantrell RP, Christopher R, Dowswell CR, Paliwal RL (1996) Maize in the third world. Westview Press, Boulder, CO
Cavanagh C, Morell M, Mackay I, Powell W (2008) From mutations to MAGIC: resources for gene discovery, validation and delivery in crop plants. Curr Opin Plant Biol 11:215–221
Chacón MI, Pickersgill SB, Debouck DG (2005) Domestication patterns in common bean (*Phaseolus vulgaris* L.) and the origin of the Mesoamerican and Andean cultivated races. Theor Appl Genet 110:432–444
Chaudhuri SK (2005) Genetic erosion of agrobiodiversity in India and intellectual property rights: interplay and some key issues. Patentmatics 5(6):1–10

Chen S, Jin W, Wang M (2003) Distribution and characterization of over 1000 T-DNA tags in rice genome. Plant J 36(1):105–113

Chessa C, Nieddu G (2005) Analysis of diversity in the fruit tree genetic resources from a Mediterranean island. Genet Resour Crop Evol 52:267–276

Cho GT, Yoon MS, Lee J, Baek HK, Kang JH, Kim TS, Paek NC (2008) Development of a core set of Korean soybean landraces [*Glycine max* (L.) Merr.]. J of Crop Sci Biotech 11(3):157–162

Choudhary G, Ranjitkumar N, Surapaneni M, Deborah DA, Vipparla A, Anuradha G, Siddiq EA, Vemireddy LR (2013) Molecular genetic diversity of major Indian rice cultivars over decadal periods. PLoS One 8:e66197

CIAT, IFPRI (2018) Crop modeling work plan 2018, pp 1–7

Cohen JI, Alcorn JB, Potter CS (1991) Utilization and conservation of genetic resources: international projects for sustainable agriculture. Econ Bot 45(2):190–199

Concibido VC, La Vallee B, Mclaird P, Pineda N, Meyer J, Hummel L, Yang J, Wu K, Delannay X (2003) Introgression of a quantitative trait locus for yield from *Glycine soja* into commercial soybean cultivars. Theor Appl Genet 106:575–582

Convention on Biological Diversity (2007) Available online at: https://www.cbd.int/doc/meetings/cop-bureau/cop-bur-2007/cop-bur-2007-10-14-en.pdf

Cooper HD, Spillane C, Hodgkin T (2001) Broadening the genetic base of crops: an overview. In: Cooper HD, Spillane C, Hodgkin T (eds) Broadening the Genetic Base of Crop Production. CAB International, Oxon, UK., pp 1–23

Corderio CMT, Morales E, Ferreira P, Rocha DMS, Costa IVA, Silva S (1995) Towards a Brazilian core collections of cassava. In: Hodgkin T, Brown AHD, van Hintum MJL, Moarles EAVA (eds) Core collections of plant genetic resources. Wiley-Sayee Publication, Chichester, UK, pp 155–168

Cox TS, Kiang YT, Gorman MB, Rodgers DM (1985) Relationship between coefficient of parentage and genetic similarity indices in soybean. Crop Sci 25:529–532

Cox TS, Murphy JP, Rodgers DM (1986) Changes in genetic diversity in the red and winter wheat regions of the United States. Proc Natl Acad Sci U S A 83:5583–5586

Crossa J, Hernandez CM, Bretting P, Eberhart SA, Taba S (1993) Statistical genetic considerations for maintaining germplasm collections. Theor Appl Genet 86:673–678

Crossa J, de Los Campos G, Pérez P, Gianola D, Burgueño J, Araus JL et al (2010) Prediction of genetic values of quantitative traits in plant breeding using pedigree and molecular markers. Genet 186:713–724

Crossa J, Beyene Y, Kassa S, Pérez P, Hickey JM, Chen C et al (2013) Genomic prediction in maize breeding populations with genotyping by sequencing. G3 (Bethesda) 3:1903–1926

da Fonseca RR, Smith BD, Wales N, Cappellini E, Skoglund P, Fumagalli M (2015) The origin and evolution of maize in the southwestern United States. Nat Plants 1:14003

Dahan Y, Rosenfeld R, Zadiranov V, Irihimovitch V (2010) A proposed conserved role for an avocado *fw2.2*-like gene as a negative regulator of fruit cell division. Planta 232:663–676

Dalrymple DG (1986) Development and spread of high-yielding wheat varieties in developing countries, 7th edn. US Agency for International Development, Washington, DC

Daniel C (2006) A 'Forever' seed bank takes root in the Arctic. Science 312(5781):1730–1731

Darlington CD (1973) Chromosome botany and the origins of cultivated plants, 3rd edn. Allen and Unwin, London

Davari A, Khoshbakht K, Ghalegolab Behbahani A, Veisi H (2013) A qualitative assessment of diversity and factors leading to genetic erosion of vegetables: a case study of Varamin (Iran). Int J Agric Sci 3(3):198–212

De Benoist, B, McLean, E, Egli, I, Cogswell, M (2008) Worldwide prevalence of anemia 1993–2005. World Health Organization. p 35

de Candolle A (1890) Origin of cultivated plants. Haffner, New York

de los Campos G, Naya H, Gianola D, Crossa J, Legarra A, Manfredi E et al (2009) Predicting quantitative traits with regression models for dense molecular markers and pedigree. Genet 182:375–385

Dempewolf H et al (2012) Reproductive isolation during domestication. Plant Cell 24:2710–2717

Dhillon BS, Chhuneja P (2014) New initiatives for precision breeding. In: Chhuneja P, Brar KS, Ghumber RK (eds) Souvenir book of national symposium on crop improvement for inclusive sustainable development. Punjab Agricultural University, Ludhiana, pp 9–167th-9th Novemeber, 2014

Dirzo R, Raven PH (2003) Global state of biodiversity and loss. Ann Rev Environ Resour 28:137–167

Diwan N, Bauchan GR, Mclintosh MS (1994) A core collection for the United States annual *Medicago* germplasm collections. Crop Sci 34:279–285

Diwan N, Mclntosh MS, Bauchan GR (1995) Methods of developing a core collection of annual *Medicago* species. Theor Appl Genet 90:755–761

Doebley JF et al (2006) The molecular genetics of crop domestication. Cell 127:1309–1321

Dvorak J, Luo MC, Akhunov ED (2011) Vavilov's theory of Centres of diversity in the light of current understanding of wheat diversity, domestication and evolution. Czech J Genet Plant Breed:47

Edwards D, Batley J (2010) Plant genome sequencing: applications for crop improvement. Plant Biotechn 8:2–9

Edwards SV, Potter S, Schmitt CJ, Bragg JG, Moritz C (2016) Reticulation, divergence, and the phylogeography–phylogenetics continuum. Proc Natl Acad Sci U S A 113:8025–8032

Ellis P, Roberts M (2016) Leveraging urbanisation in South Asia: managing spatial transformation for prosperity and liability. International Bank for Reconstruction and Development/The World Bank, Washington, DC, p 204

Ellis PR, Pink DAC, Phelps K, Jukes PL, Breeds SE, Pinnegar AE (1998) Evaluation of a core collection of *Brassica oleracea* accessions for resistance to *Brevicoryne brassicae*, the cabbage aphid. Euphytica 103:149–160

EOS [Earth and Space Science News] (2019) Which greenhouse gas does the most damage to crops? Available from https://eos.org/research-spotlights/which-greenhouse-gas-does-the-most-damage-to-crops. Accessed on 17 June 2019

Eticha F, Sinebo W, Grausgruber H (2010) On-farm diversity and characterization of barley (*Hordeum vulgare* L.) landraces in the highlands of West Shewa Ethiopia. Ethnobot Res Appl 8:25–34

ETMarkets (2018) Agriculture sector to grow 2.1%: can it double farm income by 2022? Available from https://economictimes.indiatimes.com/markets/stocks/news/agriculture-sector-to-grow-2-1-can-it-double-farm-income-by-022/articleshow/62692884.cms?from=mdr. Accessed on 17 June 2019

Evans LT (1993) Crop evolution, adaptation and yield. Cambridge University Press, Cambridge, UK

Evenson RE, Gollin D (2003) Assessing the impact of the green revolution, 1960 to 2000. Science 300:758

FAO (1996a) Report on the state of the World's plant genetic resources for food and agriculture, prepared for the international technical conference on plant genetic resources, Leipzig, Germany, 17–23 June 1996. Food and Agricultural Organization of the United Nations, Rome

FAO (1996b) Global action plan for the conservation and sustainable utilization of plant genetic resources for food and agriculture: Leipzig declaration International Technical Conference on plant genetic resources Jure Leipzig, Germany

FAO (1996c) World food summit 1996. Final report Part 1. Available from www.fao.org/docrep/003/w3548e/w3548e00.htm. Accessed on 17 June 2019

FAO (1980) Assessment and Collection of Data on Postharvest Food Grain Losses. FAO Economic and Social Development Paper 13. Rome

FAO (2004) Rice is life. Available from http://www.fao.org/rice2004/. Accessed on 17 June 2019

FAO (2011) Labour. Available from http://www.fao.org/3/i2490e/i2490e01b.pdf. Accessed on 17 June 2019

FAO (2012) Food and Agriculture Organization of the United Nations, Italy

FAO (2017) Food security and nutrition around the World. Available from http://www.fao.org/state-of-food-security-nutrition/en/. Accessed on 17 June 2019

FAO/IAEA (1977) Induced mutations against plant diseases, STI/Pub/462. IAEA, Vienna
Fischer I, Steige KA, Stephan W, Mboup M (2013) Sequence evolution and expression regulation of stress-responsive genes in natural populations of wild tomato. PLoS One 8:e78182
Flint-Garcia SA (2013) Genetics and consequences of domestication. J Agri Food Chem 61:8267–8276
Forsbach A, Schubert D, Lechtenberg B, Gils M, Schmidt R (2003) A comprehensive characterization of single-copy T-DNA insertions in the *Arabidopsis thaliana* genome. Plant Mol Biol 52(1):161–176
Frankel OH, Bennett E (1970) Genetic resources – introduction. In: Frankel OH, Bennett E (eds) Genetic resources in plants – their exploration and conservation, IPB handbook, vol 11. International Biological Programme, London, pp 7–18
Frankel OH, Brown AHD (1984) Current plant genetic resources – a critical appraisal. In: Genetics: new Frontiers, vol 4. Oxford and IBH publ. Co., New Delhi, pp 1–11
Frey KJ (1981) Capabilities and limitations of conventional plant breeding. In: Rachie KO, Lyman JM (eds) Genetic Engineering for Crop Improvement – Working papers. Rockefeller Foundation, New York, pp 15–65
Friis-Hansen E (1999) Erosion of plant genetic resources: causes and effect. Danish J Geo Spec Iss 1:61–68
Fu YB (2015) Understanding crop genetic diversity under modern plant breeding. Theor Appl Genet 128:2131–2142
Fu YB, Peterson GW, Scoles G, Rossnagel B, Schoen DJ, Richards KW (2003) Allelic diversity changes in 96 Canadian oat cultivars released from 1886 to 2001. Crop Sci 43:1989–1995
Fu YB (2006) Impact of plant breeding on genetic diversity of agricultural crops: searching for molecular evidence. Plan Gen Resour 4:71–78
Fu YB, Peterson GW, Morrison MJ (2007) Genetic diversity of Canadian soybean cultivars and exotic germplasm revealed by simple sequence repeat markers. Crop Sci 47:1947–1954
Funk WC, Mckay JK, Hohenlohe PA, Allendorf FW (2012) Harnessing genomics for delineating conservation units. Trends Ecol Evol 27:489–496
Galwey NW (1995) Verifying and validating the representativeness of a core collection. In: Hodgkin T, Brown AHD, van Hintum MJL, Morales EAV (eds) A Core Collections of Plant Genetic Resources. Wiley-Sayee Publications, Chichester, UK, pp 187–198
Gao JP, Lin HX (2013) Methods in molecular biology. Humana Press, New York
Gaut BS (2014) Evolution is an experiment: assessing parallelism in crop domestication and experimental evolution. Mol Biol Evol 32:1661–1671
Gaut BS (2015) Evolution is an experiment: assessing parallelism in crop domestication and experimental evolution: (Nei Lecture, SMBE 2014, Puerto Rico). Mol Biol Evol 32:1661–1671. https://doi.org/10.1093/molbev/msv105
Gautam PL, Singh BB, Sexana S, Sharma RK (2004) Collection, conservation and utilization of plant genetic resources. In: Jain HK, Kharkwal MC (eds) Plant breeding – Mendelian to molecular approaches. Narosa Publishing House, New Delhi, pp 657–690
Gelvin SB (2003) Agrobacterium-mediated plant transformation: the biology behind the 'gene-jockeying' tool. Microbiol Mol Biol Rev 67(1):16–37
Gelvin SB (2017) Integration of agrobacterium T-DNA into the plant genome. Annual Rev Genet 51:195–217
Gheysen G, Van Montagu M, Zambryski P (1987) Integration of *Agrobacterium tumefaciens* transfer DNA (T-DNA) involves rearrangements of target plant DNA sequences. Proc Nat Acad Sci USA 84(17):6169–6173
Gill KS, Sandhu TS, Singh K, Brar JS (1975) Crop Improv 2:99
Glémin S, Bataillon T (2009) A comparative view of the evolution of grasses under domestication. New Phytol 183:273–290
Golicz AA, Batley J, Edwards D (2016a) Towards plant pangenomics. Plant Biotech J 14:1099–1105

Golicz AA, Bayer PE, Barker GC, Edger PP, Kim H, Martinez PA, Chan CKK, Severn-Ellis A, McCombie WR, Parkin IAP et al (2016b) The pangenome of an agronomically important crop plant *Brassica Oleracea*. Nat Commun 7:13390

González-Camacho JM, de Los Campos G, Pérez P, Gianola D, Cairns JE, Mahuku G et al (2012) Genome-enabled prediction of genetic values using radial basis function neural networks. Theor Appl Genet 125:759–771

Good Food World (2011) Where have all the farmers gone? November 16th, 2011. Available from http://www.goodfoodworld.com/2011/11/where-have-all-the-farmers-gone/. Accessed on 17 June 2019

Gordon SP, Contreras-Moreira B, Woods DP, Marais DLD, Burgess D, Shu SQ, Stritt C, Roulin AC, Schackwitz W, Tyler L et al (2017) Extensive gene content variation in the *Brachypodium distachyon* pan-genome correlates with population structure. Nat Commun 8:2184

Gowda J, Rekha D, Bharti S, Krishnappa M (2007) Methods of constructing core set using agro-morphological traits in foxtail millet. J Plant Genet Res 20(1):38–42

Gowda J, Rekha D, Somu G, Bharti S, Krishnappa M (2008) Development of core set in little millet (*Panicum sumatrense*) germplasm using data on twenty one morpho-agronomic characters. Environ Ecol 26(3):1055–1060

Gowda J, Bharati S, Somu G, Krishnappa M, Rekha D (2009) Formation of core set in barnyard millet (*Echenocloa frumentaces* (Rox b.) Link) germplasm using data on twenty four morpho-agronomic characters. Int J Plant Sci 4:1): 1–1): 5

Grauke LJ, Thompson TE (1995) Evolution of pecan [*Carya illinoinensis* (Wangenh.) K. Koch] germplasm collections and designation of a core subset. Hort Sci 30(5):50–54

Greene T (2019) Genome editing in crops. In: XIV Agricultural Science Congress, 20th–23rd February, 2019, NASC, Pusa, New Delhi, India. p 9

Grenier C, Bramel PJ, Hamon P (2001) Core collection of the genetic resources of sorghum: 1. Stratification based on eco-geographical data. Crop Sci 41:234–240

Grenier C, Cao TV, Ospina Y, Quintero C, Châtel MH, Tohme J et al (2015) Accuracy of genomic selection in a rice synthetic population developed for recurrent selection breeding. PLoS One 10:e0136594. https://doi.org/10.1371/journal.pone.0136594

Gressel J (2004) Major heretofore intractable biotic constraints to African food security that may be amenable to novel biotechnological solutions. Crop Protect 23:661–689

Groombridge B (1992) Global biodiversity:status of the Earth's living resources. A report compiled by the world conservation monitoring Centre. Chapman and Hall, London

Gross BL, Olsen KM (2010) Genetic perspectives on crop domestication. Trends Plant Sci 15:529–537

Guan JC, Koch KE, Suzuki M, Wu S, Latshaw S, Petruff T et al (2012) Diverse roles of strigolactone signaling in maize architecture and the uncoupling of a branching-specific subnetwork. Plant Physiol 160:1303–1317. https://doi.org/10.1104/pp.112.204503

Gustavasson J, Cederberg C, Sonesson U, van Otterdijk R, Meybeck A (2011) Global food losses and food waste: extent causes and prevention. Food and Agriculture Organization (FAO) of the United Nations, p 23

Hallauer AR (2011) Evolution of plant breeding. Crop Breed Appl Biotechn 11:197–206

Hammer K, Knuepffer H, Xhuveli L, Perrino P (1996) Estimating genetic erosion in landraces—two case studies. Genet Res Crop Evol 43:329–336

Hamon S, Noirot M (1990) Some proposed procedures for obtaining a core collection using quantitative plant characterization data. In: Proceedings of international workshop on okra. NBPGR, New Delhi, pp 8–12

Hamon S, Van Sloten DH (1989) Characterization and evaluation of okra. In: Brown AHD, Frankel OH, Marshall DR, Williams JT (eds) The use of plant genetic resources. Cambridge University Press, Cambridge, UK, pp 173–196

Hamon S, Noirot M, Anthony F (1995) Developing a coffee core collection using the principal components score strategy with quantitative data. In: Hodgkin T, Brown AHD, van Hintum MJL, Morales EAV (eds) Core collections of plant genetic resources. A Wiley-Sayee Publications, Chichester, UK, pp 117–126

Hao C, Wang L, Zhang X, You G, Dong Y, Jia J, Liu X, Shang X, Liu S, Cao Y (2006) Genetic diversity in Chinese modern wheat varieties revealed by microsatellite markers. Science China C Life Sci 49:218–226
Haradari CB (2009) Formation of core set in African and Indian finger millet (*Eleusine coracana* (L.) Gaertn) accessions. M.Sc. (Ag.) Thesis, University of Agricultural Sciences, Bangalore
Harlan HV (1957) One man's life with barley. Exposition Press, New York
Harlan JR (1965) Plant introduction and biosynthesis. In: Plant breeding symposium, pp 55–83
Harlan JR (1970) Evolution of cultivated plants. In: Frankel OH, Bennett E (eds) Genetic resources in plants – IBP handbook no 11. International Biological Programme, London, pp 19–32
Harlan JR (1975) Our vanishing genetic resources. Science 188:618–621
Harlan JR (1987) Gene centres and gene utilization in American agriculture. In: Yeotman CW, Xaflon D, Wilkes G (eds) Plant genetic resources: a conservation imperative, AAAS selected symposium 87. Westview Press, Boulde Colorado, pp 111–129
Harlan JR (1992) Crops and man, 2nd edn. American Society of Agronomy and Crop Science Society of America, San Antonio
Harlan JR (1995) The living fields: our agricultural heritage. Cambridge University Press, Cambridge, UK
Harlan JR, de Wet JMJ (1971) Toward a rational classification of cultivated plants. Taxon 20:509–517
Haussmann BIG, Parzies HK, Presterl T, Susic Z, Miedaner T (2004) Plant genetic resources in crop improvement. Plant Genet Res 2(1):3–21
Hawkes JG (1977) The importance of wild germplasm in plant breeding. Euphytica 26:615–621
Heppner MJ (1923) The factor for bitterness in the sweet almond. Genet 8:390–391
Hernández M, Pla M, Esteve T, Prat S, Puigdomènech P, Ferrando A (2003) A specific real-time quantitative PCR detection system for event MON810 in maize YieldGard based on the 3′-transgene integration sequence. Transgenic Res 12(2):179–189
Hernandez-Verdugo S, Luna-Reyes R, Oyama K (2001) Genetic structure and differentiation of wild and domesticated populations of *Capsicum annuum* (Solanaceae) from Mexico. Plant Syst Evol 226:129–142
Heslot N, Rutkoski J, Poland J, Jannink JL, Sorrells ME (2013) Impact of marker ascertainment bias on genomic selection accuracy and estimates of genetic diversity. PLoS One 8:e74612. https://doi.org/10.1371/journal.pone.0074612
Heszkyl L (2008) Challenges of plant breeding early in 21th century. In: Hungarian agricultural research. Institute of Genetics and Biotechnology, St. Istvan University, Godollo, pp 4–8
Heun M, Schaefer-Pregl R, Klawan D, Castagna R, Accerbi M, Borghi B, Salamini F (1997) Site of einkorn wheat domestication identified by DNA fingerprinting. Science 278:1312–1314
Heywood VH (1990) Botanic gardens and the conservation of plant resources. Impact Sci Soc 158:121–132
Heywood VH (1995) Global biodiversity assessment. UNEP. Cambridge University Press, Cambridge, UK
Hirsch CN, Foerster JM, Johnson JM, Sekhon RS, Muttoni G, Vaillancourt B, Penagaricano F et al (2014) Insights into the maize pangenome and pan-transcriptome. Plant Cell 26:121–135
Hodkinson TR, Parnell JAN (2007) Reconstructing the tree of life: taxonomy and systematic of species rich taxa, Systematics association, special series, vol 72. Taylor and Francis, CRC Press, Boca Raton
Holbrook CC (1999) Testing and utilization of a core collection for the US germplasm collection of peanut. In: Johnson RC, Hodgkin T (eds) Core collections for today and tomorrow. International Plant Genetic Resources Institute, Rome, pp 68–73
Holbrook CC, Anderson WF, Pittman RN (1993) Selection of a core collection from the US germplasms collection of peanuts. Crop Sci 33:859–861
Holbrook CC, Timper P, Xue HQ (2000) Evaluation of the core collection approach for identifying resistance to *Meloidogyne arenaria* in peanut. Crop Sci 40:1172–1175
Holden JHW (1984) The second ten years. In: Holden JHW, Williams JT (eds) Crop genetic resources: conservation and evaluation. George Allen and Unwin, London, UK, pp 277–285

Holland W (2018) Kerala flooding: agricultural impacts and environmental degradation. The Cabi Blog. Available at https://blog.cabi.org/2018/09/26/kerala-flooding-agricultural-impacts-and-environmental-degradation/. Accessed on 10 June 2019

Hoyt E (1988) Conserving the wild relatives of crops. IBPGR – IUCN – WWF, IBPGR, Rome

Huaman Z, Aguilar C, Ortiz R (1999) Selecting a Peruvian sweet potato core collection on the basis of morphological, eco-geographical, disease and pest reaction data. Theor Appl Genet 98:840–844

Huang H (2011) Plant diversity and conservation in China: planning a strategic bioresource for a sustainable future. Bot J Linn Soc 166:282–300

Huang XQ, Coster H, Ganal MW, Roder MS (2003) Advanced backcross QTL analysis for the identification of quantitative trait loci alleles from wild relatives of wheat (*Triticum aestivum* L.). Theort Appl Genet 106:1379–1389

Huang X, Paulo MJ, Boer M, Effgen S, Keizer P, Koornneef M, van Eeuwijk FA (2011) Analysis of natural allelic variation in *Arabidopsis* using a multiparent recombinant inbred line population. Proc Nat Acad Sci USA 108:4088–4093

Huang H, Oldfield S, Qian H (2013) Chapter 2: Global significance of plant diversity in China. In: De-Yuan H, Blackmore S (eds) Plants of China. Science Press, Beijing, pp 7–34

Hufford MB et al (2012) Comparative population genomics of maize domestication and improvement. Nat Genet 44:808–811

Hummer KE, Hancock JF (2015) Vavilovian centers of plant diversity: implications and impacts. HortSci 50(6):780–783

Hurgobin B, Golicz AA, Bayer PE, Chan CKK, Tirnaz S, Dolatabadian A, Schiessl SV, Samans B, Montenegro JD, Parkin IA (2018) Homeologous exchange is a major cause of gene presence/absence variation in the amphidiploid *Brassica napus*. Plant Biotech J 16:1265–1271

Ichikawa T, Nakazawa M, Kawashima M (2003) Sequence database of 1172 T-DNA insertion sites in *Arabidopsis* activation-tagging lines that showed phenotypes in T1 generation. Plant J 36(3):421–429

ICRISAT (1976) Chickpea breeding. Annual Report, 1975–1976. ICRISAT, Hyderabad

ICRISAT (2004) Success story 22 Jan 2004. Patancheru, Hyderabad, India. Online website www.icrisat.org

Igartua E, Gracia MP, Lasa JM, Medina B, Molina-Cano JL, Romagosa I (1998) The Spanish barley core collection. Genet Res Crop Evol 45(5):475–481

IME [Institution of Mechanical Engineers] (2013) Global food waste not, want not. Available at: https://www.imeche.org/policy-and-press/reports/detail/global-food-waste-not-want-not. Accessed on 17 June 2019

IPCC [Intergovernmental Panel on Climate Change] (2013) Climate change 2013: the physical science basis, Working group I contribution to the IPCC fifth assessment report. Cambridge University Press, Cambridge, UK. Available at: www.ipcc.ch/report/ar5/wg1. Accessed on 17 June 2019

IPGRI (1993) Diversity for development. International Plant Genetic Resources Institute, Rome, p 136

ISAAA (2017) Global status of commercialized biotech/GM crops in 2017: biotech crop adoption surges as economic benefits accumulate in 22 years, ISAAA Brief No. 53. ISAAA, Ithaca, NY

Jacquemin J, Bhatia D, Singh K, Wing RA (2013) The international *Oryza*, map alignment project: development of a genus-wide comparative genomics platform to help solve the 9 billion-people question. Curr Opin Plant Biol 16:147–156

Jaganathan D, Ramasamy K, Sellamuthu G, Jayabalan S, Venkataraman G (2018) CRISPR for crop improvement: an update review. Front Plant Sci 9:985

Jain SM (2001) Tissue culture-derived variation in crop improvement. Euphytica 118(2):153–166

Jain SK, Omprakash (2019) Pre-breeding: a bridge between genetic resources and crop improvement. Int J Curr Microbiol Appl Sci 8(2):1998–2007

Jana R, David VLK (2016) Application of next-generation sequencing in plant breeding. Czech J Genet Plant Breed. https://doi.org/10.17221/192/2016-CJGPB

Jana S, Pietrzak LN (1988) Comparative assessment of genetic diversity in wild and primitive cultivated barley in a center of diversity. Genet 119:981–990
Jannink JL, Lorenz AJ, Iwata H (2010) Genomic selection in plant breeding: from theory to practice. Brief Funct Genomics 9:166–177
Jauhar A, Santiaguel FA (2011) Greener rice. Rice Today 10:27–29
Jia L, Agrama H, Yeater KM, Yan W, Mc Clung AM, Wu D (2009) Evaluation of the USDA rice core collection for sheath blight disease using micro-chamber. In: Proceedings of 2009 international annual meetings of ASA-CSSA-SSSA, November 1–5, 2009, Pittsburgh, Pennsylvania
Jiang HF, Ren XP, Liao BS, Huang JQ, Lei Y, Chen BY, Guo B, Holbrook CC, Upadhyaya HD (2008) Peanut core collection established in China and compared with ICRISAT mini core collection. Acta Agron Sinica 34(1):25–30
Jones MR, Good JM (2016) Targeted capture in evolutionary and ecological genomics. Mol Ecol 25:185–202
Jorger KM, Schrödl M (2013) How to describe a cryptic species? Practical challenges of molecular taxonomy. Front Zool 10:59
Jyoti K, Ruchi B, Sundeep K (2017) Plant genetic resources: effective utilization for sustainability. Adv Biotech Micro 3(2):555609. https://doi.org/10.19080/AIBM.2017.03.555609
Kaga A et al (2008) The genetics of domestication of the azuki bean (*Vigna angularis*). Genet 178:1013–1036
Kaneda T, Haub C (2018) How many people have ever lived on earth? Available at: https://www.prb.org/howmanypeoplehaveeverlivedonearth/. Accessed on 2 Feb 2019
Kang LZ et al (2014) The 3000 Rice genome project. Giga Science. https://doi.org/10.1186/2047-217X-3-7
Kaya H, Sato S, Tabata S, Kobayashi Y, Iwabuchi M, Araki T (2000) *hosoba toge toge,* a syndrome caused by a large chromosomal deletion associated with a T-DNA insertion in *Arabidopsis*. Plant Cell Physiol 41(9):1055–1066
Kiambi D, Ford-Lloyd B, Jackson MT, Guarino L, Maxted N, Newbury HJ (2005) Collection of wild rice (*Oryza* L.) in east and southern Africa in response to genetic erosion. Plant Genet Res Newslett 142:10–20
Kijima Y, Sserunkuuma D, Otsuka K (2006) How revolutionary is the "Nerica revolution"? Evidence from Uganda. Develop Econ 44:252–267
Kim SR, Lee J, Jun SH (2003) Transgene structures in T-DNA-inserted rice plants. Plant Mol Biol 52(4):761–773
Kim S, Veena, Gelvin S (2007) Genome-wide analysis of *Agrobacterium* T-DNA integration sites in the *Arabidopsis* genome generated under non-selective conditions. Plant J 51:779–791
Kistler L, Yoshi Maezumi S, Gregorio de Souza J, Przelomska NAS, Malaquias Costa F, Smith O, Loiselle H, Ramos-Madrigal J, Wales N, Ribeiro ER, Morrison RR, Grimaldo C, Prous AP, Bernardo Arriaza M, Gilbert TP, de Oliveira Freitas F, Allaby RG (2018) Multiproxy evidence highlights a complex evolutionary legacy of maize in South America. Science 362:1309–1313
Knupffer H, van Hintum MJL (1995) The barley core collection- an international effort. In: Hodgkin T, Brown AHD, van Hintum MJL, Morales EAV (eds) Core collections of plant genetic resources. A Wiley-Sayee Publications, Chichester, pp 171–178
Kombo GR, Dansi A, Loko LY et al (2012) Diversity of cassava (*Manihot esculenta* Crantz) cultivars and its management in the department of Bouenza in the Republic of Congo. Genet Res Crop Evol 59(8):1789–1803
Kottapalli P, Upadhyaya HD, Kottapalli KR, Payton PR Puppala N (2010) Molecular characterisation and assessment of genetic diversity in Valencia mini core using SSR markers. In: Proceedings of the American Peanut Research and Education Society, Inc. July 14th–17th, 2009, Releigh, North Carolina, p 38
Kover PX, Valdar W, Trakalo J, Scarcelli N, Ehrenreich IM, Purugganan MD, Durrant C, Mott R (2009) A multiparent advanced generation inter-cross to fine-map quantitative traits in *Arabidopsis thaliana*. PLoS Genet 5:e1000551

Krysan PJ, Young JC, Sussman MR (1999) T-DNA as an insertional mutagen in Arabidopsis. Plant Cell 11(12):2283–2290

Kumar S, Fladung M (2002) Transgene integration in aspen: structures of integration sites and mechanism of T-DNA integration. Plant J 31(4):543–551

Kusaba M, Miyahara K, Iida S (2003) *Low glutelin content1*: a dominant mutation that suppresses the *glutelin* multigene family via RNA silencing in rice. Plant Cell 15(6):1455–1467

Ladizinsky G (1987) Pulse domestication before cultivation. Econ Bot 41:60–65

Ladizinsky G (1993) Lentil domestication: on the quality of evidence and arguments. Econ Bot 47:60–64

Ladizinsky G (1998) Plant evolution under domestication. Kluwer Academic Publishers, New York

Ladizinsky G (1999) On the origin of almond. Genet Res Crop Evol 46:143–147

Lado B, Barrios PG, Quincke M, Silva P, Gutiérrez L (2016) Modeling genotype-environment interaction for genomic selection with unbalanced data from a wheat breeding program. Crop Sci 56:1–15. https://doi.org/10.2135/cropsci2015.04.0207

Laghetti G, Fiorentin G, Hammer K, Pignone D (2009) On the trail of the last autochthonous Italian einkorn (*Triticum monococcum* L.) and emmer (*Triticum dicoccon* Schrank) populations: a mission impossible? Genet Res Crop Evol 56(8):1163–1170

Larson G, Burger J (2013) A population genetics view of animal domestication. Trends Genet 29:197–205

Leache AD, Oaks JR (2017) The utility of single nucleotide polymorphism (SNP) data in phylogenetics. Ann Rev Ecol Syst 48:69–84

Lelley T, Stachel M, Grausgruber H, Vollmann J (2000) Analysis of relationships between *Aegilops tauschii* and the D genome of wheat utilizing microsatellites. Genome 43(4):661–668

Lenser T, Theißen G (2013) Molecular mechanisms involved in convergent crop domestication. Trends Plant Sci 18:704–714

Lester RN (1989) Evolution under domestication involving disturbance of genic balance. Euphytica 44:125–132

Li DJ, Sun CQ, Fu YC, Li C, Zhu ZF, Chen L, Cai HW, Wang XK (2002) Identification and mapping of genes for improving yield from Chinese common wild rice (*O. rufipogon* Griff.) using advanced backcross QTL analysis. Cjin Sci Bull 47:1053–1059

Li Y, Rosso MG, Ulker B, Weisshaar B (2006) Analysis of TDNA insertion site distribution patterns in *Arabidopsis thaliana* reveals special features of genes without insertions. Genomics 87:645–652

Li YH, Zhou G, Ma J, Jiang W, Jin LG, Zhang Z, al GY (2014) *De novo* assembly of soybean wild relatives for pan-genome analysis of diversity and agronomic traits. Nat Biotech 32:1045–1052

Liang HZ, Zi-Chao L, Deng-Quan L, Ya-Wen Z, Shi-Quan S, Ping M, Zhong-Yi Y, Xiang-Kun W (2004) Microsatellite analysis of landrace rice core collection in Yunan, China. Chinese J Agril Biotech 1:23–30

Lin K, Zhang N, Severing E, Nijveen H, Cheng F, Visser R, Wang X et al (2014) Beyond genomic variation – comparison and functional annotation of three *Brassica rapa* genomes: a turnip, a rapid cycling and a Chinese cabbage. BMC Genomics 15:250

Ling LX, Yong-gen LU, Jin-quen LI, Hai-ming XU, Muhammad Q (2011) Strategies on sample size determination and qualitative and quantitative traits integration to construct core collection of rice (*Oryza sativa* L.). J Rice Sci 18(1):46–55

Litt M, Luty JA (1989) A hypervariable microsatellite revealed by in vitro amplification of a dinucleotide repeat within the cardiac muscle actin gene. Amer J Human Genet 44(3):397–401

Lokanathan TR, Phundan Singh, Agarwal DK., Punit Mohan, Suman Bala Singh, Vinita Gotmare, Singh VV (2003) Genetic enhancement in cotton. Technical Bulletin from CICR Report No. 26

Londo JP, Chiang YC, Hung KH, Chiang TY, Schaal BA (2006) Phylogeography of Asian wild rice, *Oryza rufipogon*, reveals multiple independent domestications of cultivated rice, *Oryza sativa*. Proc Nat Acad Sci USA 103:9578–9583

Louwaars NP (2018) Plant breeding and diversity: a troubled relationship? Euphytica 214:114–122

Lu XM, Hu XJ, Zhao YZ, Song WB, Zhang M, Chen ZL, Chen W, Dong YB, Wang ZH, Lai JS (2012) Map-based cloning of *zb7* encoding an IPP and DMAPP synthase in the MEP pathway of maize. Mol Plant 5(5):1100–1112

Lucht JM, Mauch-Mani B, Steiner HY, Metraux JP, Ryals J, Hohn B (2002) Pathogen stress increases somatic recombination frequency in *Arabidopsis*. Nat Genet 30(3):311–314

Luciano LN, Peterinain E (2000) Prebreeding: a link between genetic resources and maize breeding. Sci Agric 57:3):1–3)10

Mackay MC (1990) Strategic planning for effective evaluation of plant germplasm. In: Srivastava IP, Damania AB (eds) Wheat genetic resources: meeting diverse needs. Wiley, New York, pp 49–53

Madlung A, Comai L (2004) The effect of stress on genome regulation and structure. Ann Bot 94(4):481–495

Mahajan RK, Bisht IS, Agarwal RC, Rana RS (1996) Studies on South Asian okra collection: methodology for establishing a representative core set using characterization data. Genet Res Crop Evol 43:249–255

Majeský L, Vasut RJ, Kitner M, Trávnicek B (2012) The pattern of genetic variability in apomictic clones of *Taraxacum officinale* indicates the alternation of asexual and sexual histories of apomicts. PLoS One 7:e41868

Makarevitch I, Svitashev SK, Somers DA (2003) Complete sequence analysis of transgene loci from plants transformed via microprojectile bombardment. Plant Mol Biol 52(2):421–432

Mallet J (2007) Hybrid speciation. Nature 446:279–283. https://doi.org/10.1038/nature05706

Malvar RA, Butron A, Alvarez A, Ordas B, Soengas P, Revilla P, Ordas A (2004) Evaluation of the European Union maize landrace core collection for resistance to *Sesamia nonagrioides* (Lepidoptera: Noctuidae) and *Ostrinia nubilalis* (Lepidoptera:Crambidae). J Econ Entomol 97:628–634

Malysheva-Otto L, Ganal MW, Law JR, Reeves JC, Roder MS (2007) Temporal trends of genetic diversity in European barley cultivars (*Hordeum vulgare* L.). Mol Breed 20:309–322

Mammadov J, Buyyarapu R, Guttikonda SK, Parliament K, Abdurakhmonov IY, Kumpatla SP (2018) Wild relatives of maize, rice, cotton, and soybean: treasure troves for tolerance to biotic and abiotic stresses. Front Plant Sci 9:886. https://doi.org/10.3389/fpls.2018.00886

Mangelsdorf PC (1966) Genetic potential for increasing yields of food crops and animals. Proc Nat Acad Sci USA 56:370–375

Mao L, Begum D, Chuang HW, Budiman MA, Szymkowiak E, Irish EJ et al (2000) *JOINTLESS* is a *MADS-box* gene controlling flower abscission zone development. Nature 406:910–913

Martinez-Ainsworth NE, Tenaillon MI (2016) Superheroes and masterminds of plant domestication. C R Biol 339:268–273

Martinville B d, Wyman AR, White R, Francke U (1982) Assignment of the first random restriction fragment length polymorphism (RFLP) locus (D14S1) to a region of human chromosome 14. Amer J Human Genet 34(2):216–226

Mathews P, Ambrose MJ (1994) Development and use of a core collection for the John Inns Pisum collection. In: Proceedings of the Eucarpia meeting: evaluation and exploitation of genetic resources. March 15th–18th, Clermont Ferrand, France. pp 99–102

Matsuoka Y, Mitchell SE, Kresovich S, Goodman M, Doebley J (2002) Microsatellites in Zea—variability, patterns of mutations, and use for evolutionary studies. Theor Appl Genet 104(2–3):436–450

McCallum CM, Comai L, Greene EA, Henikoff S (2000) Targeted screening for induced mutations. Nat Biotech 18:455–457

McClung AM, Chen M, Bockelman HE, Bryant RJ, Yan W, Fjellstrom RG (2004) Characterization of a core collection of rice germplasm and elite breeding lines in the US with genetic markers associated with cooking quality. In: Proceedings of the 2nd international Rice functional genomics conference, Tuscon, Arizona. p 127

Meyer RS, Purugganan MD (2013) Evolution of crop species: genetics of domestication and diversification. Nat Rev Genet 14:840–852

Miller AJ (2007) Crop plants: evolution. In: Hetherington AM (ed) eLS. Wiley, New York, pp 1–7

Mohan M, Nair S, Bhagwat A (1997) Genome mapping, molecular markers and marker-assisted selection in crop plants. Mol Breed 3(2):87–103

Mongkolporn O, Suntree Hanyong S, Julapark Chunwongse J, Sirikul Wasee S (2015) Establishment of a core collection of chilli germplasm using microsatellite analysis. Plant Genet Res 13(2):104–110

Montenegro JD, Golicz AA, Bayer PE, Hurgobin B, Lee H, Chan CKK, Visendi P, Lai K, Dolezel J, Batley J et al (2017) The pangenome of hexaploid bread wheat. Plant J 90:1007–1013

Moody DB, Rathjen AJ, Cartwright B, Paull JF, Lewis J (1988) Genetic diversity and geographical distribution of tolerance to high levels of soil boron. In: Miller TE, Koebner RMD (eds) Proceedings of the 7th international wheat genetic symposium. Cambridge University Press, Cambridge, UK, pp 859–865

Muller HJ (1927) Artificial transmutation of the gene. Science 66:84–87

Nass LL, Paterniani E (2000) Pre-breeding: a link between genetic resources and maize breeding. Sci Agric 57:581–587

Nater A, Mattle-Greminger MP, Nurcahyo A, Nowak MG, de Manual M, Desai T et al (2017) Morphometric, behavioral, and genomic evidence for a new orangutan species. Curr Cell Biol 27:3487–3498

NIDM [National Institute of Disaster Management] (2015) Impact and damage assessment. In: Uttarakhand disaster 2013. National Institute of Disaster Management (Ministry of Home Affairs, Government of India), pp 89–114

Ochoa CM (1975) Potato collecting expeditions in Chile, Bolivia and Peru, and the genetic erosion of indigenous cultivars. In: Frankel OH, Hawkes JG (eds) Crop genetic resources for today and tomorrow. International biological Programme 2. Cambridge University Press, Cambridge, UK, pp 167–173

Oliveira MF, Randall LN, Isaias OG, Cosme DC, José Francisco F, Toledo d (2010) Establishing a soybean germplasm core collection. Field Crops Res 119(2–3):277–289

Olsen KM, Wendel JF (2013) A bountiful harvest: genomics insights into crop domestication phenotypes. Annu Rev Plant Biol 64:47–70

Ortiz R, Lund B, Anderson SB (2003) Breeding gain and changes in morphotype of Nordic spring wheat (1901–1993) under contrasting environments. Genet Res Crop Evol 50:455–459

Paris HS, Brown RN (2005) The genes of pumpkin and squash. HortSci 40:1620–1630

Paun O, Hörandl E (2006) Evolution of hypervariable microsatellites in apomictic polyploid lineages of *Ranunculus carpaticola*: directional bias at dinucleotide loci. Genet 174:387–398

Paun O, Greilhuber J, Temsch E, Hörandl E (2006) Patterns, sources and ecological implications of clonal diversity in apomictic *Ranunculus carpaticola* (*Ranunculus auricomus* complex, Ranunculaceae). Mol Ecol 15:897–910

Peach C, Velten J (1991) Transgene expression variability (position effect) of CAT and GUS reporter genes driven by linked divergent T-DNA promoters. Plant Mol Biol 17(1):49–60

Pickersgill B (2009) Domestication of plants revisited – Darwin to the present day. Bot J Linn Soc 161:203–212

Pickersgill B (2018) Parallel vs. convergent evolution in domestication and diversification of crops in the Americas. Front Ecol Evol. https://doi.org/10.3389/fevo.2018.00056

Pierre CS, Burgueño J, Crossa J, Dávila GF, López PF, Moya ES et al (2016) Genomic prediction models for grain yield of spring bread wheat in diverse agro-ecological zones. Sci Rep 6:1–11

Pimm SL, Gittlaman JL, McCracken GF, Gilpin M (1989) Genetic bottlenecks: alternative explanations for low genetic variability. Trends Ecol Evol 4:176–177

Pingali PL (2012) Green revolution: impacts, limits, and the path ahead. Proc Nat Acad Sci USA 109:12302–12308

Pkania CK, Jian-Guo W, Hai-Ming X, Chun-Hai S, Chang-Tao L (2006) Addressing rice germplasm genetic potential using genotypic value to develop quality core. J Sci Food Agric 87(2):326–333

Plunkett DL, Smith NJH, Williams JT, Anishetty NM (1987) Gene banks and the worlds food. Princton University Press, Princeton

Poland J, Endelman J, Dawson J, Rutkoski J, Wu S, Manes Y et al (2012) Genomic selection in wheat breeding using genotyping-by-sequencing. Plant Genome 5:103–113

Poncet V, Robert T, Sarr A, Gepts P (2004) Quantitative trait locus analysis of the domestication syndrome and domestication process. In: Goodman RM (ed) Encyclopedia of plant and crop science. M. Dekker, New York, pp 1069–1074

Purugganan MD, Fuller DQ (2009) The nature of selection during plant domestication. Nature 457:843–848

Qi Y, Zhang D, Zhang H, Wang M, Sun J, Wei X, Qiu Z, Tang S, Cao Y, Wang X, Li Z (2006) Genetic diversity of rice cultivars (*Oryza sativa* L.) in China and the temporal trends in recent fifty years. Chin Sci Bull 51:681–688

Qualset CO, McGuire PE, Warbuton ML (1995) Agrobiodiversity – key to agricultural productivity. California Agri 49:45–49

Qualset CO, Damania AB, Zanatta ACA, Brush SB (1997) Locally based crop plant conservation. In: Maxted N, Ford-Lloyd BV, Hawkes JG (eds) Plant genetic conservation: the in situ approach. Chapman and Hall, London, pp 160–175

Radovic G, Jelovac D (1994) The possible approach in maize core collection development. In: Proceedings of Eucarpia meeting: Evaluation and exploitation of genetic resources. March 15th–18th, Clermont Ferrand, France. pp 109–115

Ram T, Majumder ND, Krishnaveni D, Ansari MM (2007) Rice variety Dhanrasi, an example of improving yield potential and disease resistance by introgressing gene(s) from wild species (*Oryza rufipogon*). Curr Sci 92:987–992

Ramachandran P (2013) Food and nutrition security: challenges in the new millennium. Indian J Medical Res 138(3):373–382

Ramesh Babu (2019) Genomic selection in crops: status and prospects. In: XIV agricultural science congress, 20th–23rd February, 2019, NASC, Pusa, New Delhi, India. p 7

Rao KEP, Rao VR (1995) The use of characterization data in developing a core collection of sorghum. In: Hodgkin T, Brown AHD, van Hintum MJL, Morales EAV (eds) Core collections of plant genetic resources. A Wiley-Sayee Publications, Chichester, UK, pp 109–116

Rao SLV, Chaitanya U, Gautam NK, Viraktamath BC (2012) Development of rice mini core. In: Proceedings of International symposium on 100 years of Rice Science and Looking Beyond, Jan 9th–12th 2012, TNAU, Coimbatore, Tamil Nadu. p 24

Rauf S, da Silva JT, Khan AA, Naveed A (2010) Consequences of plant breeding on genetic diversity. Int J Plant Breed 41:1–21

Reddy LJ, Upadhyaya HD, Gowda CLL, Singh S (2005) Development of core collection in pigeon pea [*Cajanus cajan* (L.) Millspaugh] using geographic and qualitative morphological descriptors. Genet Res Crop Evol 52:1049–1056

Reif JC, Hamrit S, Heckenberger M, Schipprack W, Maurer HP, Bohn M, Melchinger AE (2005) Trends in genetic diversity among European maize cultivars and their parental components during the past 50 years. Theor Appl Genet 111:838–845

Rick CM (1979) Potential improvement of tomatos by controlled introgression of genes from wild species. In: Proceedingds of conference on broadening Genetic Base of crops. 1978. Wageningen Pudoc. Wageningen, The Netherlands

Rick CM (1984) Plant germplasm resources. In: Evans DA, Sharp WR, Ammirato PV, Yamada Y (eds) Hand book of cell culture. Mac Milan, New York, pp 9–37

Rieger R, Michaelis A, Green M (1976) Glossary of genetics and cytogenetics. Springer, New York

Rieseberg LH, Willis JH (2007) Plant speciation. Science 317:910–914

Rodgers DM, Murphy JP, Frey KJ (1983) Impact of plant breeding on the grain yield and genetic diversity of spring oats. Crop Sci 23:737–740

Ronen G, Carmel GL, Zamir D, Hirchberg J (2000) An alternative pathway to carotene formation in plant chromoplasts discovered by map-based cloning and old gold color mutation in tomato. Proc Nat Acad Sci USA 97:11102–11107

Ronfort J, Glemin S (2013) Mating system, Haldene's sieve and the domestication process. Evolution 67:1518–1526

Ross H (1979) Wild species and primitive cultivars as ancestors of potato varieties. In: Proceedings Conference on Broadening Genetic Base of Crops. 1978. Wageningen Pudoc. Wageningen, The Netherlands

Ross-Ibarra J, Morrell PL, Gaut BS (2007) Plant domestication, a unique opportunity to identify the genetic basis of adaptation. Proc Natl Acad Sci U S A 104(Suppl 1):8641–8648

Roussel V, Koenig J, Beckert M, Balfourier F (2004) Molecular diversity in French bread wheat accessions related to temporal trends and breeding programmes. Theor Appl Genet 108:920–930

Roux C, Fraïsse C, Romiguier J, Anciaux Y, Galtier N, Bierne N (2016) Shedding light on the grey zone of speciation along a continuum of genomic divergence. PLoS Biol 14:e2000234. https://doi.org/10.1371/journal.pbio.2000234

Rutkoski JE, Poland JA, Singh RP, Huerta-Espino J, Bhavani S, Barbier H et al (2014) Genomic selection for quantitative adult plant stem rust resistance in wheat. Plant Genome 7:1–10

Ryu CH, You JH, Kang HG (2004) Generation of T-DNA tagging lines with a bidirectional gene trap vector and the establishment of an insertion-site database. Plant Mol Biol 54(4):489–502

SADC [Southern African Development Community] (2014) Food and nutrition security strategy 2015–2025. Southern African Development Community p 46

Saiki RK, Gelfand DH, Stoffel S, Scharf SJ, Higuchi R, Horn GT, Mullis KB, Erlich HA (1987) Primer-directed enzymatic amplification of DNA with a thermostable DNA polymerase. Science 239(4839):487–491

Saini P (2012) Formation of core set in rice (*Oryza sativa* L.) short duration germplasm accessions. M.Sc. (Ag.) Thesis, Kerala Agricultural University, Vellanikkara, Thrissur, Kerala

Sakuma S et al (2011) The domestication syndrome genes responsible for the major changes in plant form in the *Triticeae* crops. Plant Cell Physiol 52:738–749

Saltzman A, Birol E, Bouis HE, Boy E, De Moura FF, Islam Y, Pfeiffer WH (2013) Biofortification: Progress toward a more nourishing future. Glob Food Sec 2(1):9–17

Saltzman A, Birol E, Oparinde A, Andersson MS, Asare-Marfo D, Diressie MT, Gonzalez C, Lividini K, Moursi M, Zeller M (2017) Availability, production, and consumption of crops biofortified by plant breeding: current evidence and future potential. Ann New York Acad Sci 1390:104–114

Samra JS and Singh G (2003) Cold wave of 2002–2003: impact on agriculture. Indian Council Agril Res p 49

Sang T (2009) Genes and mutations underlying domestication transitions in grasses. Plant Physiol 149:63–70

Sang T, Ge S (2013) Understanding rice domestication and implications for cultivar improvement. Curr Opin Plant Biol 16:139–146

Sanjur OI, Piperno DR, Andres TC, Wessel-Beaver L (2002) Phylogenetic relationships among domesticated and wild species of *Cucurbita* (Cucurbitaceae) inferred from a mitochondrial gene: implications for crop plant evolution and areas of origin. Proc Nat Acad Sci USA 99:535–540

Satterthwaite D, McGranahan G, Tacoli C (2010) Urbanisation and its implications for food and farming. Philos Trans R Soc B 365:2809–2820

Schatz M, Maron L, Stein J, Wences A, Gurtowski J, Biggers E, Lee H et al (2014) Whole genome de novo assemblies of three divergent strains of rice, *Oryza sativa*, document novel gene space of aus and indica. Genome Biol 15:506

Schmutz J, Cannon SB, Schlueter J, Ma J, Mitros T, Nelson W, Hyten DL, Song Q, Thelen JJ, Cheng J et al (2010) Genome sequence of the palaeopolyploid soybean. Nature 463:178–183

Schneeberger K (2014) Using next-generation sequencing to isolate mutant genes from forward genetic screens. Nat Rev Genet 15:662–676

Sebolt AM, Shoemaker RC, Diers BW (2000) Analysis of a quantitative trait locus allele from wild soya bean that increases seed protein concentration in soy bean. Crop Sci 40:1438–1444

Seetharam A (2007) Pre-breeding: An important step in the effective utilization of conserved germplasm. National workshop on utilization of wild mulberry genetic resources 2nd & 3rd Nov. 2007. 9–16

Sharma JR (1994) Principles and practice of plant breeding. Tata McGraw-Hill Publishing Company Limited, New Delhi
Sharma VP (2015) Dynamics of land use competition in India: perceptions and realities. Indian Institute of Management, Ahmadabad, Working paper no 2015-06-02
Sharma RK, Jagshoran (2000) Contribution of exotic germplasm in wheat breeding in India. Indian J Plant Genet Res 13:75–79
Sharma S, Upadhyaya HD, Varshney RK, Gowda CLL (2013) Pre-breeding for diversification of primary gene pool and genetic enhancement of grain legumes. Front Plant Sci. https://doi.org/10.3389/fpls.2013.00309
Shimizu K, Takahashi M, Goshima N, Kawakami S, Irifune K, Morikawa H (2001) Presence of an SAR-like sequence in junction regions between an introduced transgene and genomic DNA of cultured tobacco cells: its effect on transformation frequency. Plant J 26(4):375–384
Shoemaker RC (1986) Chloroplast DNA variation in the genus *Glycine* subgenus *soya*. J Hered 77:26–30
Siddiq EA, Vemireddy LR, Nagaraju J (2012) Basmati rices: genetics, breeding and trade. Agril Res 1:25
Siebert C (2011) Food ark. Natl Geogr 220(1):108–131
Simmonds NW (1970) Genetic resources in plants—their exploration and conservation. In: Frankel OH, Bennett E (eds) Experimental agriculture. Blackwell Scientific Publications, Oxford, p 554
Singh BP, Srivastava U (2013) Plant genetic resources in Indian perspective. Indian Council of Agricultural Research, New Delhi. ISBN: 978-81-7164-148-2
Skinner DZ, Bauchan GR, Auricht G, Hughes S (1999) Developing core collection from a large annual Medicago germplasm collection. In: Johnson RC, Hodgkin T (eds) Core collections for today and tomorrow. International Plant Genetic Resources Institute, Rome, pp 61–67
Slabbert R (2004) Drought tolerance, traditional crops and biotechnology: breeding towards sustainable development. South African J Bot 70:116–123
Smith NJH (1986) Botanic gardens and Germplasm conservation. Harold L. Lyon Arboretum, University of Hawaii Press, Honolulu
Smith JSC, Duvick DN, Smith OS, Cooper M, Feng L (2004) Changes in pedigree backgrounds of Pioneer brand maize hybrids widely grown from 1930 to 1999. Crop Sci 44:1935–1946
Sneyers A (2017) Food, drought and conflict: evidence from a case study on Somalia. HiCN Working Paper, 252
SOFI [The State of Food Security and Nutrition in the World] (2018) Food security and butrition around the World. Available at: http://www.fao.org/publications/sofi. Accessed on 17 June 2019
Somssich M, Je BI, Simon R, Jackson D (2016) *CLAVATA WUSCHEL* signaling in the shoot meristem. Development 143:3238–3248. https://doi.org/10.1242/dev.133645
Song QJ, Hyten DL, Jia GF, Quigley CV, Fickus EW, Nelson RL, Cregan PB (2015) Fingerprinting soybean germplasm and its utility in genomic research. G3: Genes Genomes Genet 5:1999–2006
Spielman D, Kolady D, Ward P (2013) The prospects for hybrid rice in India. Food Security 5:651–665
Spindel J, Begum H, Akdemir D, Virk P, Collard B, Redoña E et al (2015) Genomic selection and association mapping in rice (*Oryza sativa*): effect of trait genetic architecture, training population composition, marker number and statistical model on accuracy of rice genomic selection in elite, tropical rice breeding lines. PLoS Genet 11:e1004982. https://doi.org/10.1371/journal.pgen.1004982
Stabbins GL (1957) Genetics, evolution and plant breeding. In: Proceedings of symposium on genetics and plant breeding in South-East Asia, January, 1957, New Delhi, India. Indian J Genet 17:129–141
State Forestry Administration of China (2012) The Saving and Conservation Program on Extremely Small Populations in China. https://www-bcf.usc.edu/~forsburg/A
Strigens A, Freitag NM, Gilbert X, Grieder C, Riedelsheimer C, Schrag TA, Messmer R, Melchinger AE (2013) Association mapping for chilling tolerance in elite flint and dent

maize inbred lines evaluated in growth chamber and field experiments. Plant Cell Environ 36(10):1871–1887

Struck TH, Feder JL, Bendiksby M, Birkeland S, Cerca J, Gusarov VI (2017) Finding evolutionary processes hidden in cryptic species. Trends Ecol Evol 33:153–163. https://doi.org/10.1016/j.tree.2017.11.007

Suprunova T, Krugman T, Distelfeld A, Fahima T, Nevo E, Korol A (2007) Identification of a novel gene (*Hsdr4*) involved in water-stress tolerance in wild barley. Plant Mol Biol 64:17–34

Swaminathan MS (1978) Inaugural address. In: National Symposiu on plant and animal genetic resources, December 1978. IARI, New Delhi

Swamy MBP, Upadhyaya HD, Goudar PVK (2006) Characterization of Asian core collection of groundnut for morphological traits. Indian J Crop Sci 1(1–2):129–134

Szabados L, Kovács I, Oberschall A (2002a) Distribution of 1000 sequenced T-DNA tags in the *Arabidopsis* genome. Plant J 32(2):233–242

Szabados L, Kovacs I, Oberschall A, Abraham E, Kerekes I, Zsigmond L, Nagy R, Alvarado M, Krasovskaja I, Gal M (2002b) Distribution of 1000 sequenced T-DNA tags in the Arabidopsis genome. Plant J 32:233–242

Taba S, Diaz J, Franco J, Crossa J (1998) Evaluation of Caribbean maize accessions to develop a core subset. Crop Sci 38:1378–1386

Tabare A, Magalhaes JR, Parentoni SN, Corderio C, de Andrade RV (1999) The core collection of maize germplasm of Brazil. Plant Genet Res Newslett 117:55–56

Tabashnik BE (2010) Communal benefits of transgenic corn. Science 330:189–190

Takahashi R, Hayashi J (1964) Linkage study of two complementary genes for brittle rachis in barley. Bericht des Ohara Instituts fu¨r Landwirtschaftliche Biologie, Okayama 12: 99–105

Tamura Y, Hattori M, Yoshioka H, Yoshioka M, Takahashi A, Wu J, Sentoku N, Yasui H (2014) Map-based cloning and characterization of a brown plant hopper resistance gene *BPH26* from *Oryza sativa L.* ssp. *indica* cultivar ADR52. Scientific Report pp 1–4

Tanksley SD, McCouch SR (1997) Seed banks and molecular maps: unlocking genetic potential from the wild. Science 277(5329):1063–1066

Tao Y, Mace ES, Tai S, Cruickshank A, Campbell BC, Zhao X, Van Oosterom EJ, Godwin ID, Botella JR, Jordan DR (2017) Whole-genome analysis of candidate genes associated with seed size and weight in *Sorghum bicolor* reveals signatures of artificial selection and insights into parallel domestication in cereal crops. Front Plant Sci 8:1237

Tao Y, Mace ES, George-Jaeggli B, Hunt C, Cruickshank A, Henzell R, Jordan D (2018) Novel grain weight loci revealed in a cross between cultivated and wild sorghum. Plant Genome 11:1–10

Tavakol E, Bretani G, Rossini L (2017) Natural genetic diversity and crop improvement. In: Pilu R, Gavazzi G (eds) More food: road to survival. Bentham Science Publishers, Sharjah, pp 218–254

Teklu Y, Hammer K (2006) Farmers' perception and genetic erosion of Ethiopian tetraploid wheat landraces. Genet Res Crop Evol 53:1099–1113

Tettelin H, Masignani V, Cieslewicz MJ, Donati C, Medini D, Ward NL, Angiuoli SV et al (2005) Genome analysis of multiple pathogenic isolates of *Streptococcus agalactiae*: implications for the microbial "pan-genome". Proc Nat Acad Sci USA 102:13950–13955

Thalapati S, Batchu AK, Neelamraju S, Ramanan R (2012) *Os11Gsk* gene from a wild rice, *Oryza rufipogon* improves yield in rice. Funct Integrat Genomics 12:277–289

Tikader A, Ananda Rao A (2003) Inter and intra-specific hybridization studies in mulberry (*Morus* spp.) germplasm. Bull Indian Acad Sericulture (6):17–22

Tikader A, Dandin SB (2001) Breeding behaviour of some wild mulberry. Indian Silk 40(1):9–10

Tikader A, Kamble CK (2007) Survival potential of different mulberry (*Morus* Spp.) accessions. Geobios 34:282–286

Tikader A, Thangavelu K (2002) Breeding performance of some wild mulberry (*Morus* Spp.). In: Proceedings of National Conference on recent trends in plant science research 14th–15th November 2002, pp 106–111

Tohme J, Jones P, Beebe S, Iwanga M (1995) The combined use of agro-ecological and characterization data to establish the CIAT *Phaseolus vulgaris* core collection. In: Hodgkin T, Brown

AHD, van Hintum MJL, Morales EAV (eds) Core collections of plant genetic resources. A Wiley-Sayee Publications, Chichester, UK, pp 95–107

Tomooka N, Lairunhreang C, Nakeeraks P, Egawa Y, Thavarasook C (1992) Center of genetic diversity and dissemination pathways in mungbean deduced from seed protein electrophoresis. Theor Appl Genet 83:289–293

Tsegaye B, Berg T (2007) Genetic erosion of Ethiopian tetraploid wheat landraces in Eastern Shewa, Central Ethiopia. Genet Res Crop Evol 54(4):715–726

Tyler GM (2004) Sustaining the earth: an integrated approach. Thomson/Brooks/Cole, p 349

UCDP [Uppsala Conflict Data Program] (2018). Available at http://ucdp.uu.se/. Accessed 17 June 2019

UNEP (2011) Convention on biological diversity. http://hdr.undp.org/en/statistics/

UNHCR [United Nation Higher Commission for Refugees] (2018) Global appeal 2018–2019. Available https://www.unhcr.org/publications/fundraising/5a0c05027/unhcr-global-appeal-2018-2019-full-report.html. Accessed 17 June 2019

United Nations, Department of Economic and Social Affairs, Population Division (2017) World population prospects: the 2017 revision, key findings and advance tables. Working paper no.ESA/P/WP/248

Upadhyaya HD, Ortiz R (2001) A mini core subset for capturing diversity and promoting utilization of chickpea genetic resources in crop improvement. Theor Appl Genet 102:1292–1298

Upadhyaya HD, Ortiz R, Bramel PJ, Singh S (2002a) Phenotypic diversity for morphological and agronomic characteristics in chickpea core collection. Euphytica 123(3):333–342

Upadhyaya HD, Bramel PJ, Ortiz R, Singh S (2002b) Developing a minicore of peanut for utilization of genetic resources. Crop Sci 42:2150–2156

Upadhyaya HD, Ortiz R, Bramel PJ, Singh S (2003) Development of a groundnut core collection using taxonomical, geographical, and morphological descriptors. Genet Res Crop Evol 50:139–148

Upadhyaya HD, Sharma NDRK, Ravishankar CR, Albrecht T, Narasimhudu Y, Singh SK, Varshney SK, Reddy VG, Singh S, Dwivedi SL, Wanyera N, Oduri COA, Mgonja MA, Kisandu DB, Parzies HK, Gowda CLL (2010) Developing minicore collection in finger millet using multi-location data. Crop Sci 50:1924–1931

Upadhyaya HD, Sharma S, Gowda CLL, Reddy VG, Singh S (2011) Developing proso millet (*Panicum miliaceum* L.) core collection using geographical and morpho-agronomic data. Crop Pasture Sci 62(5):383–389

USEPA [United States Environmental Protection Agency] (2017) Climate change indicators: greenhouse gases. Available at https://www.epa.gov/climate-indicators/greenhouse-gases. Accessed on 17 June 2019

Valkoun JJ (2001) Wheat pre-breeding using wild progenitors. Euphytica 119:17–23

van Beuningen LT, Busch RH (1997) Genetic diversity among North American spring wheat cultivars: I. Analysis of the coefficient of parentage matrix. Crop Sci 37:570–579

van de Wouw M, van Hintum T, Kik C, van Treuren R, Visser B (2010a) Genetic erosion in crops: concept, research results and challenges. Plant genetic resources characterization and utilization. https://doi.org/10.1017/S1479262109990062

van de Wouw M, van Hintum T, Kik C, van Treuren R, Visser B (2010b) Genetic diversity trends in twentieth century crop cultivars: a meta analysis. Theor Appl Genet 120:1241–1252

van der Knaap E, Chakrabarti M, Chu YH, Clevenger JP, Illa-Berenguer E, Huang Z et al (2014) What lies beyond the eye: the molecular mechanisms regulating tomato fruit weight and shape. Front Plant Sci 5:227

Vaughan DA, Balazs E, Heslop-Harrison JS (2007) From crop domestication to super domestication. Annals Bot 100:893–901

Vavilov NI (1926) Studies on the origin of cultivated plants. Bulletin of Applied Botany, Genet and Plant Breed, State Press, Leningrad, USSR 16:1–248

Vavilov NI (1935) The phytogeographical basis of plant breeding. In: University of California, Division of Agriculture and Natural Resources. http://pba.ucdavis.edu/files/198376.pdf

Vavilov NI (1951) The origin, variation, immunity and breeding of cultivated plants. Chron Bot 13:1–366

Vavilov NI, Dorofeev FV (1992) Origin and geography of cultivated plants. Cambridge University Press

Vellve R (1992) Saving the seeds: genetic diversity and European agriculture. Earthscan Publication, London

Verma MM, Kumar J (1974) Breeding for yield in self-pollinators – an introspection and reorientation. Crop Improv 1:15–31

Vernikos G, Medini D, Riley DR, Tettelin H (2015) Ten years of pan-genome analyses. Curr Opin Microbiol 23:148–154

Vos P, Hogers R, Bleeker M, Reijans M, van de Lee T, Hornes M, Frijters A, Pot J, Peleman J, Kuiper M (1995) AFLP: a new technique for DNA fingerprinting. Nucleic Acids Res 23(21):4407–4414

Wang WS, Mauleon R, Hu ZQ, Chebotarov D, Tai SS, Wu ZC, Li M, Zheng TQ, Fuentes RR, Zhang F et al (2018) Genomic variation in 3010 diverse accessions of Asian cultivated rice. Nature 557:43–49

Weiguo Z, Cho GT, Ma KH, Chung JW, Gwag JG, Park YJ (2010) Development of an allele mining set in rice using heuristic algorithm and SSR genotype data with least redundancy for the post genomic era. Mol Breed 26(4):639–651

Weltzein E (1989) Differentiation among barley landrace populations from the near east. Euphytica 43:29–39

White J, Law JR, MacKay I, Chalmers KJ, Smith JSC, Kilian A, Powell W (2008) The genetic diversity of UK, US and Australian cultivars of *Triticum aestivum* measured by DArT markers and considered by genome. Theor Appl Genet 116:439–453

Williams JG, Kubelik AR, Livak KJ, Rafalski JA, Tingey SV (1990) DNA polymorphisms amplified by arbitrary primers are useful as genetic markers. Nucleic Acids Res 18(22):6531–6535

Wilson EO (1992) The diversity of life. Penguin, London

Wilson EO, Frances MP (1988) Biodiversity. National Academy Press, Washington, DC

Windels P, Taverniers I, Depicker A, Van Bockstaele E, De Loose M (2001) Characterisation of the roundup ready soybean insert. Eur Food Res Tech 213(2):107–112

Witcombe JR (1999) Does plant breeding lead to a loss of genetic diversity? In: Wood D, Lenni JM (eds) Agrobiodiversity – characterization, utilization and management. CABI Publishing, London, pp 245–272

World Bank (2018a) Life expectancy at birth, total (years). Available at https://data.worldbank.org/indicator/SP.DYN.LE00.IN?view=chart. Accessed on 17 June 2019

World Bank (2018b) Population growth (annual %). Available at https://data.worldbank.org/indicator/SP.POP.GROW?view=chart. Accessed on 17 June 2019

World Bank (2018c) Agricultural land (% of land area). Available at https://data.worldbank.org/indicator/AG.LND.AGRI.ZS?view=chart. Accessed on 17 June 2019

World Bank (2018d) Employment in agriculture (% of total employment) (modeled ILO estimate). Available at https://data.worldbank.org/indicator/sl.agr.empl.zs. Accessed on 17 June 2019

Xiao JH, Li JM, Grandillo S, Ahn SN, Yuan LP, Tanksley SD, McCouch SR (1998) Identification of trait-improving quantitative trait loci alleles from a wild rice relative, *Oryza rufipogon*. Genetics 150:899–909

Xu X, Liu X, Ge S, Jensen JD, Hu FY, Li X, Wang W (2012) Resequencing 50 accessions of cultivated and wild rice yields markers for identifying agronomically important genes. Nat Biotech 30:105–111

Xu Y, Li P, Zou C, Lu Y, Xie C, Zhang X, Prasanna BM, Olsen MS (2017) Enhancing genetic gain in the era of molecular breeding. J Exp Bot 68(11):2641–2666

Yan W, Rutger N, Bryant RJ, Bockelman HE, Fjellstrom RG, Chen MH, Tai TH, McClung AM (2007) Development and evaluation of a core subset of the USDA rice germplasm collection. Crop Sci 47:869–878

Yang WZ (2017) A sophisticated species conservation strategy for *Nyssa yunnanensis*, a species with extremely small populations in China. Biodivers Conserv 26:967–981

Yao QL, Fan P, Zou SX (2008) Constructing a core collection for maize (*Zea mays* L.) landrace from wuling mountain region in China. Agril Sci China 7(12):1423–1432
Ye X, Al-Babili S, Kloti A, Zhang J, Lucca P, Beyer P, Potrykus I (2000) Engineering the Provitamin a (beta-carotene) biosynthetic pathway into (carotenoid free) rice endosperm. Science 287:303–305
Zeuli SPL, Qualset CO (1993) Evaluation of five strategies for obtaining a core subset from large genetic resources collection of *Triticum durum*. Theor Appl Genet 87:295–304
Zeven AC, de Wit JMJ (1982) Dictionary for cultivated plants and their regions of diversity excluding most ornamentals, forest trees and lower plants. Center for Agricultural Publishing and Documentary, Wageningen
Zhang H, Zhang D, Wang M, Sun J, Qi Y, Li J, Wei X, Qiu Z, Tang S, Li Z (2011) A core collection and mini core collection of *Oryza sativa* L. in China. Theor Appl Genet 122(1):49–61
Zhang S et al (2013) Localization of a new gene for bitterness in cucumber. J Heredity 104:134–139
Zhao K, Tung CW, Eizenga GC, Wright MH, Ali ML et al (2011) Genome-wide association mapping reveals a rich genetic architecture of complex traits in oryza sativa. Nat Commun 2:467
Zhao Q, Feng Q, Lu HY, Li Y, Wang A, Tian QL, Zhan QL, Lu YQ, Huang T, Wang TC et al (2018) Pan-genome analysis highlights the extent of genomic variation in cultivated and wild rice. Nat Genet 50:278–284
Zhou ZK, Jiang Y, Wang Z, Gou ZH, Lyu J, Li WY, Tian ZX (2015) Resequencing 302 wild and cultivated accessions identifies genes related to domestication and improvement in soybean. Nat Biotech 33:408–414
Zhu W, Ausin I, Seleznev A, Méndez-Vigo B, Picó FX, Sureshkumar S et al (2015) Natural variation identifies *ICARUS1*, a universal gene required for cell proliferation and growth at high temperatures in *Arabidopsis thaliana*. PLoS Genet 11(5):e1005085. https://doi.org/10.1371/journal.pgen.1005085
Zhukovsky PM (1965) Genetic and botanical irregularities in the evolution of cultivated plants. Genetika Mosc 1:9–41

Omics and Plant Genetic Resources: Towards Mining Potential Genes

4

Nagesha N and Praveen L. Bennur

Contents

Abstract

Conservation of plant genetic resources is critical for increasing the resilience and agricultural production system to meet the increasing demand for future world food security. Ex situ and in situ conservation of plant genetic resources are particularly important for their efficient utilization in crop improvement programmes. Different types of gene pool techniques are important for restoration of gene pool of crops. Omics can enable further expansion of agricultural research in food, health, energy, chemical feedstock, and specialty chemicals while helping to preserve, enhance, and remediate the environment. Recent advances in genomic technologies especially omics-based technologies play a major role in improving the utilization of genetic resources and mining of potential genes. This chapter mainly emphasizes on importance of plant genetic resources and intervention of different omics-based technologies including

Nagesha N (✉) · P. L. Bennur
Department of Plant Bioechnology, College of Agriculture, UAS, GKVK, Bangalore, Karnataka, India
e-mail: nagesha.n@uasbangalore.edu.in; praveenlbennur@gmail.com

R. K. Salgotra, S. M. Zargar (eds.), *Rediscovery of Genetic and Genomic Resources for Future Food Security*,
https://doi.org/10.1007/978-981-15-0156-2_4

genomics, transcriptomics, proteomics, and metagenomics for the identification, characterization, mining and introgression of the identified beneficial genes for disease and pest resistance, qualitative traits, drought tolerance traits, and their introgression into the plants for increasing the future world food security.

Acronyms

cDNA	Complementary deoxyribonucleic acid
CRISPR	Clustered regularly interspaced short palindromic repeats
DNA	Deoxyribonucleic acid
DREB	Dehydration-responsive element binding
DT	Drought tolerance
EST	Expressed sequence tag
FAO	Food and Agriculture Organization
GWAS	Genome-wide association studies
GBS	Genotyping-by-sequencing
GolS	Galactinol synthase
LEA	Late embryogenesis abundant proteins
MAS	Marker-assisted selection
MS	Mass spectrometry
NGS	Next-generation sequencing
PAMP	Pathogen-associated molecular pattern
PGR	Plant genetic resources
PTM	Post-translational modifications
QTL	Quantitative trait loci
RGPs	Restoration of gene pools
RNA	Ribonucleic acid
RNAi	RNA interference
SGS	Second-generation sequencing
SNPs	Single-nucleotide polymorphism
SMRT	Single-Molecule Real-Time Sequencer
TFs	Transcription factors
TGS	Third-generation sequencing technologies
TILLING	Targeting Induced Local Lesions in Genomes

4.1 Introduction

4.1.1 Plant Genetic Resources and Its Conservation

Plant genetic resources are the material derived from plants, which is of value as a resource for present and future generations of people. The Food and Agriculture Organization (FAO) defined plant genetic resources as the entire generative and vegetative reproductive material of species with economical and/or social value, especially for the agriculture of the present and the future, with special emphasis on nutritional plants. Genetic resources can be defined as all materials that are available for improvement of a cultivated plant species (Becker 1993). Hallauer and Miranda (1981) reported that exotics for pre-breeding purposes include any germplasm that does not have immediate usefulness without selection for adaptation for a given area. In this sense, exotic germplasms are represented by races, populations, inbred lines, etc.

The quality of these resources is decreasing in faster rate due to increasing population and unsustainable practices including modern technologies eradicating these natural resources. As many reported, the exponential loss of biodiversity is well documented; but a major concerning factor is that process shows no signs of slowing down and threatens to become even more severe in the future, creating huge problems to the human beings (Maxted and Kell 2003).

The demand for food is increasing every year. In order to feed this growing population, there is a need for increasing the production of food with available land and resources. The breeding programmes mainly involved in the development of new varieties of crops with increased yield and maintaining nutritional qualities require a range of diversified genetic resources of the plant species. In order to access this range of genetic diversity by the plant breeders, there is a need for global effort to assemble, document, and utilize these resources to fight against hunger (Hoisington et al. 1999).

Plant genetic resources (PGRs) form the natural variations that have supported human kind for several millennia and are the basis for food security in addition to being sources of energy, animal feed, fibre as well as other ecosystem services. Plant genetic resources play an important role in addressing the global challenges that are currently facing the human population, particularly the twin challenge of climate change and food scarcity. Therefore, effective conservation and sustainable utilization of these resources are critically important. There are huge numbers of accessions that are conserved in gene banks for various species. Promoting the sustainable utilization of biodiversity is a key goal of various global and regional efforts and initiatives as well as international agreements and treaties governing genetic resources (Wambugu et al. 2018).

The only authoritative account of agricultural biodiversity status at the global level is represented by the First and Second Reports on the State of the World's Plant Genetic Resources for Food and Agriculture published by the Food and Agriculture Organization (FAO) of the United Nations (FAO 1998, 2010). The Second Report mentions that there are about 7.4 million accessions conserved in over 1750 gene banks around the world in either seed banks, field collections, or in vitro and cryopreservation conditions (FAO 2010). This represents an increase of more than 1.4 million accessions added to ex situ collection since the publication of the First Report on the State of the World's Plant Genetic Resources for Food and Agriculture. Although reportedly over-represented, a large part of the genetic diversity of major food crops is stored in ex situ collections (ex situ conservation – the technique of conservation of all levels of biological diversity outside their natural habitats through different techniques like zoo, captive breeding, aquarium, botanical garden, and gene bank). The exact proportion is still uncertain, but estimates suggest that more than 70% of the genetic diversity of some 200–300 crops is already conserved in gene banks. In addition, there are over 2500 botanic gardens maintaining samples of some 80,000 plant species (FAO 2010). However, regeneration of gene bank accessions remains a major problem, threatening collections (FAO 1998).

Jones (2003) reported that for restoration of gene pool, he used the concept of Harlan and DeWet's classification of gene pool, whereas they defined the primary gene pool as the biological species, that is, all materials that easily crossed, generating offspring with approximately normal fertility and segregation in succeeding generations. Harlan and DeWet's secondary gene pool included all other biological species that have significant genetic incompatibility barriers to crossing but may cross with the primary gene pool under natural, albeit exceptional, circumstances. Harlan and DeWet's tertiary gene pool includes taxa that may be crossed with the species of interest but only through extreme measures that would probably occur at most rarely in nature. The tertiary gene pool is not a taxonomic unit but defines the extreme outer limits of the gene pool potentially useful to the plant breeder, albeit only with extraordinary artificial effort. Harlan and DeWet's concept can be adapted for restoration. However, for restoration the primary and secondary RGPs encompass the same taxon as the target population, whereas taxa represented in the tertiary and quaternary RGPs are distinct from the primary RGP taxon.

Some of the reports indicated that in maize, for example, genetic variation in the primary gene pool is so large that the secondary or tertiary gene pools are rarely used. In rapeseed, on the other hand, genetic variation in the primary gene pool is small, and breeders have to transfer important traits from *Brassica* species of the secondary and tertiary gene pool into the cultivated species (Hu et al. 2002).

4.1.2 Conservation of Plant Genetic Resources

The major objective of plant genetic resource conservation is to ensure maintenance of the maximum possible genetic diversity of a taxon and their availability for utilization. Plant genetic resource conservation acts as a link between the genetic diversity of a plant and its utilization or exploitation by the plant breeders in the development of improved crop varieties. The overall concept of plant genetic resource conservation is represented in Fig. 4.1 (Maxted and Kell 2003).

A major idea in conservation genetics is that small, isolated populations can be threatened for the following genetic reasons (Frankham et al. 2009; Ouborg et al. 2006). The dynamics of genetic variation over space and time in a population containing only few individuals are expected to be strongly dominated by the random genetic drift and inbreeding. Genetic drift will lead to random loss and fixation of alleles; owing to the random nature of genetic drift, adaptive alleles can be lost, and deleterious alleles can become fixed in the population. Inbreeding, in the context of conservation genetics, is often invoked as biparental inbreeding and leads to increased frequency of homozygotes in the population.

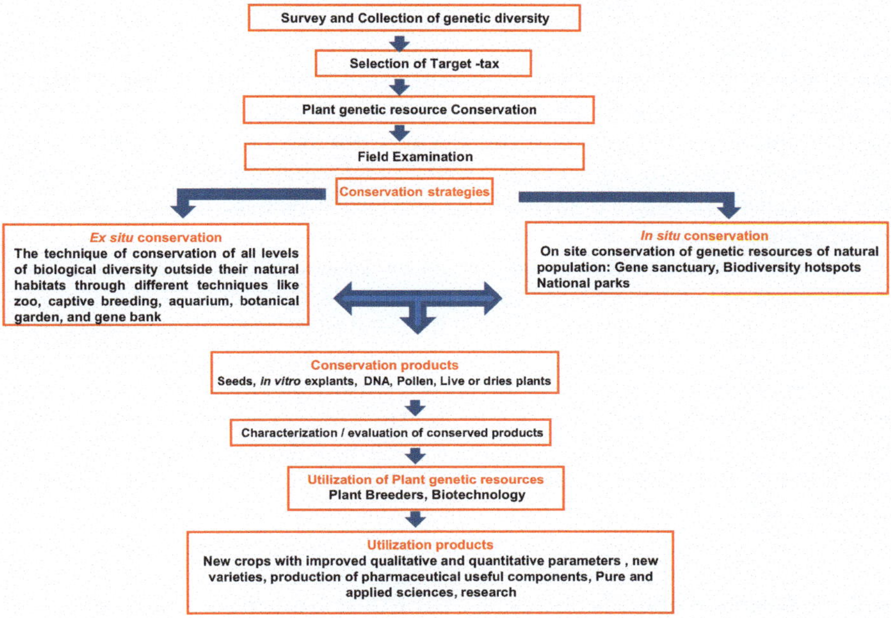

Fig. 4.1 Plant genetic resources, conservation, and their utilization

4.2 The Significance of Genes and Genotypes as Reported by the FAO 1998

The different genetic characteristics or genes used for the various agronomic qualities are:

- Adaptations to resistance against biotic stress like pests and other diseases caused by bacteria, fungi, and virus
- Adaptations to abiotic stresses such as drought and salinity tolerance
- Plant height along with other factors that affect productivity
- Improved qualitative parameters such as higher oil or protein content
- Culinary and other factors of cultural importance

The above-mentioned characteristics are important not only to farmers but are also of major global importance as they are introduced into large number of cultivated crops. Particular genotypes are also especially important to farmers in resource-poor areas, as they tend to be well adapted to local conditions. The particular combination of genes in a well-adapted landrace, for example, may be difficult or impossible to reconstruct (FAO 1998).

The introgression of genes that are required for improving the qualitative and quantitative characters of the crop plants is very critical for improved breeding programmes. For example, the introgression of genes for reduced plant height and increased disease and viral resistance in wheat provided major avenues for the foundation for the "Green Revolution" and demonstrated the essential role of genetic resources on improving the production. The wheat hybrids and synthetics developed (synthetics-reproducing the hybridization event that leads to creation of hexaploid wheat from a cross of tetraploid with diploid wheat which represents source of novel genetic variation) from these may provide the yield increases needed in the future (Hoisington et al. 1999).

The wild relatives of crop plants contain the genetic resources required for the development of biotic and abiotic resistance crop plants (Pritham 2015). The conservation of these wild relatives is critical for the plant breeders. Some of the examples are, a wild relative of maize '*Tripsacum*' belongs to grass family will have the genetic resource for abiotic and biotic stress resistance and for apomixes. This trait helps in the development of maize hybrids. The application of modern molecular biology and genetic engineering technologies enhances the use of genetic resources. The potential and complementary use of all the modern technological tools and resources will be helpful in meeting the increasing global demand for food.

4.3 Omics-Based Technology for Plant Genetic Resource Conservation and Mining of the Beneficial Genes for Increased Crop Productivity

4.3.1 Omics-Based Technologies

Omics: The term "omic" is derived from the Latin suffix "ome" meaning mass or many. Thus, Omics involves a mass (large number) of measurements per endpoint (Jackson et al. 2006). In biological context, the suffix – omics – is used to refer to the study of large sets of biological molecules (Smith et al. 2005). New technologies that permit simultaneous monitoring of many hundreds or thousands of macro- and small molecules promise to allow functional monitoring of multiple key cellular pathways simultaneously. The new "global" methods of measuring families of cellular molecules, such as RNA, proteins, and intermediary metabolites, have been termed "Omics" technologies, based on their ability to characterize all, or most, members of a family of molecules in a single analysis. With these new tools, we can now obtain complete assessments of the functional activity of biochemical pathways and of the structural genetic (sequence) differences among individuals and species that were previously unattainable. Fundamental biological processes can now be studied by applying the full range of Omics technologies (genomics, transcriptomics, proteomics, metabolomics, and beyond) to the same biological sample (Fig. 4.2) (Morrison et al. 2006).

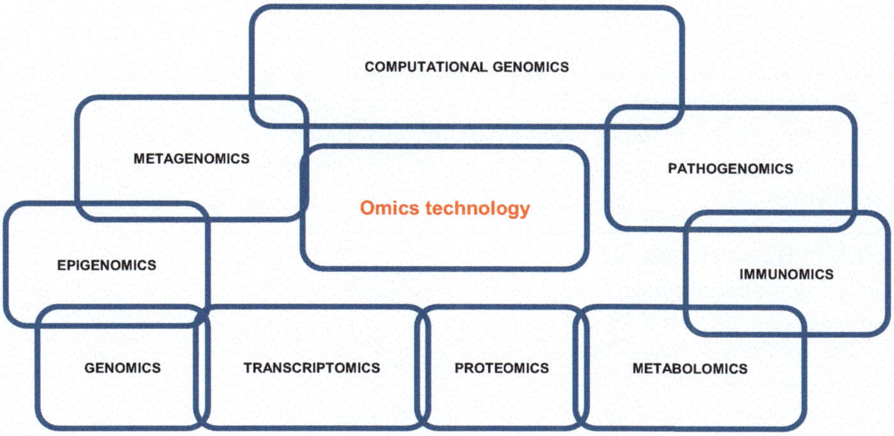

Fig. 4.2 Types of omics technologies

4.3.2 Genomics

The field of genomics has been divided into three major categories.

1. *Genotyping* (focused on the genome sequence): the physiological function of genes and the elucidation of the role of specific genes in disease susceptibility (Syvanen 2001).
2. *Transcriptomics* (focused on genomic expression): the abundance of specific mRNA transcripts in a biological sample is a reflection of the expression levels of the corresponding genes (Lowe et al. 2017).
3. *Epigenomics* (focused on epigenetic regulation of genome expression): study of the complete set of epigenetic modifications on the genetic material of a cell, known as the epigenome on a large (ultimately genome-wide) scale (Feinberg 2007).

4.3.2.1 Genotyping

- Identification of the physiological function of genes
- Role of specific genes in disease susceptibility (Syvanen 2001)

Common parameters used

- Among different variations (insertions, deletions, SNPs, etc.), single-nucleotide polymorphisms (SNPs) are the most commonly investigated and can be used as markers for diseases.
- Tag SNPs (informative subset of SNPs) and fine mapping are further used to identify true cause of phenotype.

Application

- Identification of genes associated with disease
- Recent improvement in genotyping and array-based genotyping techniques, allowing the simultaneous assessment (up to 1 million SNPs) per assay, leads to the genotyping of entire genome known as genome-wide association studies (GWAS).

4.3.2.2 Transcriptomics

Gene expression profiling

- The identification and characterization of the mixture of mRNA that is present in a specific sample.

Principle

- The abundance of specific mRNA transcripts in a biological sample is a reflection of the expression levels of the corresponding genes (Lowe et al. 2017).

Application

- To associate differences in mRNA mixtures originating from different groups of individuals to phenotypic differences between the groups

Challenge

- The transcriptome in contrast to the genome is highly variable over time, between cell types and environmental changes.

4.3.2.3 Epigenomics

Epigenetic processes

- Mechanisms other than changes in DNA sequence that cause effect in gene transcription and gene silencing.
- Number of mechanisms of epigenomics but is mainly based on two mechanisms, DNA methylation and histone modification.
- Recently RNAi has acquired considerable attention.

Goal

- The focus of epigenomics is to study epigenetic processes on a large (ultimately genome-wide) scale to assess the effect on diseases.

Association with disease

- Hypermethylation of CpG islands located in promoter regions of genes is related to gene silencing. Altered gene silencing plays a causal role in human disease.
- Histone proteins are involved in the structural packaging of DNA in the chromatin complex. Post-translational histone modifications such as acetylation and methylation are believed to regulate chromatin structure and therefore gene expression

4.3.3 Computational Genomics

Computational genomics (often referred to as computational genetics) refers to the use of computational and statistical analysis to decipher biology from genome sequences and related data, including both DNA and RNA sequences as well as other "post-genomic" data (i.e. experimental data obtained with technologies that require the genome sequence, such as genomic DNA microarrays). Compuational genomics in combination with computational and statistical approaches to understanding the function of the genes and statistical association analysis, this field is also often referred to as Computational and Statistical Genetics/genomics.

4.3.4 Proteomics

- Proteomics provides insights into the role of proteins in biological systems. The proteome consists of all proteins present in specific cell types or tissue and highly variable over time and between cell types and will change in response to changes in its environment, which is a major challenge.
- The overall function of cells can be described by the proteins (intra- and intercellular) and the abundance of these proteins.
- Although all proteins are directly correlated to mRNA (transcriptome), post-translational modifications (PTM) and environmental interactions impede to predict from gene expression analysis alone (Hanash et al. 2008).

  Tools for proteomics

  Mainly two different approaches that are based on detection

1. Mass spectrometry (MS)
2. Protein microarrays using capturing agents such as antibodies

  Major focuses

- The identification of proteins and protein interaction.
- Then the quantification of the protein abundance. The abundance of a specific protein is related to its role in cell function (Fliser et al. 2007).

Applications of proteomics

- Protein Mining: cataloging all the proteins present in a tissue, cell, organelle, etc.
- Differential Expression Profiling: Identification of proteins in a sample as a function of a particular state: differentiation, stage of development, disease state, and response to stimulus or environments.
- Network Mapping: Identification of proteins in functional networks: biosynthetic and signal transduction pathways and multi-protein complexes.
- Mapping Protein Modifications: Characterization of post-translational modifications: phosphorylation, glycosylation, oxidation, etc.

4.3.5 Metabolomics

Metabolomics refers to the systematic identification and quantification of the small-molecule metabolic products (metabolome) of a biological system (cell, tissue, organ, biological fluid, or organism) at a specific point in time. Mass spectrometry and NMR spectroscopy are the techniques most often used for metabolome profiling.

Metabolomics is defined as the measurement of the amounts (concentrations) and locations of all the metabolites in a cell, the metabolites being the small molecules.

- The metabolome consists of small molecules (e.g. lipids or vitamins) that are also known as metabolites (Claudino et al. 2007).
- Metabolites are involved in the energy transmission in cells (metabolism) by interacting with other biological molecules following metabolic pathways.
- Metabolic phenotypes are the by-products of interactions between genetic, environmental, lifestyle, and other factors.
- The metabolome is highly variable and time dependent, and it consists of a wide range of chemical structures.
- An important challenge of metabolomics is to acquire qualitative and quantitative information with perturbance of environment.

Terminologies related to metabolomics

1. *Metabolites*: are the intermediates and products of metabolism. Within the context of metabolomics, a metabolite is usually defined as any molecule less than 1 kDa in size.
2. *Metabolome*: refers to the complete set of small-molecule metabolites (such as metabolic intermediates, hormones, and other signalling molecules and secondary metabolites) found within a biological sample. The word was coined in analogy with transcriptomics and proteomics; like the transcriptome and the proteome, the metabolome is dynamic, i.e. changing from second to second.
3. *Metabonomics*: is defined as "the quantitative measurement of the dynamic multiparametric metabolic response of living systems to pathophysiological stimuli or genetic modification".

4. *Metabonomics*: is a subset of metabolomics and is defined as the quantitative measurement of the multiparametric metabolic responses of living systems to pathophysiological stimuli or genetic modification, with particular emphasis on the elucidation of differences in population groups due to genetic modification, disease, and environmental (including nutritional) stress (Ramsden 2009).

Applications of metabolomics

- Characterization of metabolism
- Identification of regulated key sites in network
- Biofortification and genetic modification
- Investigation of gene function under stress conditions

4.3.6 Metagenomics

Metagenomics is the study of metagenomes, genetic material recovered directly from environmental samples. The broad field may also be referred to as environmental genomics, ecogenomics, or community genomics.

4.3.7 Functional Genomics

Functional genomics is a field of molecular biology that attempts to make use of the vast wealth of data produced by genomic projects (such as genome sequencing projects) to describe gene (and protein) functions and interactions.

- Unlike genomics, functional genomics focuses on the dynamic aspects of genome such as transcription, translation, and protein-protein interactions, as opposed to the static aspects of the genomic information such as DNA sequence or structures.
- Functional genomics attempts to answer questions about the function of DNA at the levels of genes, RNA transcripts, and protein products. A key characteristic of functional genomics studies is their genome-wide approach to these questions, generally involving high-throughput methods rather than a more traditional "gene-by-gene" approach.

4.3.8 Immunomics

Immunomics is the study of immune system regulation and response to pathogens using genome-wide approaches. With the rise of genomic and proteomic technologies, scientists have been able to visualize biological networks and infer interrelationships between genes and proteins; recently, these technologies have been used to help better understand how the immune system functions and how it is regulated.

4.3.9 Pathogenomics

Pathogen infections are among the leading causes of infirmity and mortality among humans and other animals in the world. It has been difficult to compile information to understand the generation of pathogen virulence factors as well as pathogen behaviour in a host environment.

- The study of pathogenomics attempts to utilize genomic and metagenomics data gathered from high-throughput technologies (e.g. sequencing or DNA microarrays), to understand microbe diversity and interaction as well as host-microbe interactions involved in disease states. The bulk of pathogenomics research concerns itself with pathogens that affect human health; however, studies also exist for plant- and animal-infecting microbes.

4.3.10 Regenomics

Regenomics represents the merger of two fields of scientific endeavour: regenerative medicine and genomics. New technologies to reprogramme aged somatic cells back to pluripotency and to restore telomere length are currently used in research in regenerative medicine, though FDA-approved cellular therapies using reprogrammed cells are currently not available in the United States.

- The culture and banking of somatic cells also allows the parallel sequencing of their nuclear DNA to provide individuals with potentially valuable information for guiding them in lifestyle choices, but also 1 day, potentially in preventative strategies where cell types are made in advance for high-risk categories of disease, i.e. preparing cardiac progenitor cells for individuals at high risk for heart disease.

4.3.11 Personal Genomics

Personal genomics is the branch of genomics concerned with the sequencing and analysis of the genome of an individual.

- The genotyping stage employs different techniques, including single-nucleotide polymorphism (SNP) analysis chips (typically 0.02% of the genome) or partial or full genome sequencing.
- Once the genotypes are known, the individual's genotype can be compared with the published literature to determine likelihood of trait expression and disease risk.

4.4 Next-Generation Sequencing (NGS) Technologies

A number of next-generation sequencing (NGS) technologies have recently become available which are capable of generating hundreds of thousands or tens of millions of short DNA sequence reads at a relatively low cost. At present, these NGS technologies, referred to as second-generation sequencing (SGS) technologies, are being utilized for de novo sequencing, genome re-sequencing, and whole genome and transcriptome analysis. Now, new generation of sequencers, based on the "next-next" or third-generation sequencing (TGS) technologies like the Single-Molecule Real-Time (SMRT™) Sequencer, the Heliscope™ Single-Molecule Sequencer, and the Ion Personal Genome Machine™, are becoming available that are capable of generating longer sequence reads in a shorter time and at even lower costs per instrument run. Ever-declining sequencing costs and increased data output and sample throughput for NGS and TGS sequencing technologies enable the plant genomics and breeding community to undertake genotyping-by-sequencing (GBS). Data analysis, storage, and management of large-scale second or TGS projects, however, are essential, and they have reported the different sequencing technologies with an emphasis on TGS technologies and bioinformatics tools required for the latest evolution of DNA sequencing platforms (Thudi 2012).

Omics can enable further expansion of agricultural research in food, health, energy, chemical feedstock, and specialty chemicals while helping to preserve, enhance, and remediate the environment. Omics technologies focus on key traits of interest with precision. Omics can lead to enhancement of the nutritional properties of food for consumer benefit, such as a tomato that is high in lycopene, fruit with delayed ripening characteristics, and produce with potent antioxidant capabilities (Ahmad et al. 2012).

"The tools of genome research may finally unleash the genetic potential of our wild and cultivated Germplasm resources for the benefit of the society" (Tanksley and McCouch 1997). The utility of molecular markers and genome research in the context of using PGR for crop improvement includes:

- Diversity studies to identify genetically similar or distinct accessions and to determine individual degrees of heterozygosity and heterogeneity within populations of PGR
- Genetic mapping to identify simply inherited markers in close proximity to genetic factors affecting quantitative traits (QTL), followed by marker-assisted selection (MAS) of desired genotypes in segregating populations
- Exploitation of valuable QTL from PGR by advanced backcross QTL analysis to combine QTL analysis with the development of superior genotypes or by marker-assisted, controlled introgression of PGR into breeding materials through the development of introgression libraries
- Association studies to mine directly the allelic diversity of PGR and to identify those alleles

As outlined by the Food and Agriculture Organization (FAO) of the United Nations in the Global Plan of Action for the Conservation and Sustainable Utilization of Plant Genetic Resources, a more efficient use of plant genetic diversity is a prerequisite to meeting the challenges of development, food security, and poverty alleviation (FAO 1996b).

Concrete aims of FAO for conservation and sustainable utilization of plant genetic resources are:

- To develop cultivars that are specifically adapted to marginal or stress environments
- To assure sustainable production in high-yielding environments through better input-output relations, i.e. through reduced application of agrochemicals and increased nutrient and water efficiency
- To open production alternatives for farmers through development of industrial or pharmaceutical crops

To achieve these aims, extensive ex situ and in situ conservation of PGR must be assured. Evaluation of conserved accessions and their use by plant breeders or farmers needs to be supported and facilitated. The aim should not only be to exploit intraspecific variation within a crop but also to increase interspecific diversity in agriculture through genetic improvement and promotion of less popular, neglected, or underutilized crop species (Padulosi et al. 2002).

Many underutilized species are particularly useful in marginal lands where they have been selected to tolerate stress conditions and contribute to sustainable production. These genetic resources need to be evaluated for their outcrossing rates, yield potential, response to inputs, agronomic value, and the amount of genetic variation for specific traits, to allow more efficient genetic improvement and promotion. The genetic improvement of PGR for specific traits, followed by successful cultivation and marketing or consumption of the improved materials, is probably one of the most sustainable ways to "conserve" valuable genetic resources for the future.

4.5 Applications of Omics and Plant Genetic Resources for Mining of the Beneficial Genes with Suitable Examples

4.5.1 Methods of Using Genetic Resources in Plant Breeding

As per the published reports, there are three ways of using PGR in plant breeding (Simmonds 1993; Cooper et al. 2001):

1. Introgression involves the transfer of one or few genes or gene complexes (chromosome segments) from the PGR into breeding materials.
2. Incorporation (also named genetic enhancement or base broadening) describes the development of new, genetically broad, adapted populations with large variation and acceptable performance level.

3. Pre-breeding refers to more basic research activities with the goal of facilitating use of "difficult" materials.

Nonetheless, the three categories cannot be clearly separated from each other.

4.5.1.1 Introgression

The major objective of introgression is improving highly heritable qualitative traits that are governed by single or few major genes or gene complexes. The backcrossing method is conventionally used to introgress traits like resistance or restorer genes from wild relatives into breeding materials for the development of new hybrids possessing the desirable traits. The genetic problems play a minor role when introgressing major genes from PGR into high-yielding genotypes. Searching for specific traits, breeders would principally consider PGR of the primary gene pool, followed by the secondary gene pool and eventually the tertiary gene pool (Becker 1993).

4.5.1.2 Incorporation

Incorporation, genetic enhancement, or base broadening mainly aims at increasing the genetic variation for quantitative traits in breeding materials. Various methods of population improvement can be used. The methods vary depending on the crop species (self- or cross-pollinating) and the available time frame. Initially, selection may concentrate on adaptation traits that are highly heritable; performance traits are selected at a later stage. Diversity and recombination are maximized in the initial phase, with minimal selection intensities.

According to the available time frame, Cooper et al. (2001) identified three methods:

1. Development of synthetic or composite cross populations (long term)
2. Incorporation of PGR in a region's breeding materials to reduce the effects of historical bottlenecks during the evolutional spread of the crop (medium term)
3. Genetic enhancement to increase the actual variation in breeding populations (short term)

4.5.2 Application of Omics in Plant Abiotic Stress Tolerance and Biotic Stress Resistance

4.5.2.1 Application of Omics in Plant Abiotic Stress Tolerance

Climate change and food production are becoming major concerns in the present era which lead to abiotic stresses affecting crop production. Abiotic stress is defined as the negative impact of nonliving factors on the living organisms in a specific environment. The nonliving variable must influence the environment beyond its normal range of variation to adversely affect the population performance or individual physiology of the organism in a significant way (Vinebrooke et al. 2004). Abiotic stress is the main factor negatively affecting crop growth and productivity worldwide. The advances in physiology, genetics, and molecular biology have greatly improved our understanding of plant responses to stresses (Gao and Lin 2007).

Extreme environmental conditions such as drought, salinity, and freezing temperature cause adverse effect on the growth and productivity of crop plants. The temperature (heat, chilling, and freezing), drought, and salinity stress together represent abiotic stress. Exposure of plants in general to these abiotic stresses is inevitable in nature.

4.5.2.1.1 Drought

Drought is a devastating factor for global agronomic production. Drought being a yield-limiting factor has become a major threat to international food security. It is a complex trait, and drought tolerance response is carried out by various genes, transcription factors (TFs), microRNAs (miRNAs), hormones, proteins, cofactors, ions, and metabolites (Budak et al. 2015). Drought tolerance is an important quantitative trait with multipart phenotypes that are often further complicated by plant phenology. The identification of the candidate genes responsible for plant tolerance under different abiotic stresses, along with the use of the most suitable promoters associated with these events, is essential to develop transgenic crops with enhanced drought stress tolerance (Oosten et al. 2016).

Genome annotation, functional genomics, and molecular physiology studies have been conducted in several model and major crops to identify candidate genes involved in drought tolerance. These candidate genes include a large family of genes expressed under drought stress. Different proteins expressed by drought stress-associated candidate genes play significant roles in (i) cellular protection, such as osmotic adjustment, structural adaptation, repair, degradation, and detoxification, and (ii) positive interactions with other proteins and transcription factors, such as protein kinases and bZIP, MYB, and DREB, which are involved in plant drought responses by regulating other responsive genes, such as those involved in cell protection, to cope with drought stress in plants. Identifying drought tolerance (DT) genes from different model plants and major crops is vital to understanding the functional basis of the DT mechanism and its downstream use, including validation via MAS through molecular breeding. The transcriptomics responses of some candidate DT genes identified from different plant species have been characterized and evaluated.

The tools of genomics offer the means to produce comprehensive datasets on changes in gene expression, protein profiles, and metabolites that result from exposure to drought. Abiotic stress tolerance involves similar transcription factors in both dicotyledonous and monocotyledonous plants, and some molecular mechanisms of drought tolerance have been extensively described. It includes signal transduction cascade and activation/regulation of transcription, functional protection of proteins by late-embryogenesis abundant proteins (e.g. dehydrins) and chaperone proteins (e.g. heat shock proteins), accumulation of osmolytes (proline, glycine betaine, trehalose, mannitol, *myo*-inositol), induction of chemical antioxidants (ascorbic acid and glutathione), and enzymes reducing the toxicity of reactive oxygen species (superoxide dismutase, glutathione *S*-transferase) (Shinozaki and Yamaguchi-Shinozaki 2007).

QTLs, miRNAs, and genome-editing systems (e.g. CRISPR/Cas system) are major genomics-based methods applied to discover and manipulate related genomic regions. Candidate genes should be validated via approaches, such as expression analysis, qRT-PCR, incorporation into QTL maps, linkage mapping, TILLING, and allele mining, and applications of these approaches have been reviewed previously. In recent years, many transcriptomics and functional genomics studies have been conducted to understand the stress mechanisms in different crop plants. One common approach that effectively isolates the candidate genes responsible for drought stress in drought-resistant genotypes is the generation of expressed sequence tags (ESTs) from cDNA libraries of tissues collected under drought. To date, many drought-responsive genes have been identified from several crop species. Normalized cDNA libraries from rice seedlings led to the identification of many genes responsible for drought tolerance that were highly expressed under drought (Sahebi et al. 2018).

Introduction by gene transfer of several stress-inducible genes has demonstrated enhanced abiotic stress tolerance in transgenic plants (Zhang et al. 2004). These particular genes encode key enzymes regulating biosynthesis of compatible solutes such as amino acids (e.g. proline), quaternary and other amines (e.g. glycine betaine and polyamines), and a variety of sugars and sugar alcohols (e.g. mannitol, trehalose, galactinol, and raffinose). Genes encoding LEA proteins and heat shock proteins have also been used to improve drought tolerance in transgenic plants. A gene encoding galactinol synthase (GolS), a key enzyme involved in raffinose family oligosaccharide biosynthesis, was introduced to improve drought stress tolerance in transgenic *Arabidopsis* (Taji et al. 2002). Prior analyses demonstrate that GolS genes are induced by drought, cold, and ABA. Moreover, expression of the gene encoding raffinose synthase is also induced by drought stress. Additionally, recent metabolome analysis indicated significant accumulation of both galactinol and raffinose under drought stress. Not only metabolites, but also some stress-responsive proteins such as LEAs, have also been implicated in detoxification and alleviation of cellular damage during dehydration. Other studies demonstrate that overexpression of some LEA class genes results in enhanced tolerance to dehydration, although the precise mechanism is still unknown. LEA proteins may also function as chaperone-like protective molecules to combat cellular damage (Umezawa et al. 2006).

4.5.2.2 Application of Omics in Biotic Stress Resistance Specific to Plant Disease Resistance

The term Omics, which look into the global profiling and analysis of various cellular molecules, has gained new insight with the advancement of next-generation sequencing and mass spectrometry technologies (Bhadauria et al. 2016). It has broader implication in genetic improvement of crops for resistance against various diseases of economic significance, and some important contributions are summarized as follows.

Crop production is greatly affected by biotic stresses which are greater constraints to plant growth and development. To withstand such stresses, plants have developed stress-specific adaptations as well as simultaneous responses. The efficacy and magnitude of inducible adaptive responses are dependent on activation of signalling pathways and intracellular networks by modulating expression, abundance, and/or post-translational modification of proteins associated with defence mechanisms. In this regard, proteomics plays an important role in elucidating plant defence mechanisms by mining the differential regulation of proteins to various biotic stresses. Rice, most widely cultivated as staple food crops in world, is greatly affected by a variety of biotic stresses, and high-throughput proteomics approaches have been employed to unravel the molecular mechanism of the biotic stresses-response in rice. Latest advances of proteomic studies on defence responses discuss the potential relevance of the proteins identified by proteomic means in rice defence mechanism (Chen et al. 2016).

Lentil (*Lens culinaris*) crop is strongly affected by fungal diseases reducing the yield as well as quality which is a grain crop and an important source of dietary proteins and fibre. Lentil has a narrow genetic base presumably due to a bottleneck during domestication, and as a result, any resistance to fungal diseases in the cultivated gene pool is gradually eroded and overcome by pathogens. So there is a need for identification of new sources of resistance from wild lentil (*Lens ervoides*) which provides resistance genes to be transferred to lentil cultivars using next-generation sequencing-based genotyping, comparative genomics, and marker-assisted selection breeding (Bhadauria 2016).

In nature, plants are continuously exposed to the attack of various pathogens, such as fungi or bacteria, and also viruses. Naturally plants develop different metabolic and genetic responses whose final outcome is the production of either toxic compounds that kill the pathogen or deter its growth and/or semiotic molecules that alert other individuals from the same plant species. These metabolites are derived from the secondary metabolism, and their production is induced upon detection of a pathogen-associated molecular pattern (PAMP). These PAMPs are different molecules that are perceived by the host cell triggering defence responses. Plant metabolite profiling techniques have allowed the identification of novel antimicrobial molecules that are induced upon elicitation (Arbona and Cadenas 2016).

The *Brassica* genus involves a variety of economically important species and cultivars used as oil seeds, vegetable source, forage, and ornamental. During the infection of pathogens, plant growth and development and crop productivity are greatly affected. Advancement in the field of plant physiology, genetics, and molecular biology has greatly improved our understanding of plant responses to biotic stress conditions which includes "omics-based technologies" which enables the qualitative and quantitative monitoring of the abundance of various biological molecules in a high-throughput manner and thus allows determination of their variation between different biological states on a genomic scale. Modern approaches are being used to elucidate the molecular mechanisms that underlie *Brassica* disease resistance (Francisco et al. 2016).

Rice (*Oryza sativa*) is one of the most important cereal grain crops in the world, especially in Asian countries. Virus infection/diseases of rice are considered as the most serious threat to rice yields. Most of the rice viruses are transmitted by the vector insects belonging to hemipteran insects such as plant hoppers and leafhoppers. These plant hoppers transmit five rice viruses in Asian areas, viz., Rice stripe virus, Rice black-streaked dwarf virus, Rice ragged stunt virus, Rice grassy stunt virus, and Southern rice black-streaked dwarf virus. Several antiviral genes have shown to provide resistance to viruses. For the isolation of virus resistance genes, omics studies are required to better understand their responses including relationships between plant hoppers and viruses by using various omics techniques (Cui et al. 2016).

The attack of different pathogens, such as bacteria, fungi, and viruses, has a negative impact on crop production. To counter such attacks, plants have developed different strategies involving the modification of gene expression, activation of several metabolic pathways, and post-translational modification of proteins, which culminate into the accumulation of primary and secondary metabolites implicated in plant defence responses. The recent advancement in omics techniques allows the increased coverage of plant transcriptomics, proteomes, and metabolomes during pathogen attack and the modulation of the response after the infection. Omics techniques also allow us to learn more about the biological cycle of the pathogens in addition to the identification of novel virulence factors in pathogens and their host targets. Both approaches become important to decipher the mechanism underlying pathogen attacks and to develop strategies for improving disease-resistant plants. Some of the contribution of genomics, transcriptomics, proteomics, metabolomics, and metallomics in devising the strategies to obtain plants with is increased resistance to pathogens. These approaches constitute important research tools in the development of new technologies for the protection against diseases and increased plant production (Gomez-Casati et al. 2016b).

Plant diseases are responsible for important losses in crops and cause serious impacts in agricultural production. In the last years, proteomics has been used to examine plant defence responses against pathogens. Such studies may be pioneer in the generation of crops with enhanced resistance. Proteomics advances in the understanding of host and non-host resistance against pathogens (Grandellis et al. 2016).

The study of the association between the traits and biomarkers will dramatically decrease the time and costs required to bring new improved disease resistance lines to market. The field of omics has an enormous potential to assess diseases more precisely, including the identification and understanding of pathogenic mechanisms in legume crops, and has been exemplified by a relatively large number of studies. Recently, molecular genetic studies have accumulated a huge amount of genotypic data, through a more affordable next-generation sequencing (NGS) technology, causing the omics approaches to fall behind. They have reported overview of genomics and proteomics and their use in legume crops, including the use of comparative genomics to identify homologous markers within legume crops (Diapari 2016).

4.6 Conclusion

The demand for food production increases every year; thus there is a need to develop and improve production of plant crops with qualitative and quantitative characteristics. Many of the naturally available wild relatives of crop plants and close relatives of domesticated plants possess diversified genes in their gene pool that will provide resistance to both biotic and abiotic stresses by increasing the quantity of crop production. Plant breeders require access to new genetic diversity to satisfy the demands of a growing human population for more food that can be produced in a variable or changing climate and to deliver the high-quality food with nutritional and health benefits demanded by consumers. In conservation of plant genetic resources, genetics is mainly focused on determining the relationship between species and population, studying the cross-species variation, and describing the interactions between species and their threatening processes. Besides preventing the threatened species, diversity can benefit from looking beyond these and considering the genome of rare species and others that share a common environment. By identifying the factors or processes that influence the genomic composition of the threatened or extinction species, we can predict and identify the ecologically and genetically unique species. At present omics-based technologies play a major role in the identification of diversified genes in different sources that will be utilized for transferring those genes into crop plants for increased production of food. In the future, we may effectively utilize these technologies for crop improvement programmes for conservation as well as introgression of these genes into crop plants. This is likely to result in increased agricultural productivity, thereby having a positive impact on global food and nutritional security.

References

Ahmad P, Ashraf M, Younis M, Hu X, Kumar A, Akram N, Al-Qurainy F (2012) Role of transgenic plants in agriculture and biopharming. Biotech Adv 30:524–540

Arbona V, Cadenas AG (2016) Metabolomics of disease resistance in crops. Omics in Plant Disease Resistance, Curr Issues Mol Biol:13–30

Becker H (1993) Pflanzenzüchtung. Verlag Eugen Ulmer, Stuttgart

Bhadauria V (2016) Omics in plant disease resistance. Curr Issues Mol Biol Book 19:1–2

Bhadauria V, Melissa ML, Wong Kirstin E, Bett SB (2016) Wild help for enhancing genetic resistance in lentil against fungal diseases. Omics in Plant Disease Resistance, Curr Issues Mol Biol:3–6

Budak H, Hussain B, Khan Z, Ozturk NZ, Ullah N (2015) From genetics to functional genomics: improvement in drought signaling and tolerance in wheat. Front Plant Sci 6:1–13

Chen X, Bhadauria V, Ma B (2016) Current status of proteomic studies on Defence responses in rice. Omics in Plant Disease Resistance, Curr Issues Mol Biol:7–12

Claudino WM, Quattrone A, Biganzoli L, Pestrin M, Bertini I, Leo AD (2007) Metabolomics: available results, current research projects in breast Cancer, and future applications. J Clin Oncol 25:2840–2846

Clouse S (2002) Brassinosteroids. In: Somerville CR, Meyerowitz EM (eds) The Arabidopsis book. Rockville, American Society of Plant Biologists, pp 1–23

Cooper HD, Spillane C, Hodgkin T (2001) Broadening the genetic base of crops: an overview. In: Cooper HD, Spillane C, Hodgkin T (eds) Broadening the genetic base of crop production. CABI Publishing in cooperation with FAO and IPGRI, CAB International, Wallingford, pp 1–23

Cui F, Zhao W, Luo L, Kang L (2016) Rice responses and resistance to plant hopper-borne viruses at transcriptomic and proteomic levels. Omics in Plant Disease Resistance, Curr Issues Mol Biol 19:43–52

Deshmukh R, Sonah H, Patil G, Chen W, Prince S, Mutava R, Vuong T, Valliyodan B, Nguyen HT (2014) Integrating omics approaches for abiotic stress tolerance in soybean. Front Plant Sci 5:1–12

Diapari M (2016) Linking biomarker and comparative omics to pathogens in legumes. Omics in Plant Disease Resistance, Curr Issues Mol Biol pp:137–144

Diego F Gomez-Casati, Pagani MA, Busi MV, Vijai Bhadauria (2016a) Omics approaches for the engineering of pathogen resistant plants. Omics in plant disease resistance, Curr Issues Mol Biol 89–98

FAO (1989) Annex to the international undertaking on plant genetic resources, conference resolution 5/89. Twenty- fifth sessions of the FAO conference, Rome, 11–29 November 1989. Food and Agricultural Organization of the United Nations, Rome

FAO (1996a) Report on the state of the World's plant genetic resources for food and agriculture, prepared for the international technical conference on plant genetic resources, Leipzig, Germany, 17–23 June 1996. Food and Agricultural Organization of the United Nations, Rome

FAO (1996b) Global plan of action for the conservation and sustainable utilization of plant genetic resources and the Leipzig declaration, adopted by the international technical conference on plant genetic resources, Leipzig, Germany, 17–23 June 1996. Food and Agricultural Organization of the United Nations, Rome

FAO (1998) The state of the World's plant genetic resources for food and agriculture. FAO, Rome

FAO (2010) Second report on the state of the World's plant genetic resources for food and agriculture. Food and Agriculture Organization of the United Nations, Rome, Italy

Feinberg AP (2007) Phenotypic plasticity and the epigenetics of human disease. Nature 447:433–440

Fliser D, Novak J, Thongboonkerd V, Argiles A, Jankowski V, Girolami MA, Jankowski J, Mischak H (2007) Advances in urinary proteome analysis and biomarker discovery. J Am Soc Nephrol 18:1057–1071

Francisco M, Soengas P, Velasco P, Bhadauria V, Cartea ME, Rodríguez VM (2016) Omics approach to identify factors involved in *Brassica* disease resistance. Omics in Plant Disease Resistance, Curr Issues Mol Biol 19:31–42

Frankham R, Ballou JD, Briscoe DA, Karina H, McInnes KM (2009) Introduction to conservation genetics, 2nd edn. Cambridge University Press, Cambridge

Gao JP, Lin HX (2007) Understanding abiotic stress tolerance mechanisms: recent studies on stress response in rice. J Integr Plant Biol 49(6):742–750

Gomez-Casati DF, Pagani MA, Busi MV, Bhadauria V (2016b) Omics approaches for the engineering of pathogen resistant plants. Curr Issues Mol Biol 19:89–98

Grandellis C, Vranych CV, Piazza A, Garavaglia BS, Gottig N, Ottado J (2016) An overview of proteomics tools for understanding plant Defence against pathogens. Omics in Plant Disease Resistance, Curr Issues Mol Biol pp:129–136

Hallauer AR, Miranda JB (1981) Quantitative genetics in maize breeding. Iowa State University Press, Ames

Hanash SM, Pitteri SJ, Faca AM (2008) Mining the plasma proteome for cancer biomarkers. Nature 452:571–579

Harlan JR, de JMJ W (1971) Toward a rational classification of cultivated plants. Taxon 20:509–517

Hoisington D, Khairallah M, Reeves T, Ribaut J-M, Bent S, Skovmand B, Taba S, Warburton M (1999) Plant genetic resources: what can they contribute toward increased crop productivity? Proc Natl Acad Sci U S A 96:5937–5943

Hu Q, Hansen LN, Laursen J, Dixelius C, Andersen SB (2002) Intergeneric hybrids between Brassica napus and *Orychophragmus violaceus* containing traits of agronomic importance for oilseed rape breeding. Theor Appl Genet 105:834–840

Jackson O, Lay JR, Borgmann S, Liyanage R, Charles L, Wilkins (2006) Problems with the "omics". Trends Anal Chem 25(11):1046–1056

Jones TA (2003) Transcriptomics technologies. PloS Comput Biol:1–23

Lowe R, Shirley N, Bleackley M, Dolan S, Shafee T (2017) Transcriptomics technologies. PLoS Comput Biol 13(5):e1005457

Maxted N, Kell SP (2003). Biodiversity and conservation: Plant diversity, conservation and use. Encyclopedia of Appl Plant Sci

Morrison N, Cochrane G, Faruque N, Tatusova T, Tateno Y, Hancock D, Field D (2006) Concept of sample in OMICS technology. OMICS A J Integr Biol 10(2):127–137

Oosten MJ, Van CA, Punzo P, Landi S, Ruggiero A, Batelli G, Grillo S (2016) In: Hossain MA, Wani SH, Bhattacharjee S, Burritt DJ, Phan Tran L-S (eds) Drought stress tolerance in plants, vol Vol 2. Springer, Berlin, pp 39–70

Ouborg NJ, Vergeer P, Mix C (2006) The rough edges of the conservation genetics paradigm for plants. J Ecol 94:1233–1248

Padulosi S, Hodgkin T, Williams JT, Haq N (2002) Underutilized crops: trends, challenges and opportunities in the 21st century. In: Engels JMM, Rao VR, Brown AHD, Jackson MT (eds) Managing plant genetic diversity. IPGRI/CABI Publishing, Wallingford, pp 323–338

Pritham OB (2015) Crop abiotic and biotic stress monitoring by using remote sensing. Trends Biosci 8(9):2319–2320

Ramsden JJ (2009) Metabolomics and metabonomics. Bioinformatics:1–6

Sahebi M, Hanafi MM, Rafii MY, Mahmud T, Azizi P, Osman M, Atabaki N (2018) Improvement of drought tolerance in rice (*Oryza sativa* L.): genetics, genomic tools, and the WRKY gene family. BioMed Res Intl 1–20

Shinozaki K, Yamaguchi-Shinozaki K (2007) Gene networks involved in drought stress response and tolerance. J Exp Bot 58:221–227

Simmonds NW (1993) Introgression and incorporation. Strategies for the use of crop genetic resources. Biol Rev 68:539–562

Smith B, Ceusters W, Klagges B (2005) Relations in biomedical ontologies. Genome Biol:6–46

Syvanen AC (2001) Accessing genetic variation: genotyping single nucleotide polymorphisms. Nat Rev Genet 2(12):930–942

Taji T, Ohsumi C, Iuchi S, Seki M, Kasuga M, Kobayashi M, Yamaguchi-Shinozaki K, Shinozaki K (2002) Important roles of drought- and cold-inducible genes for Galactinol synthase in stress tolerance in *Arabidopsis thaliana*. Plant J 29(4):417–426

Tanksley SD, McCouch S (1997) Seed banks and molecular maps: unlocking genetic potential from the wild. Science 277:1063–1066

Thudi M, Li Y, Jackson SA, May GD, Varshney RK (2012) Current state-of-art of sequencing technologies for plant genomics research. Brief Funct Genom 11(1):3–11

Umezawa T, Fujita M, Fujita Y, Yamaguchi-Shinozaki K, Shinozaki K (2006) Engineering drought tolerance in plants: discovering and tailoring genes unlock the future. Curr Opin Biotech 17:113–122

Vinebrooke RD, Cottingham KL, Norberg J, Scheffer M, Dodson SI, Maberly SC, Sommer U (2004) Impacts of multiple stressors on biodiversity and ecosystem functioning: the role of species co-tolerance. Oikos 104:451–457

Wambugu PW, Ndjiondjop M-N, Robert J, Henry RJ (2018) Role of genomics in promoting the utilization of plant genetic resources in gene banks. Brief Funct Genom 17(3):198–206

Zhang JZ, Creelman RA, Zhu JK (2004) From laboratory to field. Using information from *Arabidopsis* to engineer salt, cold, and drought tolerance in crops. Plant Physiol 135:615–621

Genetic and Genomic Resources and Their Exploitation for Unlocking Genetic Potential from the Wild Relatives

5

Mehfuz Hasan, Abu Sayeed Md. Hasibuzzaman, Hasan Muhammad Abdullah, and Md. Mehboob Hasan Kallol

Contents

Abstract

There is a need for the plant breeders to exploit novel genetic variations to meet up with increasing food needs by the human population which can be cultivated in different climatic conditions and to provide top-quality and beneficial foods as demanded by the people. Plant breeders can utilize the crop wild relatives (CWRs), which are the closest relatives to the domesticated plants, as a practical gene pool. The expansion of genetic diversity of crop plants has been supported

M. Hasan (✉) · A. S. Md. Hasibuzzaman
Department of Genetics and Plant Breeding, Bangabandhu Sheikh Mujibur Rahman Agricultural University, Gazipur, Bangladesh

H. M. Abdullah
GIS and Remote Sensing Lab, Department of Agroforestry and Environment, Bangabandhu Sheikh Mujibur Rahman Agricultural University, Gazipur, Bangladesh
e-mail: hasan.abdullah@bsmrau.edu.bd

Md. M. H. Kallol
Senior Management Counsellor and Head, Computer Services Division, Bangladesh Institute of Management (BIM), Dhaka, Bangladesh

R. K. Salgotra, S. M. Zargar (eds.), *Rediscovery of Genetic and Genomic Resources for Future Food Security*,
https://doi.org/10.1007/978-981-15-0156-2_5

by the data revealed by the genomics of CWR. Advanced DNA sequencing technology has provided opportunities for sequencing of CWR and its enhanced application in the development of the crop. After the completion of the genomic sequencing of significant crop species, the focus has been placed in the broader gene pool analysis of the important farm produce including CWR. In order to perform an efficient exploration of the beneficial genetic variation in CWR, de novo sequencing and resequencing must collectively be applied. According to the results of genome analysis, beneficial alleles present in CWR were discovered, and genome regions where the variation is missing in domestication bottlenecks have been identified. Maximizing what the genome sequencing of CWR offers can be achieved by targeting top priority CWR for sequencing. By coordinating the worldwide efforts to use genomics, it is possible to speed up biodiversity exploitation and conservation needed to sustain agriculture and availability of foods.

Acronyms

BLAST	Basic Local Alignment Search Tool
CAPS	Cleaved amplified polymorphic sequence
CSSLS	Chromosome segment substitution lines
CWR	Crop wild relatives
DH	Doubled haploid
DNA	Deoxyribonucleic acid
FAnGR	Farm animal genetic resources
GIS	Geographic information system
GWAS	Genome-wide association studies
GBS	Genotyping-by-sequencing
GenoSIS	Genome Spatial Information System
GMOs	Genetically modified organisms
GPS	Global positioning system
HTPs	High-throughput phenotyping
HTPPs	High-throughput phenotyping platforms
InDels	Insertions or deletions
KASP	Kompetitive allele specific PCR
MAGIC	Multi-parent advanced generation intercross
MABC	Marker-assisted backcrossing
MAS	Marker-assisted selection
NAM	Nested association mapping
NILs	Near-isogenic lines
NGS	Next-generation sequencing
QTL	Quantitative trait loci
RILs	Recombinant inbred lines
RAD-seq	Restriction-site-associated DNA sequencing

SNPs	Single nucleotide polymorphism
SNVs	Single nucleotide variations
SSD	Single-seed descent
UNIX	Uniplexed Information and Computing System
WGS	Whole-genome shotgun sequence

5.1 Introduction

The increase in food demand across the globe and the likely influence of climate change have resulted into the higher demand for effective plant breeding programs to offer better yield and resilience to influence better biotic and abiotic stress in the production of crops. Furthermore, dietetic support for higher human residents has to be supplied by agricultural produce. Hence, generally, plant breeding has to concentrate on the significance of improving mandate, which is increased production, adopting environment changeability, and getting food nutrition. Existing plant genetic and genomic resources have to be explored as their effective exploitation is crucial for the enhancement of crop development programs.

While a significant amount of plant species has been explored in crop production, the ultimate human food materials came from a comparatively small proportion of plant species. Based on the assessment, 95% of food material is gotten from 30 different species of plant. Food such as rice, maize, potato, and wheat falls among these 30 species, and they constitute beyond 60% of food (Alercia et al. 2015). It is possible to improve food security by emphasizing on researches to enlarge agriculture through domestication and the use of a higher number of plant species.

The wild relatives are the progenies of every crop plant. The fundamental sources of divergence which help in plant breeding program come from the crop wild relatives (CWRs). On some occasions, the wild species which developed from a crop plant may be clearly seen in current wild plant residents. Also, there is a chance not to recognize it as the wild species may be nonextant. Mostly, plant species that happens to be a hybrid or polyploidy basis may have contributed to the adopted crop's genome. Several other species associated with the instant wild relatives may collectively use a large gene pool offering a means of divergence for the crop. A signal of the primary gene pool for plant improvement is seen by wild relatives of plants which are cross-compatible with contemporary cultivars. The more distantly related species where genes can be saved with a better degree of difficulties shows the secondary or tertiary gene pools (Harlan and de Wet 1971). Such gene pools encompass a significant material for agriculture.

With the tools of genomics, genetic variation present in CWR is better characterized than in old molecular approaches. Genome analysis priority should be given to the CWRs which have been recognized as an important species. A technology avenue that enhances the efficient application of available crop diversity is provided by the genomic analysis. Likewise, it supports the identification of beneficial genes in crop wild relatives to improve farm yield. Improvement in DNA sequencing allows whole-genome sequencing to become an enticing option to characterize the plant genetic resources. Genome sequencing of plant genetic resources offers a high potential to support better control of in situ and ex situ resources. Studies on genomics can assist in the development of new plants which can survive in several environments. In a different climate, wild relatives can be found as they supply a means of genotypes that thrive in different climatic conditions of crop production. It is possible to get insight into plant adaptation techniques under natural selection by analyzing the genome variation in wild populations; it can also create alternative applications in plant breeding for climate resilience since it offers food security for people in climate vulnerable regions of the world. One approach which has been employed is the selection of wild germplasm sampled from the environment similar to the focused area of agricultural production. By characterizing the genome of the adapted genotypes, it will serve as a tool to dissect the genetic basis of performance in those locations. The gene variations in wild populations relative to various ecological situations may signify approaches for use in the development of crop genotypes that have adapted to new or changing environments.

5.2 Scheme for the Application of CWR in Crop Enhancement

The vital processes in the use of CWR for the improvement of crops are as follows:

- Collection of targeted germplasm.
- Preservation of germplasm.
- Development of structured core samples.
- Phenotyping (characterization) based on stress tolerance, i.e., pest, diseases, nutritional values, and environmental variations.
- Performing genotyping to assess CWR.
- Molecular marker design and use of high-throughput sequencing technologies.
- Evaluate divergence by applying markers to concentrate on conservation and breeding approaches.
- The crossing of crop variety with wild germplasm.
- Segregate mapping population from the cross, for instance, retrogression line population (advanced backcross approaches), F_2.
- Genotype population – structuring linkage map.
- Trait mapping to connect with specific locus/loci – parent differs for the trait of concern – and, consequently, segregating within the population.
- Selection of lines possessing QTL of concern.
- Attempt introgress back into the crop by marker-assisted backcrossing, and purify the region of concern.
- Selection of trait to evaluate performance.

5.2.1 Characterization of the Germplasm

After the collection of CWR germplasm and saved as samples in the gene bank, then, potentially, they become obtainable for the enhancement of crop. Nevertheless, it is tasking to characterize such a significant amount of germplasm, regardless of the enthusiasm level of the gene bank staff. Several ways have been used to cross this hurdle, but, basically, the adopted idea is to create a lesser representative subset or core samples of the fundamental collection. A combination of markers along with phenotypic data can be used to create an organized archetypal sampling of the current variations for a specific plant species.

The "phenotype" of the CWR constitutes its ecological deviation response – variations in temperature, nutrient, the presence of water, etc. – stress tolerance level, growth resilience, and response to pest as well as diseases, among others. Phenotypic classification of the CWR at the fundamental level includes cultivating the CWR in a defined trial and evaluating attributes like performance and development. Afterward, further expert phenotyping may be performed to determine the experimental difficulties, for instance, an infection caused by a pathogen, or the different availability levels of water or nutrient. In terms of productivity, the CWR samples may not perform at the same level as the target crop; however, they may display significant differences in a specific trait, which may be useful to plants. Also, accurate phenotype information will add to the records of the samples used. It will be helpful for identification in future studies and, consequently, contribute to preservation endeavors.

5.2.2 Genetic Marker Design

The genome, gene expression, and the relation between the two and the environment drive the phenotype of a plant. It is pertinent to introgress genes of the CWR into the plant genome to move beneficial qualities from CWR into a crop. This can be done by knowing the aspect of the genome that drives the characteristics; however, since plant genome is broad, a recommendable method is to recognize the vast area of the plant genome responsible for the trait and then introgress the area. A genetic marker has to be created to identify relevant areas. Every marker is considered as a distinct DNA sequence, which is merely found in the plant genome and varies among polymorphic samples. Ordinarily, a bunch of markers has to be created which will enable the plant breeding operators to accurately recognize a specific locus as well as variants in the locus and, consequently, allow distinction between the target and recurrent genome; such markers are polymorphic markers. Then, the polymorphic markers will be applied to genotype offspring of the cross between CWR and crop and also assign the right parental stage at every marker. Based on the user's preference, there are various marker systems. Modern genetic panels of marker oriented from next-generation sequencing are created by analyzing some aspects or the whole genomes of the plant and that of CWR. Basically, crop genomes are better known and have more tendency to be sequenced compared to the CWR genomes; thus, the tasks may include using the identified genome or transcriptome of the crop acquired from the web hub, genome sequencing of the CWR, and the use of a computational alignment approach to equal the CWR sequence to the crop sequence. After the first alignment program, BLAST, a series of alignment technology like Tophat, Burrows-Wheelers alignment, and Bowtie are becoming the standards in this active aspect of bioinformatics study. After the alignment of the sequences, the variations between the crop and the CWR, in terms of single nucleotide polymorphism (SNPs), insertions or deletions (InDels), and single nucleotide variations (SNVs), may be recognized computationally and the suitable polymorphism confirmed by the genotyping. Then, the resultant markers are applied as a method for the assessment of the difference between more CWR and crop samples without a need to sequence their complete genomes.

5.2.3 Performing Genotyping

A specific combination of alleles present in an organism's genome constitutes the genotype of the organism. Immediately the molecular marker panel has to be recognized for a category of gene bank samples of both the CWR and the plants. In summary, it is possible to determine the genotypes of the samples. Each plant is grown, and extraction of DNA is done. Afterward, the markers are applied alongside template from the chosen plants to rate the availability of variations. There are several commercially available high-throughput methods that may evaluate a bunch of markers collectively in a test, involving Illumina Infinium arrays and KASP markers. With these approaches, a "signal" can be identified with every accession, which reveals specific variants that they contain for every marker in the bunch. An alternate means is to sequence samples directly through the use of next-generation techniques to find their genotype; this approach is increasingly affordable.

5.2.4 Evaluation of Diversity

The diversity between samples can be assessed by genotypic differences between samples. Closely related samples tend to possess a highly similar set of marker variants; on the other hand, diverse samples will have a broader diversity of genotypic variants. In order to assess the population structure, computational methods have been created. Software known as STRUCTURE is commonly used to assess population structure as it uses maximum likelihood to establish population structure among samples (Huda et al. 2019). For the recreation of evolutionary relationships, an active research approach is available which deploys statistical techniques such as maximum likelihood for further diverse samples, evolutionary parameters such as mutation rates, Bayesian techniques to jointly analyze close relationships between samples, and maximum parsimony for closely related samples. Once the genetic diversity among samples is understood, it will assist in developing a collection of samples to optimize genetic diversity and, likewise, to recognize specific samples closely or more distantly associated to plants of concern.

5.2.5 The Crossing of Crop Variety with Wild Germplasm

It is imperative to check if a cross between CWR and its domesticated crop relative can be performed before a crossing program can be initiated. In a situation where the CWR is not cross-compatible with the target crop, it may become necessary to cross the CWR to a bridging species. After the crossing, which involves transferring the pollen to the crop from the CWR (it is advisable to perform the reciprocal cross), the F_1 will be collected.

5.2.6 Segregate Mapping Population from a Cross

While several kinds of the population can be established, among the simplest designs, two of them are the F_1-selfed F_2 derivatives and, also, the backcross generation obtained from backcrossing the F_1 to the female crop. A mosaic of the two-parental genotypes can be shown by the progeny (100–300), and hopefully, it will represent an even-distributed recombination set. With these populations, it is possible to get the fastest means to a segregating population, but they possess high heterozygosity, and consequently, they do not stand for a stable resource which is ready to propagate. With this feature, these are beneficial for primary map development. A better recommendable alternative is to establish a stable mapping population which can be sustained indefinitely and, consequently, allows the screening of several traits as demanded. Some of the most used methods include single-seed descent (SSD) for self-pollinated crops to develop recombinant inbred lines (RILs). Since further recombination situations are possible in the process of RIL production, they provide excellent opportunities for mapping trait of interest. The doubled haploids (DHs) obtained from the F_1 enable a quicker means to fixation of traits. Plantlets obtained in the course of this technique have to be critically assessed to check the ploidy level to ascertain that the regenerated plantlets are truly DHs. Near-isogenic lines (NILs) can be established to get a comprehensive resolution of the segmented areas from the CWR in crop's background. Basically, NILs are obtained by backcrossing followed by selfing for some generations to obtain homozygous lines. Collections from the original BC_1 obtained from it can be improved by applying the marker-assisted backcrossing. By applying this technique, there is a chance to develop introgression line populations as every line will contain a donor species segment in the current background, altogether offering coverage for the genome of the donor. Generally, the target is the homozygosity, i.e., fixation. Fixed lines provide a chance to execute replicated experiments in several environments and throughout the year. Doing this will allow a precise approximation of the genetic constituent of the target character and, consequently, alleviate the variations due to environmental issues.

5.3 Underpinning Technologies

5.3.1 Technological Advances in Using Crop Wild Relative (CWR)

After the introduction of the genetic material from CWR into a domesticated background, populations are mostly created to assess the inserted genes of interest and identify the position in the genome, and through the application of marker-assisted selection (MAS), they will be further deployed. MAS enhances the breeding process efficiency as a significant amount of offspring can be assessed in an organized manner. The trait will have to be genetically mapped, and the genotype impacts must be confirmed before the MAS can be conducted. Through genetic mapping, the genotypes will be correlated with phenotypes and, also, tag the alleles operating each trait with the aid of molecular marker(s).

Genome-wide association studies (GWAS) or QTL mapping techniques are mostly utilized in molecular breeding. Currently, extra mapping techniques are being used to improve the precision of mapping. For instance, nested association mapping (NAM) populations (Guo et al. 2010) and multi-parent advanced generation intercross (MAGIC) populations are being employed (Huynh et al. 2018). The target of some population development programs is to develop specific kinds of recombinants. The series of introgression lines form the chromosome segment substitution lines (CSSLS), and each line possesses chromosome fragments obtained from the donor parent within a collective recipient context, which altogether denotes the complete genome of the donor species. With these CSSLs, it becomes possible to detect and map QTLs throughout the entire genome and assess the interaction between QTLs. Within the early generation, desired recombinants can be created, provided it is a sufficiently large population. For instance, marker-assisted backcrossing (MABC) was used for below sea-level rice cultivation to incorporate a significant QTL into a new genetic background in merely two sets of backcrossing (Iftekharuddaula et al. 2011). Additionally, genomic tools allow a direct assessment of the CWR contribution in contemporary varieties (Baute et al. 2015), while various software tools are being created to assist such studies.

CWR diversity and landrace systematic investigation, also known as indigenous trait research, is witnessing a rejuvenation in some aspects of the private organizations. Sometimes, this is chosen as an option to the application of transgenic approaches that run at a high cost of regulation and prevalence of end user's concern. Nevertheless, transgenic techniques remain beneficial as an experimental approach to understand the performance of candidate genes from CWR. Immediately the performance of an allele is verified; incorporation of alleles can be done with MAS without incurring regulatory expenses related to GMOs. Mostly, genetic engineering is used in the private industry, and due to our lack of accessibility toward such commercially important data, it is possible to have underestimated their actual use here.

The frequency of molecular marker development and utilization has been revolutionized by genomics. Currently, many crops have SNP chip platforms – while the price of DNA sequencing has progressively declined with time. *Helianthus annuus*

L. (Livaja et al. 2016) and *Brassica napus* L. (Clarke et al. 2016) are among the most recently developed single nucleotide polymorphism platforms. In crops which are deficient in SNP genotyping platforms or situations where the novel, divergence resource is being utilized – for instance, crop wild relatives – high-throughput sequencing provides an inexpensive and faster means to deploy huge number of markers for mapping analysis. Genotyping-by-sequencing, NimbleGen exome capture, and restriction-site-associated DNA sequencing (RAD-seq) are all reduced representation techniques, which have been applied in many CWRs. The solidity of marker provided through such technologies enables for an excellent quick mapping and is capable of saturating mapping populations regarding detection of every recombination event.

The application of high-throughput sequence data can be significantly enhanced by the presence of a reference genome sequence. While most of the early sequencing approaches aimed basically on grown materials, large-scale CWR genomic (re) sequencing is currently being performed. Characterization of CWR germplasm can be performed by using the whole-genome shotgun sequence (WGS) techniques. For instance, rice possesses a relatively little genome (430 Mbp), and resequencing of many wild genomes has been done with the use of WGS (Li et al. 2014; Alexandrov et al. 2014). Wheat (Sehgal et al. 2015), soybean (Song et al. 2015), and chickpea (Varshney 2016) are among several germplasm that have been genotyped or sequenced. These details will improve the knowledge and information concerning their gene pools.

Another means to find putatively useful alleles in CWR is through landscape genomics and environmental analysis (Anderson et al. 2016). Therefore, genomics may provide support for the collection, conservation, and use of CWR. In contrast to the collection information like the environment and local ecology, genomic data can be produced from all plants. Indeed, sequence data shows the diversity in samples and could be applied as a typical currency that can be used for sample comparison. Nevertheless, it is worthy to note that the development and use of genomic resources remain as a non-priority to every community of crops.

The establishment of genome editing is expected to change modern breeding. With these applications, researchers can modify an organism's genome with unprecedented accuracy and with no need for DNA introduction from another organism. While these techniques are still young, their progressive precision tends to lead to significant efficiency gains in improvement processes of crops. Nevertheless, due to the continuous evolution of the regulatory landscape, it is still unclear if they will be a useful option to conventional and transgenic breeding methods. However, gene editing works on a detailed knowledge of the molecular background of a particular trait which is oftentimes unavailable for a significant amount of traits of concern that are usually quantitative and difficult. But, modifications of bigger fractions of sequence data and various genes simultaneously such as haplotype blocks, including structural variants – for instance, chromosome segments – are emerging as suitable options. Tools for gene editing can offer more enhancement to the application of CWRs in the future.

5.3.2 Genomics and "Next-Generation Sequencing"

The knowledge of the CWR genome content and that of the associated crop can be particularly useful in understanding and finding the cryptic potential of the CWRs. The genome of every organism consists of some chromosomes which house DNA that contains the genes which the organism's cell utilizes to produce the macromolecules required for growth, to react to the environment, and for production of plant parts that ultimately develop to the crops that we consume. Within the DNA, genes are coded with a four-lettered chemical structure, and every letter is known as a single base; typically, a gene is denoted by about 1000 different bases. Generally, a plant's genome possesses several thousands of different genes set up in many chromosomes. The difference between a plant and a weed as well as the disparities between a plant and an advanced plant depends on their genic interaction and their reaction to environmental issues.

The present technique for sequencing plant genomes includes splitting the DNA of an organism into a larger amount of small sections, sequencing such small DNA fragments, and, then, trying to put them back bioinformatically through the use of overlapping segments to comprehend where each piece fits in. This technique is known as shotgun sequencing which was obtained from the technique created at the Sanger Institute. Shotgun sequencing was applied to sequence the human genome. In fact, most of the practices for experimenting crops along with CWR started from studies which were meant to apprehend and enhance human well-being.

The present "next-generation" shotgun sequencing (NGS) techniques have been used to create sequencing machines commercially from several technology companies such as Illumina, Roche Diagnostics, and Life Technologies. Primarily, the machines apply a series of reagents and electronic and photographic techniques to find the DNA sequences by sequencing several millions of nitrogenous bases within a very short time. Table 5.1 shows the improvement in sequencing CWR. There are several significant species which are yet to be sequenced. De novo assembly sequence of data produces the reference genome sequences which offer the ultimate resources for efficient application of a genetic resource in a CWR.

5.3.3 Geographical Information System for Natural Resource Management

The procedure and methodology advancements to assess genetic information have increased over the years, whereas the cost has declined significantly, and as data are mostly analyzed quickly, results are readily more obtainable by a larger number of researchers and organizations. Consequently, there has been an influx of genetic information across several fields such as epidemiology, ecology, and molecular biology, among others. To understand species distributions and range shifts as well as management of invasive species, pests, and diseases, the novel applications of such genetic information are vital. The integration of the genetic data with GIS and remote sensing shows a promising platform to assess the influence of biotic and

Table 5.1 Improvement on crop wild relatives' genome sequencing

CWR	Technology for sequencing	References
O. nivara, O. barthii (wild rice)	Illumina GAII, HiSeq2000	Zhang et al. (2014)
Musa balbisiana (wild banana)	Illumina HiSeq2000	Davey et al. (2013)
M. acuminata (wild banana)	Sanger, Illumina GAIIx, Roche 454	D'Hont et al. (2012)
Manihot esculenta ssp. *flabellifolia* (wild cassava)	Illumina GAII, HiSeq2000, Roche 454	Wang et al. (2014)
Glycine soja (wild soya bean)	Illumina GAII	Qi et al. (2014)
Cucumis sativus var. hardwickii (wild cucumber)	Illumina GAII, HiSeq2000	Qi et al. (2013)
Raphanus raphanistrum (wild radish)	Illumina GAII, Roche 454	Moghe et al. (2014)
Solanum pennellii (wild tomato)	Sanger, Illumina	Bolger et al. (2014)
Capsicum annuum var. glabriusculum (wild pepper chiltepin)	Illumina HiSeq2000	Qin et al. (2014)

abiotic factors in community structure (Manel et al. 2003). Thus, recently, a geographic information system is emerging as a vital aspect of natural resource management and conservation. Currently, this tool is commonly applied for mapping and conservation of genomic resources. Across the world, initiatives have been used to alleviate erosion of genetic resources (Hasan and Abdullah 2015). For instance, the global plan of action for FAnGR (farm animal genetic resources) specifically records the location of endangered species by using the GIS. The same technique can be applied to CWR resource conservation. The listed techniques include locations of the concerned populations and concentration (density) of population based on allelic richness and species diversity. For improvement of perennials, some of the well-characterized germplasm which breeders can utilize is mostly restricted to genetic diversity. Desirable traits such as disease resistance, rootstock attributes, and fruit quality can be found in wild relatives which are still widely untapped. Although genomics will continuously be applied in combination with conventional breeding techniques, however, it is a sophisticated tool to increase the speed and alleviate the cost of breeding in the period of harvesting potentials of wild relatives for the enhancement of perennial crops. In order to adapt the GIS model to back up the spatial representation of genome features, the Genome Spatial Information System (GenoSIS) was made available (Dolan et al. 2002). GenoSIS uses GIS performances for zooming and panning as well as highlighting the qualities of concern or filtering those with certain features; also, it uses standard cartographic methods to encode variables with the use of graphic symbols. Hence, several widely used genome browsers are available, which display sequence annotations and related qualities as horizontal information tracks which can be switched on and off based on the user's preference. Majority of genome browsers apply only simple keyword searches and restrict the show of detailed annotations to a single chromosomal region of the genome per time. Compared to the horizontally stacked data tracks paradigm utilized by several genome browsers, GenoSIS applies the idea of registered spatial layers made up of spatial objects to display diverse data. Combined with the basic keyword searches, GenoSIS gives support to complicated queries such as spatial queries and produces genome maps dynamically. Thirty percent of the Earth's surface is filled with forest trees, which are studied by scientists globally for conservation and economic benefits. Due to the onset of high-throughput technologies, massive phenotypic and genomic information have been produced for more than hundreds of species. These stagnant and long-lived individuals are the perfect models to examine population structure and environmental adaptability. Even though the comprehensive data is present, scientists are faced with integrating genotype, environment, and phenotype in a single location. To achieve this objective, CartograTree has been designed as an open repository and used as an open-source analytic framework for environment, genotypic, and phenotypic data for forest trees. Among its vital elements is the integration of geospatial data which permits the display of environmental layers and collation of environmental factors relative to the locations of georeferenced individuals. Presently, the Google Map API is used by the CartograTree to generate environmental data. Inherent restriction to this API is moving the new improvement with emphasis on performance to supply effective queries of several environmental factors. Improvements on genomic analysis will enable a complete genome sequencing of broader sets of germplasm and, consequently, leads to the examination of the complete genome, specific genes, or biosynthetic pathways, whereas the genome sequence and marker data will ensure the development of the NILs for the assessment of the impacts of particular areas/QTL from wild relative (Porch et al. 2013). Through genome-wide association studies, a combination of the phenotypic and allelic data points is generating the discovery of genetic loci which are connected with essential agronomic traits (D'Agostino and Tripodi 2017).

5.3.3.1 High-Throughput Phenotyping

There is an expectation on genomics and phenomics to revolutionize the plant breeding sector. Recently, a combination of these two concepts has started and is being facilitated via big data by improving next-generation sequencing and enhancements of artificial machine learning algorithms, artificial intelligence, and field-based high-throughput phenotyping (HTP) platforms. Recently, high-throughput phenotyping and genomic selections have been enticing crop breeding community interest from the private and public sectors globally. The two approaches are expected to change the prediction of complex traits such as growth, stress adaption, and yield. Scientists' ability to dissect the genetics of quantitative traits, most especially those concerned with yield and stress tolerance (such as heat tolerance, yield potential, increased drought, and nutrient efficiency, among others), has been limited by constraints in field phenotyping capability. The creation of impactful field-based high-throughput phenotyping platforms (HTPPs) is still a bottleneck for impending breeding advances. Nevertheless, improvement in sensors, high-performance computing, and aeronautics are creating solutions. Since wild relatives are an excellent source of breeding materials, high-throughput phenotyping/sensor-based field phenotyping may perform a significant role in unlocking the potentials of wild relatives.

5.3.3.2 GIS-Based Allelic Richness Mapping

Genetic diversity can be directly measured by allelic richness which is usually applied in experiments based on molecular markers with objectives to select population for conservation. An important tool for the recognition of the desired trait is the DNA marker. Geographic information system and integration of DNA fingerprinting enhanced allelic data, and global positioning system (GPS) technology assists us in recognition of wild relative diversity. This can be used in future breeding schemes. The application of GIS contributed considerably due to its ability to conserve, control, and incorporate molecular and spatial data and to obtain more in-depth data from pre-existing information. Also, GIS is beneficial for spatial evaluation, modeling, and data mapping and visualization. Furthermore, GIS may be applied to connect genetic data to available spatial information which is relevant to the CWR genetic resource conservation such as to show short-term challenges like accessibility as well as long-term difficulties like a change in climate.

5.3.3.3 Species Abundance Probability Mapping

The complex interactions happening over time and space between physical, socio-cultural, and biological processes shaped the existing distribution of species. By understanding the biogeographical distribution, the fundamental evolutionary processes, and the diversity patterns of living species, it is possible to conserve and manage genetic resources. Characterization of the species distribution can be performed by population density or occurrence probability, stated for every position in many spatial extents. By defining distributions using these two factors, we will be able to prevent ambiguity concerning the indices of occurrence and abundance generated by several presence-only algorithms. Possibly, abundance is the most vital ecological quantity needed to understand the population dynamics and decision-making in biological conservation and management, such as evaluating extinction chances of endangered species. While invasive species threatened species of interest, wild species often possess useful and desired traits, and therefore, they are significant genomic resources. The identification of possible species absence and presence can be assisted by GIS-based species abundance mapping. Through a combination of tools from landscape ecology, spatial statistics, and population genetics, landscape genetics will become a representative of powerful technique to assess geographic patterns of CWR genetic resources at the population level.

5.3.4 Bioinformatics for Genomic Researches

The word "bioinformatics" originates from the combination of the words "biology," "information," "technology," as well as "statistics." There are three main activities which constitute the field of bioinformatics: (1) the creation of statistical techniques and new algorithms for the evaluation of associations among massive genetic data, (2) utilization of these approaches to evaluate and interpret significant biological set of data, and (3) development of databases for an effective analysis of the concerned data. Evolution of bioinformatics can be attributed to the necessities of the new tools and methods in the handling of vast nucleotide sequence information. Most of the tools which can be applied for CWR's marker detection, development, association analysis, data storage, and gene prediction, among others, are summarized in the subsequent sections.

5.3.4.1 AutoSNP

The AutoSNP computer program recognizes SNPs and insertion/deletion (InDel) variation. SNP can be discovered in real time by using the SNPServer. With this it is possible to use BLAST, CAP3, and AutoSNP programs. The outcomes from the SNP discovery pipeline and the EST data source are saved in autoSNPdb. The database is freely available on http://autosnpdb.qfab.org.au/.

5.3.4.2 SNP2CAPS

The screens of SNP2CAPS several aligned sequences for restriction sites which are polymorphic, evaluate such sites and recognize these sites which are probably candidates for CAPS marker development. By visiting http://pgrc.ipk-gatersleben.de/snp2caps/, anyone can freely access the SNP2CAPS.

5.3.4.3 TASSEL

Population and family structure can be assessed through the software TASSEL. Linkage disequilibrium can be clearly visualized by this software. The TASSEL executables, as well as user manual, can be accessed freely by visiting http://www.maizegenetics.net/tassel (Bradbury et al. 2007).

5.3.4.4 STRUCTURE

The presence of two or more homogenous groups within one population can be detected by using the STRUCTURE software (ver. 2.3.4 in 2012) (Pritchard et al. 2000). A group of organisms which is at Hardy-Weinberg equilibrium for every random marker is considered as the homogenous group. The program is especially useful to infer the origin of the samples with unknown characteristics.

5.3.4.5 Microarray Software

A powerful precision experimental tool for the assessment of gene expression is called Microarray Software. This tool produces a massive amount of data which require well-designed and easy-to-navigate software for the collection, analysis, storage, and management. The tool can be used freely by scientists (http://www.tigr.org/software).

5.3.4.6 A C. Elegans Database (AceDB)

The system is now used to manage genomic databases of several entities. Also, it is designed to manage diverse data types as well as those related to maps (genetic and physical maps) and DNA sequences. Relevant documents of AceDB compatible with UNIX, Windows, and Macintosh environments are accessible at the website http://www.acedb.org/.

5.3.4.7 GenScan

GenScan (Burge and Karlin 1997) can predict the gene structure, poly-A signals, and promoter sides in genome sequences of target germplasm. The program server is accessible at http://genes.mit.edu/GENSCAN.html.

5.3.4.8 ClustalW

The ClustalW is user-friendly and one of the most used multiple sequence tool. The tool applies to check the similarity of the observed sequences. Every matching sequence pair is treated as one sequence, and the sequences collected again are checked two-by-two and aligned in pairs. A researcher may visit DDBJ (http://clustalw.ddbj.nig.ac.jp/) to access the ClustalW ver 2.1.

5.4 Genomic Resources and Limitations

Even though wild relatives possess immense potential for crop improvement, the first process to exploit them via genomic-based approach is to find markers connected to the beneficial phenotypes. Although the genetic divergence between the wild relatives and cultivated germplasm is exactly the reason why wild relatives provide such special characters, it tends to create challenges for the discovery of marker and breeding. In combination with the challenges potentially related to crossing more distant relatives, wild relatives tend to have greater diversity, and therefore, DNA sequencing and genotyping methods created for cultivated crops may not perform that effectively since the markers are constructed based on a reference genome.

A different option to a genotyping microarray is the next-generation DNA sequencing technologies (NGS) like genotyping-by-sequencing (GBS) (Elshire et al. 2011), and restriction-site-associated DNA (RAD) sequencing (Baird et al. 2008) which do not need markers to be found before genotyping. Nevertheless, in some situations, a reference genome is still applied for DNA sequence mapping which results from NGS and recognizes SNPs for mapping of association or GS.

There is an absence of reference genomes for wild relatives despite the proliferation of reference genome sequences. Between 2000 and 2014, more than a hundred plant genomes were sequenced; however, only 15 happened to be wild relatives (Michael and VanBuren 2015). Therefore, there is an obvious need for reference genomes in wild relatives to map sequence. It is possible that several benefits of wild germplasm are still yet to be discovered. Locations of where such germplasm can be found are one of the major challenges of improving characterization of wild germplasm. Most times, wild relatives have to be sampled from places which are difficult to reach, and therefore, the collection of new germplasm can be costly and time-consuming.

5.5 Conclusions

It is progressively important that we know how wild species are being applied to enhance crops and where to focus our efforts to surpass use constraints as the need to explore wild genetic diversity increases with challenges of climate change and reducing genetic base of crops. By looking at the present state of CWR closely, persisting challenges and considerable opportunities to exploit their values can be seen. Our ability to solve these issues and exploit the adaptive qualities present in CWR is predicated on enhanced coordination, distribution of information, and investigation in pre-breeding programs as well as the development of human capacity. Already, profits have been achieved, and momentum is rising. A clearer picture of where the bottleneck remains will help us realize the values of CWR for the enhancement of the adaptive capacity of crops.

References

Alercia A, Diulgheroff S, Mackay M (2015) FAO/bioversity multi-crop passport descriptors V. 2.1 [MCPD V. 2.1]

Alexandrov N, Tai S, Wang W, Mansueto L, Palis K, Fuentes RR, Ulat VJ, Chebotarov D, Zhang G, Li Z, Mauleon R (2014) SNP-Seek database of SNPs derived from 3000 rice genomes. Nucleic Acids Res 43(D1):D1023–D1027. https://doi.org/10.1093/nar/gku1039

Anderson JE, Kono TJ, Stupar RM, Kantar MB, Morrell PL (2016) Environmental association analyses identify candidates for abiotic stress tolerance in Glycine soja, the wild progenitor of cultivated soybeans. G3 Genes Genomes Genet 6(4):835–843

Baird NA, Etter PD, Atwood TS, Currey MC, Shiver AL, Lewis ZA, Selker EU, Cresko WA, Johnson EA (2008) Rapid SNP discovery and genetic mapping using sequenced RAD markers. PLoS One 3(10):e3376. https://doi.org/10.1371/journal.pone.0003376

Baute GJ, Kane NC, Grassa CJ, Lai Z, Rieseberg LH (2015) Genome scans reveal candidate domestication and improvement genes in cultivated sunflower, as well as post-domestication introgression with wild relatives. New Phytol 206(2):830–838

Bolger A, Scossa F, Bolger ME, Lanz C, Maumus F, Tohge T, Quesneville H, Alseekh S, Sørensen I, Lichtenstein G, Fich EA (2014) The genome of the stress-tolerant wild tomato species *Solanum pennellii*. Nat Genet 46(9):1034. https://doi.org/10.1038/ng.3046

Bradbury PJ, Zhang Z, Kroon DE, Casstevens TM, Ramdoss Y, Buckler ES (2007) TASSEL: software for association mapping of complex traits in diverse samples. Bioinformatics 23(19):2633–2635. https://doi.org/10.1093/bioinformatics/btm308

Burge C, Karlin S (1997) Prediction of complete gene structures in human genomic DNA. J Mol Biol 268(1):78–94

Clarke WE, Higgins EE, Plieske J, Wieseke R, Sidebottom C, Khedikar Y, Batley J, Edwards D, Meng J, Li R, Lawley CT (2016) A high-density SNP genotyping array for Brassica napus and its ancestral diploid species based on optimised selection of single-locus markers in the allotetraploid genome. Theor Appl Genet 129(10):1887–1899

D'Agostino N, Tripodi P (2017) NGS-based genotyping, high-throughput phenotyping and genome-wide association studies laid the foundations for next-generation breeding in horticultural crops. Diversity 9(3):38. https://doi.org/10.3390/d9030038

D'hont A, Denoeud F, Aury JM, Baurens FC, Carreel F, Garsmeur O, Noel B, Bocs S, Droc G, Rouard M, Da Silva C (2012) The banana (*Musa acuminata*) genome and the evolution of monocotyledonous plants. Nature 488(7410):213. https://doi.org/10.1038/nature11241

Davey MW, Gudimella R, Harikrishna JA, Sin LW, Khalid N, Keulemans J (2013) A draft Musa balbisiana genome sequence for molecular genetics in polyploid, inter-and intra-specific *Musa* hybrids. BMC Genom 14(1):683. https://doi.org/10.1186/1471-2164-14-683

Dolan ME, Holden C, Beard MK, Bult CJ (2002) GenoSIS: genome data interpretation using GIS. Proceedings of the Environmental Systems Research Institute

Elshire RJ, Glaubitz JC, Sun Q, Poland JA, Kawamoto K, Buckler ES, Mitchell SE (2011) A robust, simple genotyping-by-sequencing (GBS) approach for high diversity species. PLoS One 6(5):e19379. https://doi.org/10.1371/journal.pone.0019379

Guo B, Sleper DA, Beavis WD (2010) Nested association mapping for identification of functional markers. Genetics 186(1):373–383

Harlan JR, de JM W (1971) Towards a rational classification of cultivated plants. Taxon 1:509–517. https://doi.org/10.2307/1218252

Hasan M, Abdullah HM (2015) Plant genetic resources and traditional knowledge: emerging needs for conservation. In: Plant genetic resources and traditional knowledge for food security. Springer, Singapore, pp 105–120

Huda MN, Hasan M, Abdullah HM, Sarker U (2019) Spatial distribution and genetic diversity of wild date palm (Phoenix sylvestris) growing in coastal Bangladesh. Tree Genet Genom 15(1):3. https://doi.org/10.1007/s11295-018-1310-9

Huynh BL, Ehlers JD, Huang BE, Muñoz-Amatriaín M, Lonardi S, Santos JR, Ndeve A, Batieno BJ, Boukar O, Cisse N, Drabo I (2018) A multi-parent advanced generation inter-cross (MAGIC) population for genetic analysis and improvement of cowpea (Vigna unguiculata L. Walp.). Plant J 93(6):1129–1142

Iftekharuddaula KM, Newaz MA, Salam MA, Ahmed HU, Mahbub MA, Septiningsih EM, Collard BC, Sanchez DL, Pamplona AM, Mackill DJ (2011) Rapid and high-precision marker assisted backcrossing to introgress the SUB1 QTL into BR11, the rainfed lowland rice mega variety of Bangladesh. Euphytica 178(1):83–97

Li JY, Wang J, Zeigler RS (2014) The 3,000 rice genomes project: new opportunities and challenges for future rice research. GigaSci 3(1):8. https://doi.org/10.1186/2047-217X-3-8

Livaja M, Steinemann S, Schön CC (2016) Sunflower polygalacturonase-inhibiting proteins (HaPGIP) are genetically conserved in cultivated sunflower (*Helianthus annuus* L.) but diverse in wild species. Mol Breed 36(2):17. https://doi.org/10.1007/s11032-016-0444-4

Manel S, Schwartz MK, Luikart G, Taberlet P (2003) Landscape genetics: combining landscape ecology and population genetics. Trends Ecol Evol 18(4):189–197

Michael TP, VanBuren R (2015) Progress, challenges and the future of crop genomes. Curr Opin Plant Biol 24:71–81

Moghe GD, Hufnagel DE, Tang H, Xiao Y, Dworkin I, Town CD, Conner JK, Shiu SH (2014) Consequences of whole-genome triplication as revealed by comparative genomic analyses of the wild radish *Raphanus raphanistrum* and three other *Brassicaceae* species. Plant Cell 26(5):1925–1937

Porch T, Beaver J, Debouck D, Jackson S, Kelly J, Dempewolf H (2013) Use of wild relatives and closely related species to adapt common bean to climate change. Agronomy 3(2):433–461

Pritchard JK, Stephens M, Rosenberg NA, Donnelly P (2000) Association mapping in structured populations. Amer Human Genet 67(1):170–181

Qi J, Liu X, Shen D, Miao H, Xie B, Li X, Zeng P, Wang S, Shang Y, Gu X, Du Y (2013) A genomic variation map provides insights into the genetic basis of cucumber domestication and diversity. Nat Genet 45(12):1510. https://doi.org/10.1038/ng.2801

Qi X, Li MW, Xie M, Liu X, Ni M, Shao G, Song C, Yim AK, Tao Y, Wong FL, Isobe S (2014) Identification of a novel salt tolerance gene in wild soybean by whole-genome sequencing. Nat Commun 5:4340. https://doi.org/10.1038/ncomms5340

Qin C, Yu C, Shen Y, Fang X, Chen L, Min J, Cheng J, Zhao S, Xu M, Luo Y, Yang Y (2014) Whole-genome sequencing of cultivated and wild peppers provides insights into capsicum domestication and specialization. Proc Natl Acad Sci 111(14):5135–5140

Sehgal D, Vikram P, Sansaloni CP, Ortiz C, Saint Pierre C, Payne T, Ellis M, Amri A, Petroli CD, Wenzl P, Singh S (2015) Exploring and mobilizing the gene bank biodiversity for wheat improvement. PLoS One 10(7):e0132112. https://doi.org/10.1371/journal.pone.0132112

Song Q, Hyten DL, Jia G, Quigley CV, Fickus EW, Nelson RL, Cregan PB (2015) Fingerprinting soybean germplasm and its utility in genomic research. G3: genes, genomes. Genetics 5(10):1999–2006

Varshney RK (2016) Exciting journey of 10 years from genomes to fields and markets: some success stories of genomics-assisted breeding in chickpea, pigeonpea and groundnut. Plant Sci 242:98–107

Wang W, Feng B, Xiao J, Xia Z, Zhou X, Li P, Zhang W, Wang Y, Møller BL, Zhang P, Luo MC (2014) Cassava genome from a wild ancestor to cultivated varieties. Nat Commun 5:5110. https://doi.org/10.1038/ncomms6110

Zhang QJ, Zhu T, Xia EH, Shi C, Liu YL, Zhang Y, Liu Y, Jiang WK, Zhao YJ, Mao SY, Zhang LP (2014) Rapid diversification of five *Oryza* AA genomes associated with rice adaptation. Proc Natl Acad Sci 111(46):E4954–E4962. https://doi.org/10.1073/pnas.1418307111

Role of Gene Banks in Maintaining Crop Genetic Resources

6

Parvaze A. Sofi, Sajad M. Zargar, Rakeeb A. Mir, and Romesh Kumar Salgotra

Contents

Abstract

Crop genetic resources are foundations of human civilisations and are indispensable for continued existence of human race. They not only help to satiate hunger but also provide livelihood and cultural identities to humans inhabiting a specific habitat. They have shaped the cultural identities of the people as well as supported the subsistence and livelihood. They are the

P. A. Sofi (✉)
Dry Land Agricultural Research Station, SKUAST, Kashmir, Jammu and Kashmir, India

S. M. Zargar
Division of Plant Biotechnology, Sher-e-Kashmir University of Agricultural Sciences and Technology of Kashmir, Srinagar, Jammu and Kashmir, India

R. A. Mir
Department of Biotechnology, School of Biosciences and Biotechnology, BGSB University, Rajouri, Jammu and Kashmir, India

R. K. Salgotra
School of Biotechnology, Sher-e-Kashmir University of Agricultural Sciences and Technology of Jammu, Chatha, Jammu, Jammu and Kashmir, India

R. K. Salgotra, S. M. Zargar (eds.), *Rediscovery of Genetic and Genomic Resources for Future Food Security*,
https://doi.org/10.1007/978-981-15-0156-2_6

basic raw material for evolving new plant varieties and are a reservoir of genetic diversity. However, genetic resources are being lost at an increasingly alarming rate. The intensification of farming systems, change in food habits as well as emerging new breeding technologies such as genetic engineering have accelerated the pace of erosion. With the erosion of these resources, mankind loses the potential to adapt to new socio-economic and environmental conditions. Currently only 150 plant species are under extensive global cultivation, with 12 crop species providing 80% of the world's food. Given the landscape of modern agriculture in terms of its enhanced market and demand orientation, the observed changes are inevitable, notwithstanding the fact that this decline of crop diversity at both the inter- and intra-species has definite negative implications for the productivity, stability and resilience of the global farming system. Therefore, crop breeders have been collecting, maintaining and distributing crop diversity in ex situ system in fairly modern facilities called gene banks in an apparent move to prevent any catastrophic situation in the future. Currently there are about 1750 gene banks in the world maintaining millions of accessions of crops and their wild relatives. Nearly 7.4 million accessions (about 2 million of which are estimated to be unique) are now conserved ex situ in over 1750 facilities worldwide. The 16 centres of CGIAR (Consultative Group on International Agricultural Research) alone account for 30–60% of the unique accessions of the world's unique holdings. The gene banks are vital to national and international efforts to conserve and harness benefits of global crop diversity and make invaluable contributions to regional and global germplasm exchange. There are obviously distinct advantages to the crop improvement and research community as a whole in also establishing large, megadiverse, international ex situ collections, as they are maintained in fairly advanced facilities, under specialised staff, with continuous refinement of biodiversity management protocols as well as follow an efficient system for material distribution worldwide to different types of users in a safe manner.

Acronyms

CGIAR	Consultative Group for International Agricultural Research
CIAT	International Center for Tropical Agriculture
FAO	Food and Agriculture Organization
HYVs	High-yielding varieties
IPCC	Intergovernmental Panel on Climate Change
IRRI	International Rice Research Institute
PGR	Plant genetic resources
SDGs	Sustainable Development Goals

6.1 Introduction

The United Nations Sustainable Development Goal (SDG) Target 2.5

By 2020 maintain the genetic diversity of seeds, cultivated plants and farmed and domesticated animals and their related wild species, through soundly managed seed and plant banks at the national, regional and international levels. Later have been set to promote access to fair and equitable sharing of benefits arising from the utilization of genetic resources and associated traditional knowledge, as agreed by international bodies for sustainable development.

World population is expected to cross 9 billion in 2050. Feeding such huge population would require extraordinary increases in food production, that too under the scenario of deteriorating biophysical resource base as well as negative implications of climate change. The available evidence continues to signal a rise in world hunger. According to available data, the number of people who suffer from hunger has been growing over the past 3 years, returning to levels from a decade ago. The absolute number of people in the world affected by undernourishment, or chronic food deprivation, is now estimated to have increased from around 804 million in 2016 to nearly 821 million in 2017 (10.2% of global population). The situation is all the more worse in South America (8.7%) and Africa (29.8%). Another disturbing fact is that the decreasing trend recorded in undernourishment scenario in Asia until recently seems to be slowing down significantly. Without increased efforts, there is a potential risk that we may fall well short of achieving the SDGs (Sustainable Development Goals) of UN for eradicating hunger by 2030. Asia and Africa together are home to about 90% of people (686.8 million) facing severe food insecurity (Food and Agriculture Organization 2018). In addition, the worrisome picture is that the predicted climate change effects are expected to dent our efforts to be able to achieve targeted food demands, given the fact that marginal sections of society with limited adaptive capacities will be hit harder by climate change. In our efforts to be able to stand up to the challenge of sustained increase in food production and availability, crop genetic resources will be a major resource that can help us to create varieties that can yield better under pressing environmental conditions.

The term "genetic resources" was first used at a conference which took place under the International Biological Program (Hawkes 1997). According to the revised International Undertaking 1983 of the FAO, plant genetic resources were defined as the entire generative, vegetative and reproductive part of species with economical and/or social value, especially for the agriculture of the present and the future, with special emphasis on nutritional plants. Brockhaus and Oetmann (1996) defined plant genetic resource (PGR) as "plant material with a current or potential value for food, agriculture and forestry". A correlated definition that appends a value of aggregation to PGR was given by FAO (1999). According to this definition, plant genetic resources refer to the economic, scientific or societal value of the heritable materials contained within and among species. They include materials used in cytogenetic, evolutionary, physiological, biochemical, pathological or ecological research on one hand and accessions evaluated for their agronomic or breeding propensities on the other.

6.2 Problem Statement

Crop genetic resources are foundations of human civilisations and are indispensable for continued existence of human race. They are the basic raw material for evolving new plant varieties and are a reservoir of genetic diversity. Genetic adaptation and the rate of evolutionary response to selective forces give species the ability to adapt to changing environments, including new pests, diseases and new climatic conditions (Hammer and Teklu 2008). The traditional farmers have given us an invaluable bio-resource heritage of innumerable locally adapted genotypes of major and minor crops that evolved under natural and artificial selection pressures (Myers 1994). These genotypes have formed the broad genetic base of germplasm reservoir from which modern varieties have been developed. However, erosion of these adapted genetic resources along with accompanying practices and knowledge that farmers use to develop, utilise and conserve crop genetic resources has come up as a severe threat to the world's food security in the long term especially under looming threats of climate change. Such erosion has resulted in narrowing of genetic base which jeopardises the potential of species to persist in the face of environmental change as well as pathogens and weeds. With the advent of modern crop varieties bred for higher yield as well as other anthropogenic pressures on bio-resources, heterogeneity has been replaced with homogeneity, and as a result, the wealth is increasingly getting dwindled, putting our ecosystem at risk of vulnerability to climate change threats (Jarvis et al. 2008).

Crop genetic diversity is among the earth's most important bio-resources for food and agriculture. The domesticated crops and their wild relatives constitute the genetic diversity of the world's food production systems. We have and will continue to depend heavily in our efforts to ensure a sustained increase in food production for increasing human population. The situation is severely challenging in view of the fact that projected implications of climate change will significantly hamper our progress as well as our ability to achieve the Sustainable Development Goals of UN for availability and access of food for all. Crop genetic diversity is undoubtedly our key resource to identify the novel variation that can help us in developing climate smart varieties. However, genetic resources are being lost at an increasingly alarming rate. The intensification of farming systems, change in food habits as well as emerging new breeding technologies such as genetic engineering have accelerated the pace of genetic erosion. With the erosion of these resources, mankind loses the potential to adapt to new socio-economic and environmental conditions. Modern cultivars developed through conventional and molecular plant breeding in major commercial crops have an extremely narrow genetic base that potentially increases their vulnerability to environmental threats (Fu 2015). Although a complete genetic wipe out of centres of diversity has not yet been reported, modern cultivars have undoubtedly replaced many traditional varieties, thereby decreasing diversity (Van de Wouw et al. 2010). Harlan (1970), one of the early contributors to the science of plant genetic resources, cautioned as early as 1970 that the varietal diversity of the plants that feed and clothe the world is slipping away, despite the hard fact that the human race simply cannot afford to lose it (Box 6.1).

Box 6.1: Plant Breeding and Crop Diversity: Key Facts

- *Plant breeding, undeniably, has been the driving force of agricultural transformations with invariably significant impacts under all management systems.*
- *The modern crop varieties developed by plant breeding have spread throughout the world post-Green Revolution that has resulted in quantum jumps in yield and returns to the farmers.*
- *While as the achievements in crop improvement driven by plant breeding efforts have been tremendous, the food and nutritional needs of many poor farming households especially in marginal farming systems have remained unaddressed, and hunger and malnutrition are still the daunting challenges.*

6.3 Loss of Diversity: The Modernisation Bottleneck

The modernisation bottleneck refers to progressive loss of crop diversity and narrowing of genetic base due to release and large-scale adoption of modern high-yielding varieties that are fairly uniform. The rapid replacement of heterogeneous landraces with modern cultivars after Green Revolution has seen large chunks of area being cultivated by only few varieties (IR-8 and IR-36 in rice, PBW-343 in wheat as example). The replacement rate has been very fast in North America and several European nations (landraces have become absent for many crops), whereas it has been slow in less developed nations. While some studies have reported a reduction in both richness of alleles (Hao et al. 2006; Thomson et al. 2007), other studies have reported no difference in the level of diversity between landraces and cultivars (Hyten et al. 2006). This differential response is largely due to differential transition period of shift from landraces to modern varieties. On a desirable note, the observed loss of alleles of landraces when compared to modern varieties might be partly due to the elimination of deleterious or unwanted alleles through purging under plant breeding. This in turn may lead to reduction in allelic diversity on account of selection rather than erosion of possibly useful genetic variation (Allard 1996). The more visible phase of genetic erosion occurs after the complete replacement of landraces by modern varieties, especially when they are similar. Even in such case, contrasting reports have been documented that showed a decrease in diversity over time (Malysheva-Otto et al. 2007), while others reported an increase in diversity (Fu 2007; White et al. 2008). An interesting observation after Green Revolution in India witnessed initial decline in crop diversity followed by a recovery phase, possibly on account of increase in varietal development and release as well as increased use of diverse germplasm. The diversity patterns are also determined by the differential varietal turnover in various crops. In case of crops where varietal turnover is low, the diversity levels may decrease upon introduction of HYVs over large areas, whereas in case of crops with a high varietal turnover and a good adoption rates by farmers demonstrate better diversity patterns (Van de Wouw et al. 2010).

6.4 Value of Genetic Resources

Human civilisations have been greatly shaped by the pattern of the domestication, conservation and use of plant species for food, feed, clothing as well as livelihood. Farmers have used the genetic variation in wild and cultivated plants to develop crops that could potentially satiate human hunger as well as meet the requirements of a civilised life. Ever since the beginning of agriculture about 10,000 years ago, nearly about 7000 plant species out of the projected number of 300,000 have been used as crops. The Food and Agriculture Organization estimates that currently only 150 plant species are under extensive global cultivation, with 12 crop species providing 80% of the world's food (FAO 2018). Although modern agriculture feeds more people on less land than ever before, it also results in high genetic uniformity by planting large areas of the same species with genetically similar cultivars, making the entire crops highly vulnerable to biotic and abiotic stresses. Many of these threatened plant species may harbour unknown attributes that could benefit agriculture or the environment. With climate change, some plant populations may disappear completely. Expanding conservation to preserve seeds and genetic resources through gene banks has become critically important, yet it is only one component of a larger conservation plan aimed at preserving plants in situ on farms, in reserves and in their native habitats. Though the challenge is great, there is more effort and interest than ever (Boxes 6.2 and 6.3).

Box 6.2: Some Facts About Crop and Animal Diversity and Utilisation (Source: FAO 1999, World Conservation Union)

- Since the 1900s, some 75% of plant genetic diversity has been lost as farmers worldwide have left their multiple local varieties and landraces for genetically uniform, high-yielding varieties.
- 30% of livestock breeds are at risk of extinction; six breeds are lost each month.
- Today, 75% of the world's food is generated from only 12 plants and 5 animal species.

(continued)

Box 6.2 (continued)

- Only 4% of the known edible plant species, only 150 to 200 are used by humans.
- Only three – rice, maize and wheat – contribute nearly 60% of calories and proteins obtained by humans from plants.
- One in every eight plant species in the world, and one in three in the USA, is threatened with extinction, resulting largely from the development of rural land and invasion of alien species.
- Animals provide some 30% of human requirements for food and agriculture, and 12% of the world's population live almost entirely on products from ruminants.
- More than 90% of crop varieties have disappeared from farmers' fields.
- Half of the breeds of many domestic animals have been lost.
- All the world's 17 main fishing grounds are now being fished at or above their sustainable limits.

Box 6.3: Role of Genetic Resources

Sustainable Agriculture: Diversification of farming systems is key to ensuring sustainability with minimal environmental impacts. The traditional farming systems are no longer sustainable in the context of climate and environment. There is a need to find sustainable and eco-friendly cropping options to diversify the traditional cropping patterns globally.

Food Security: Food security is complex to define and involves the production, processing of nutritionally acceptable food as well as the access by individuals to maintain an active and healthy life. Crop diversity is pivotal to ensuring food security. Our resolve to ensure food security and end hunger is one of the greatest challenges that has been uncompromising but has been severely dent by unprecedented rise in human population and deteriorating bio-resources and environment. Diversity of genetic resources in gene banks and access to all helps to ensure a secure food supply as it provides the raw genetic material to breed for a more nutritious and varied food supply and improved access of the poor to more affordable and healthier food.

Climate Resilience: Climate change is more than obvious now and will hit us hard, especially the small and marginal farming systems that lack adaptive capacities, which in turn would put additional pressure on our ability to ensure food security. This assumes significant importance in view of the projections of the International Panel on Climate Change (IPCC) that indicate that warming will take place over the next several decades irrespective of any action we

(continued)

Box 6.2 (continued)

take today. The same models indicate that agricultural landscapes will be dramatically different from those which prevail today. In order to adapt to such forced changes, the value of climate-resilient varieties is widely recognised in the medium- and long-term food security imperatives. The development of crop varieties that can cope with heat, drought, flood and other weather extremes may well be the most important step we can take to adapt to climate change.

Environmental Protection: Using the global crop diversity, varieties that are resistant to pests and diseases can be developed and may help reduce the need to apply harmful pesticides. Similarly, vigorously growing crop varieties can better compete with weeds, reducing the need for applying herbicides. Drought-resistant plants can help save water by reducing the need for irrigation. Deeper rooting varieties can help stabilise soils; and varieties that are more efficient in their use of nutrients require less fertiliser. On a larger scale, productive farming systems will increase resource-use efficiency on farms (of land, labour, etc.) and progressively reduce the need to deforestation or clear fragile lands to create more farmland (FAO 2003).

Nutritional Security: Crop diversity not only helps us achieve a sustainable supply of food in sufficient quantities but also plays a major role in ensuring its quality. In fact, the observed dietary diversity is a direct result of crop diversity. New varieties have been developed with improved nutritional quality, with higher levels of essential nutritional ingredients such as proteins, vitamins, minerals and iron (golden rice, quality protein maize, iron- and zinc-biofortified crops) as well as reduced anti-nutritional or toxic factors (low phytate pulses, double zero mustard, zero aflatoxin groundnut, low HCN sorghum, low lathyrin pea).

Reducing Poverty: Farming is, and will continue to remain as, the backbone of economy, even in the most industrialised nations as it not only provides food, fodder and clothing but also a significant proportion of industrial raw materials. In fact, it is the major driver of economic progress. Growth in agricultural sector is going to ensure economic upliftment of majority of economically weaker sections as it is one of the major sources of income and livelihood for such sections. The growth in agriculture driven by harnessing genetic resources has been amply evidenced in the past and is going to be more dependent on them given the varied and complex challenges facing agriculture. There are sufficient examples from Africa where drought-tolerant crops such as maize, beans and rice have improved lives of poor farmers.

6.5 Need for Gene Banks

As farmers specialise and intensify agricultural production, there is a continued decrease not only in the number of crops grown but also in the diversity within the range of varieties in the field. Given the landscape of modern agriculture in terms of its enhanced market and demand orientation, the observed changes are inevitable, notwithstanding the fact that this decline of crop diversity at both the inter- and intra-species has definite negative implications for the productivity, stability and resilience of the global farming system (Khoury et al. 2014). In light of the above situation, crop breeders have been collecting, maintaining and distributing crop diversity in ex situ system in fairly modern facilities as gene banks in an apparent move to prevent any catastrophic situation in the future. Global and national gene banks have collected, characterised and distributed millions of accessions of traditional crop varieties and related wild species from diverse, remote, dispersed locations into gene banks. Such a framework provides a means to conserve and make accessible the unique diversity for the short- and long-term breeding goals driven by the requirements and situations imposed by changing climate. The distribution of accessions for use in national and international breeding programmes is equally imperative for harnessing the value of genetic resources. The publicly funded gene banks at national and international level under CGIAR system and NARS have been at the forefront of making available valuable genetic resources that have resulted in trait-specific cultivars in different crops.

Currently there are about 1750 gene banks in the world maintaining millions of accessions of crops and their wild relatives under short-, medium- and long-term storage (Table 6.1), with only a few countries without a national gene bank, and the value placed by governments and research organisations on crop diversity is reflected in the dramatic increase in the number of collections and of the accessions hold by gene banks in the last 30 years. As per the estimates of FAO, nearly 7.4 million accessions (about 2 million of which are estimated to be unique) are now conserved ex situ in over 1750 facilities worldwide (FAO 2010). The 16 centres of CGIAR alone account for 30–60% of the unique accessions of the world's unique holdings (Hoisington et al. 1999). The work of such gene banks is vital to national and international efforts to conserve and harness benefits of global crop diversity and makes invaluable contributions to regional and global germplasm exchange. There are obviously distinct advantages to the crop improvement and research community as a whole in also establishing large, megadiverse, international ex situ collections, as they are maintained in fairly advanced facilities, under specialised staff, with continuous refinement of biodiversity management protocols as well as follow an efficient system for material distribution worldwide to different types of users in a safe manner. Between 2012 and 2017, CGIAR gene banks distributed number of samples all over the world. Since the distribution has been highly diverse, it is challenging to trace the utilisation and subsequent economic impact of these accessions. An IRRI report in 2012 suggested that the 12 most popular IRRI gene bank accessions have been used in more than 1000 breeding crosses each, and of the 4317 released rice varieties, 90% of the non-IRRI varieties and 100% of the IRRI varieties had at least

Table 6.1 Major gene banks and their holdings

Gene bank	Year of establishment	Accessions
Svalbard Global Seed Vault, Norway	2008	983,524
IRRI, Philippines	1960	130,139
CIMMYT, Mexico	1943	182,922
IPK, Germany	1948	151,002
NORDIC, Sweden	2008	32,000
ICRISAT, India	1972	156,313
ICARDA, Syria	1975	148,000
CIAT, Columbia	1962	66,787
CIP	1971	13,911
IITA, Nigeria	1967	39,756
AVRDC, Taiwan	1984	67,817
ICRAF	1978	2448
ILRI	1994	13,470
WARDA	1971	17,440
NBPGR, New Delhi	1976	440,419
USDA NPGS	1862	27,000
ECPGR	1980	60,000
National Seed Storage Laboratory, Colorado	1958	475,000
Vavilov Institute of Plant Industry, Saint. Pitttsburg	1887	325,000
Institute of Crop Germplasm Resources, Beijing	1986	355,000
Genetic Resources Division, ABI, SUWEAN Korea	NA	120,000
EMBRAPA, Brazil	1973	150,000

one accession from the gene bank as its foundation parents (IRRI 2012). The annual rate of distribution between 1985 and 2009 for nine CGIAR gene banks was 39,970 samples. Between 2012 and 2014, the same nine gene banks reported an annual average distribution of 91,973 samples (CGIAR 2019).

6.6 Cost of Conservation in Gene Banks

The ex situ conservation and dissemination of crop genetic resources provide enormous benefits to the world's population by supplying a diverse range of germplasm to crop breeding programmes worldwide. Yet such an effort comes at a considerable cost, for which sustained funding is required. A complete understanding of operating costs is important for the effective management of the gene bank. Making the cost structure public is also important for other gene banks, as they can use it as a benchmark to compare with their own cost structure (Fig. 6.1). Using data from the five centres of the Consultative Group on International Agricultural Research (CGIAR) containing 87% of the accessions held by the

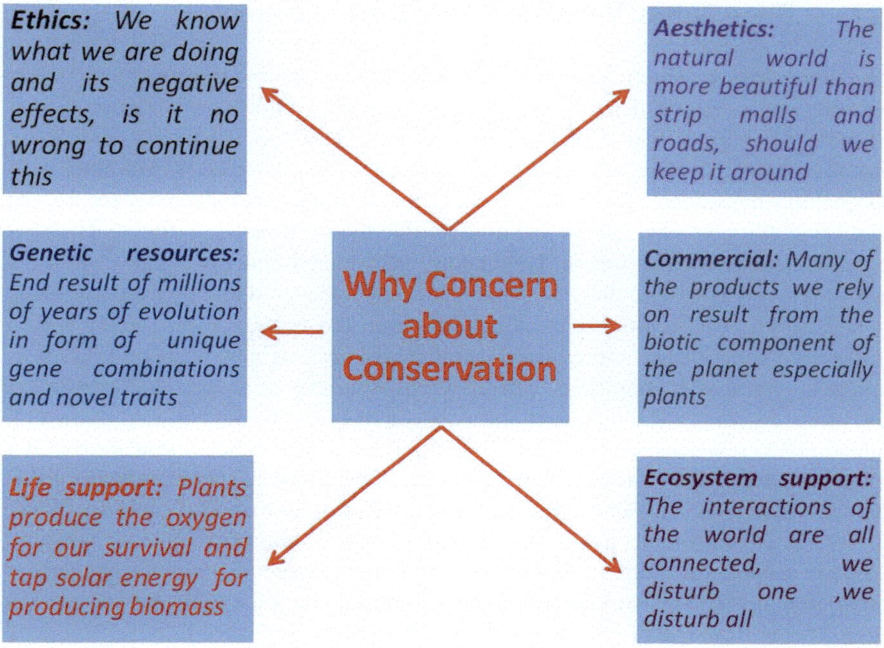

Fig. 6.1 Rationale for conservation of crop genetic resources

Table 6.2 Cost of maintaining germplasm in major gene banks

S.No	Gene bank	Cost per accession (US$)
1	IRRI	0.155
2	ICRISAT	0.131
3	ICARDA	0.103
4	CIMMYT	0.127
5	CIAT	0.150
6	AVRDC	0.103

Modified from Schreinemachers et al. (2014)

CGIAR, they estimated that the cost of storing an accession for 1 year is about US$ 1.50 for most crops. They also estimated that the cost of conserving and distributing the CGIAR collection is US$ 5.7 million per year and suggested that an endowment of US$ 149 million invested at an average annual interest rate of 4% would probably be sufficient to conserve the germplasm collections in perpetuity (Koo et al. 2004) (Table 6.2).

6.7 Gene Banks: Value of Return on Investment

The worth of crop genetic resources in terms of the economic impact is universally accepted, even though there are no concrete quantified values available on account of multiple parameters of direct and indirect impacts (Crop Trust 2015). However, two broader parameters, viz. use value and option value, have been invariably used to quantify the impact. *Use value* refers to the ability of crop diversity to provide yield (including yield stability) and non-yield (e.g. nutritional, environmental) benefits, while *option value* refers to potentially valuable but unknown genes and traits within a crop diversity collection, which may be discovered and provide use value in the future. This option value is equated with the insurance provided by crop diversity against future unpredictable challenges, such as new pests and diseases, and evolving market conditions. There are large number of documented reports on the monetary value of the current and past use of crop diversity to improve yield, nutrition and resistance to pests and diseases. Raitzer and Kelley (2008) provided a meta-analysis of the relative benefits and costs of CGIAR research investment and reported that aggregate benefit-cost ratios range from 1.94 to 17.26. Table 6.3 shows the estimated net benefits from various studies and substantiates the premise that the CGIAR research program driven by extensive use of genetic resources has been a productive investment. Similarly, Table 6.4 indicates the potential economic benefits accruing from use of crop wild relatives in various crops.

Table 6.3 Estimated benefits of crop diversity through improved varieties

S. no	Crop	Gene bank	Benefit (million US$)	References
1	Rice	IRRI	4310	Hossain et al. (2003)
2	Rice	CIAT	8280	Sanint and Wood (1998)
3	Rice	WARDA	150	Dalton and Guei (2003)
4	Wheat	Multiple	9750	Byerlee and Traxler (1995)
5	Barley	Multiple	330	Aw-Hassan et al. (2003)
6	Beans	Multiple	590	Johnson et al. (2003a)
7	Cassava	Multiple	230	Johnson et al. (2003b)
8	Maize	Multiple	440	Morris (2002)

Table 6.4 Estimated benefits of crop wild relatives through improved varieties

S. no.	Crop	Benefit (US$)	References
1	Wheat	107 million	Witt (1985)
2	Coffee	1.66 billion	Hein and Gatzweiler (2006)
3	Sunflower	23–392 million	Hunter and Heywood (2011)
4	Tomato	255 million	Hunter and Heywood (2011)

6.8 Conclusion

Genetic resources are precious and are our major armours to fight the unprecedented challenges facing agriculture including hunger, malnutrition, food and nutritional security for underprivileged classes as well as the looming threats of climate change. In our efforts to be climate change ready, we need to mine the vital traits from the gene banks to develop climate-resilient varieties. There are polarised debates over the sustainability of ex situ conservation through gene banks, but the fact of the matter is that gene banks represent institutionalised efforts to ensure safety of the genetic resources and prevent their erosion on farmlands on account of invasion of modern varieties. An integrated approach would be to link gene banks to mainstream breeding programmes as well as on farm conservation programmes to reduce the cost as well as ensure conservation by use.

References

Allard RW (1996) Genetic basis of the evolution of adaptedness in plants. Euphytica 92(1–2):1–11

Aw-Hassan A, Shideed K, Ceccarelli S et al (2003) The impact of international and national investment in barley germplasm improvement in the developing countries. Crop variety improvement and its effect on productivity: the impact of international agricultural research. CAB International, Wallingford, pp 241–246

Brockhaus R, Oetmann A (1996) Aspects of the documentation of in situ conservation measures of genetic resources. Bulletin des Ressources Phytogenetiques (IPGRI/FAO); Noticiario de Recursos Fitogeneticos (IPGRI/FAO)

Byerlee D, Traxler G (1995) National and international wheat improvement research in the post-green revolution period: evolution and impacts. Am J Agric Econ 77:268–278. https://doi.org/10.2307/1243537

CGIAR (2019). www.cgiar.org/research/program-platform/genebank-platform

Crop Trust (2015) Securing crop diversity for sustainable development. Global Crop Diversity Trust, Bonn

Dalton TJ, Guei RG (2003) Ecological diversity and rice varietal improvement in West Africa. Crop variety improvement and its effect on productivity: the impact of international agricultural research. CAB International, Wallingford, pp 109–133

FAO (1999) Agricultural biodiversity, multifunctional character of agriculture and land conference, Background Paper 1. Maastricht, Netherlands. September 1999

FAO (2003) World agriculture: towards 2015/2030. Food and Agriculture Organization of the United Nations

FAO (2010) The second report on the state of the world's plant genetic resources for food and agriculture. Rome

FAO (2018) The state of food security and nutrition in the world: building climate resilience for Food Security and Nutrition. Food and Agriculture Organization of the United Nations

Fu YB (2007) Understanding crop genetic diversity under modern plant breeding. Theor Appl Genet 128(11):2131–2142

Hammer K, Teklu Y (2008) Plant genetic resources: selected issues from genetic erosion to genetic engineering. J Agric Rural Dev Trop Subtrop (JARTS) 109(1):15–50

Hao C, Wang L, Zhang X et al (2006) Genetic diversity in Chinese modern wheat varieties revealed by microsatellite markers. Sci China, Series C: Life Sci 49:218–226

Harlan JR (1970) Evolution of cultivated plants. In: Frankel OH, Bennett E (eds) Genetic resources in plants – IBP handbook no 11. International Biological Programme, London, pp 19–32

Hawkes JG (1997) Distinguished economic botanist. Econ Bot 51(1):2–5
Hein L, Gatzweiler F (2006) The economic value of coffee (*Coffea arabica*) genetic resources. Ecol Econ 60:176–185. https://doi.org/10.1016/j.ecolecon.2005.11.022
Hoisington D, Khairallah M, Reeves T et al (1999) Plant genetic resources: what can they contribute toward increased crop productivity? Proc Natl Acad Sci 96(11):5937–5943
Hossain M, Gollin D, Cabanilla V et al (2003) International research and genetic improvement in rice: evidence from Asia and Latin America, in: crop variety improvement and its effect on productivity: the impact of international agricultural research. CABI Publishing, Wallingford
Hunter D, Heywood V (2011) Crop wild relatives: a manual of *in situ* conservation. Bioversity International, Rome, Italy
Hyten DL, Song QJ, Zhu YL et al (2006) Impacts of genetic bottlenecks on soybean genome diversity. Proc Natl Acad Sci U S A 103:16666–16671
IRRI (2012) IRRI Annual Report 2012. www.irri.org/resources/publications/annual-reports/annual-report-2012
Jarvis A, Upadhyaya HD, Gowda CLL et al (2008) Climate change and its effect on conservation and use of plant genetic resources for food and agriculture and associated biodiversity for food security. FAO document pp 26
Johnson NL, Manyong VM, Dixon AG et al (2003a) The impact of IARC genetic improvement programmes on cassava. In: Crop variety improvement and its effect on productivity: the impact of international agricultural research. CABI Publishing, Wallingford
Johnson NL, Pachico D, Voysest O (2003b) The distribution of benefits from public international germplasm banks: the case of beans in Latin America. Returns Invest Plant Genet Resour Conserv Crop Improv Res 29:277–286
Khoury CK, Bjorkman AD, Dempewolf H et al (2014) Increasing homogeneity in global food supplies and the implications for food security. Proc Natl Acad Sci 111:4001–4006
Koo B, Pardey PG, Wright BD et al (2004) Saving seeds: the economics of con- serving crop genetic resources ex-situ in the future harvest centres of the CGIAR. CABI Publishing, Oxfordshire
Malysheva-Otto L, Ganal MW, Law JR et al (2007) Temporal trends of genetic diversity in European barley cultivars (*Hordeum vulgare* L.). Mol Breed 20(4):309–322
Morris ML (2002) Impacts of international maize breeding research in developing countries, 1966–98, CIMMYT
Myers N (1994) Protected areas – protected from a greater what? Biodivers Conserv 3:411–418
Raitzer DA, Kelley TG (2008) Benefit–cost meta-analysis of investment in the International Agricultural Research Centers of the CGIAR. Agric Syst 96:108–123
Sanint LR, Wood S (1998) Impact of rice research in Latin America and the Caribbean during the past three decades. In: Impact of rice research. International Rice Research Institute, Manila
Schreinemachers P, Ebert AW, Wu M (2014) Costing the ex situ conservation of plant genetic resources at AVRDC – The World Vegetable Center. Genet Resour Crop Evol 61:757–773
Thomson MJ, Septiningsih EM, Suwardjo F et al (2007) Genetic diversity analysis of traditional and improved Indonesian rice (*Oryza sativa* L.) germplasm using microsatellite markers. Theor Appl Genet 114:559–568
Van de Wouw M, Kik C, van Hintum T et al (2010) Genetic erosion in crops: concept, research results and challenges. Plant Genet Resour 8(1):1–15
White J, Law JR, MacKay I et al (2008) The genetic diversity of UK, US and Australian cultivars of *Triticum aestivum* measured by DArT markers and considered by genome. Theor Appl Genet 116(3):439–453
Witt S (1985) Biotechnology and genetic diversity. Briefbook No. 2. California Agricultural Lands Project

Exploring Genetic Resources for Identification of Potential Novel Genes for Crop Improvement

7

Pankaj Pandotra, Parshant Bakshi, Anil Kumar Singh, and Suphla Gupta

Contents

Abstract

The emerging problem of variation in climate change with diverse environmental stress conditions has made the present situation critical for the production of food. The emitted harmful substances cause variation in the proportion of atmospheric CO_2 level, which affects the level of diversity in the biosphere. The unchecked emission of polluted harmful gases to the ecosystem also changes the

P. Pandotra (✉) · P. Bakshi
Division of Fruit Science, Sher-e-Kashmir University of Agricultural Sciences and Technology of Jammu, Chatha, Jammu, Jammu and Kashmir, India

A. K. Singh
School of Biotechnology, Sher-e-Kashmir University of Agricultural Sciences and Technology of Jammu, (SKUAST-J), Chatha, Jammu and Kashmir, India

S. Gupta
Plant Biotechnology Division, Council for Scientific and Industrial Research (CSIR), Indian Institute of Integrated Medicine (IIIM), Jammu, India

R. K. Salgotra, S. M. Zargar (eds.), *Rediscovery of Genetic and Genomic Resources for Future Food Security*,
https://doi.org/10.1007/978-981-15-0156-2_7

proportion of atmospheric CO_2 level, which is a threat to our domesticated crop species by emergence of the growth of new level of invasive weeds. Over the last three decades, a significant progress has been seen in crop genetic improvement programmes with increased tolerance/resistance to environmental stresses to enhance the crop yield by effort of the plant breeders. On the other hand, the crop wild relatives (CWRs) represent the exploring genetic resources for identification of potential novel genes for crop improvement because it has been grown in their own respective environments for many years and stored higher level of genetic diversity. In this chapter, we stress the importance of explored genetic resources in the form of CWRs for identification of potential novel genes involved in biotic and abiotic stress resistance and overall yield performance.

Acronyms

CIAT	International Centre for Tropical Agriculture
CWRs	Crop wild relatives
GP	Gene pool
GWAS	Genome-wide association studies
MAB	Marker-assisted breeding
MeAB	Metabolomics-assisted breeding
NGS	Next generation sequencing
PCR	Polymerase chain reaction
RAPD	Random amplified polymorphic DNA
RFLP	Restriction fragment length polymorphism
RILs	Recombinant inbred lines
SCAR	Sequence characterized amplified region
SNPs	Single nucleotide polymorphism
SSR	Simple sequence repeats
TG	Taxon group

7.1 Introduction

Rising global climate change, trends in human activities, exploding population growth, plant colonization and upsurge demand for land, water and energy are the key challenges confronting food production in the twenty-first century (Godfray et al. 2010). Due to global warming, we have seen that broad environmental effects such as abrupt changes in the patterns of drought and salinity and the appearance of new species of pests and outbreak of diseases directly affect the plant growth and its yield (Tester and Langridge 2010). The unchecked emission of polluted harmful gases to the ecosystem also changes the proportion of atmospheric CO_2 level, which affects our biodiversity assets and threatens our domesticated crop species by emergence of the growth of new level of invasive weeds (Raizada et al. 2009). The world population figure is estimated to touch 9 billion by the year 2050, and it has become a herculean task for the top leading agriculture production nations to meet the feeding demand for fast-growing population (Dempewolf et al. 2014; Kastner et al. 2012; Khoury et al. 2014). Over the last three decades, significant progress has been seen in crop genetic improvement programmes with increased tolerance/resistance to environmental stresses to enhance the crop yield by effort of the plant breeders. But to meet the future challenges for demand of more food security in variable climate conditions and for the growing population, attention needs to be paid to explore new genetic resources for identification of potential novel genes for crop improvement.

7.2 Irreparable Loss of Genetic Diversity in Domesticated Descendants

The beginning of the human civilization was associated with plant domestication process in which humans used wild species of different crops grown from generation to generation to fulfil their basic needs due to expression of those gene linked traits that adapted well in the local climate region. Early farmers practised limited numbers of individuals of progenitor species related to selected traits like yield, harvesting and edibility (Hua et al. 2015; Konishi et al. 2006), which later on generated strong genetic bottleneck conditions that caused reduction in the heterozygosity level of their genome (Buckler et al. 2001; Meyer et al. 2012; Miller and Gross 2011). It revealed that more than 50% of genetic variation in cultivated soybean had been lost; 2–4% of maize genes experienced artificial selection, and genetic diversity had been significantly reduced in cultivated rice compared with its wild counterpart due to reduction of the genetic diversity by the domestication process (Hyten et al. 2006; Xu et al. 2012). In addition, this domestication process also causes the reduction of diversity in closely linked loci called selective sweep that has been reported in various crops such as promoter region of axillary-branch formation-related gene *tb1* in maize (Clark et al. 2004), a 600-kb sweep at kernel-related gene *Y1* in maize (Palaisa et al. 2004) and a 260-kb sweep at amylose-related gene *waxy* in rice (Sweeney and Mccouch 2007).

Due to reduction of genetic diversity at key loci in modern crops, there will be meagre scope for developing novel varieties with improved traits due to lower content of gene pool in their genome. Besides this, the significant involvement of only 30 crop species supplied 95% of total food production, in which four main crops such as rice, wheat, maize and potato hold leading contribution. To provide solution to the problem of food security, it is better to diversify agriculture domesticating by adopting other larger number of untapped resources of plant species. But the big question is whether consumers will accept it or not. It needs perfect, optimized production systems for introducing new crops. Another way is to explore genetic resources in the form of crop wild relatives (CWRs) for identification of novel genes that could be utilized in crop improvement programmes (Warschefsky et al. 2014).

CWRs reflecting richest source of genetic diversity at both the population level and the individual level as compared to their domesticated descendants counter the challenges associated with food security problem (McKey et al. 2010).

7.3 Primary Source of Diversity

The explored genetic resource in CWRs represents the primary source of diversity in wild relatives that inherited to the domesticated crop plants. CWR is defined in two ways; one is the gene pool (GP) concept in which CWRs classify into groups (GP-1–GP-3) based on the relative transfer of gene with cultivated crops. Gene exchange occurs relatively easily between primary (GP-1) and secondary GPs (GP-2) by crossing (and fertile hybrids can be produced), whereas gene transfer between primary and tertiary (GP-3) groups is usually difficult. Even though the CWRs that have been used in crop improvement mostly belong to GP-1 (Munns et al. 2012) or GP-2 categories (Fetch et al. 2009; Saintenac et al. 2013), there are some examples where useful alleles from distant wild relatives, such as GP-3 plants, have been successfully transferred for crop improvement (Abedinia et al. 2000; Marais et al. 2003).

The second concept for CWRs is that of the taxon group (TG), a system that is based on the ranking of the taxonomic hierarchy to crops (Maxted et al. 2006). A TG may include a wide range of wild species that may be evolutionarily closely or distantly related to crop species within the same genus. With this concept, CWRs were defined in a range from TG1 (same species as the crop) to TG4 (different species within the same genus as the crop).

7.4 Distribution Pattern of Explored Genetic Resources and Their Conservation Strategy

The explored genetic resources of CWRs, approximately 50,000–60,000 species, are widely distributed in all continents, in which only 10,739 species (or even more) have been feeding the whole world as food security (Castaneda-Alvarez et al. 2016; Larson et al. 2014; Maxted and Kell 2009). The proper utilization of genetic diversity possessed in CWR resources for identification of potential novel genes in plant breeding lags behind. The alarming rise of global climate temperature and unchecked human population growth cause threat to most of the CWRs and place them at the danger of near extinction (Ford-Lloyd et al. 2011). The prioritized conservation steps should be taken in those geographical areas where the most critical collecting gaps occur. It includes the Mediterranean and the Near East, Western and Southern Europe, South-East and East Asia and South America. The joint conservation efforts of organizations like Global Crop Diversity Trust (Crop Trust), the International Centre for Tropical Agriculture (CIAT) and the Royal Botanic Gardens (Kew), with national and international agricultural research institutes, have started to fill these gaps (Khoury et al. 2010; Maxted et al. 2012). During the last decade, the number of data published shows the use of CWRs in different research programmes of crop improvement than for long-term conservation strategy because of their importance in global food security (Maxted et al. 2012). Therefore, it is necessary to feed an increasing population size by the conserved plant wild relatives to ensure sustainable agricultural development. The first step to conserve CWRs for candidate species is inventories that must be made at international, national and regional levels (Berlingeri and Crespo 2012; Khoury et al. 2013; Landucci et al. 2014). Recently, many of the deposited inventories information such as particular crop taxonomic origins, regional priority categories and proposed conservation strategies (Castaneda-Alvarez et al. 2015; Khoury et al. 2015) have entered into electronic databases that are accessible to public domain. Many national as well as local inventories targeting the major crops known as Prioritizing CWRs fall into primary or secondary GPs over other more distant ones for efficient conservation because it is comparatively easy to transfer traits between species within the primary GP (Berlingeri and Crespo 2012; Khoury et al. 2013; Landucci et al. 2014; Vincent et al. 2013). The Prioritizing CWRs are prominent in the breeding community as they play a key role in the global food supply (wheat, rice, maize, sugarcane), nutrition (quinoa) and biofuel (cassava).

Conservation of CWRs can be carried out either in situ, such as grown in natural habitats, or ex situ in the form of seed, in vitro, or field gene banks (Maxted et al. 2012). As far as ex situ conservation is concerned, it provides a convenient way to maintain the viability of seeds for long duration, but it compromises lower genetic diversity due to increased homozygosity and inbreeding depression level (Barazani et al. 2008; Maxted et al. 2012) and also ceases natural phenomenon of evolutionary processes when genetic resources are stored in seed banks. To overcome this drawback, another reliable approach, that is, in situ conservation, ensures the conservation of genetic resources without any loss of genetic diversity in their contents (Maxted et al. 2012).

Therefore, numerous international collection centres such as CIAT, Crop Trust and Kew and national gene banks such as the U.S. germplasm resource (http://www.ars-grin.gov/) have successfully conserved many CWR resource species and stored their vital linked information like prebreeding data, geographic distribution and potentially useful traits. The accessibility of CWR resources platform acquired by the researchers and breeders has utilization in the exploration of potential novel genes for future crop improvement breeding regimes.

7.5 Explore Genetic Resources for the Improvement of Quantity and Quality Value Traits

The treasure trove of genetic diversity potential present in CWRs has been unlocking for the improvement of various crop species to enhance their quantity as well as quality value traits (Foolad and Panthee 2012). Recently, numerous literatures revealed that the adoption of advanced biotechnology techniques like next generation sequencing, phenomics and genome editing have proven successful in the journey of gene discovery and transfer of superior alleles from CWRs to domesticated descendant crops (Honsdorf et al. 2014).

7.5.1 Abiotic Stress Tolerance

The environmental factors associated with the two main abiotic stress parameters, that is, salinity and drought, contribute to lower yield worldwide of primary crop species such as soybeans, tomatoes and cereals as well as of their wild relatives (Qi et al. 2014). The genetic resources for salt and drought tolerance in wild relatives for their improvement in respective domesticated crop are explored below:

Salinity stress: The production of wheat crop ranked third position after maize and rice, which is significantly affected by salinity problem in the world (FAO Stat 2015). The salt-tolerant gene, namely, *TmHKT1;5-A*, was searched in wheat wild relatives; *Triticum monococcum* has recently introgressed into commercial durum wheat variety via cross-breeding (Nevo and Chen 2010). The developed introgressed line shows 25% more yield performance in high-saline fields compared to its Tamaroi parent (Munns et al. 2012). A remarkable feature of this gene is to reduce the Na^+ level content in plant leaves that helps cease yield losses under salinity stress. The success achieved by the utilization of CWR resources for improvement of crop salt tolerance could be transferred to other commercial lines of wheat cultivars such as bread wheat. Soybean, the moderately salt-sensitive crop in which all the developmental stages are badly affected by salinity condition, results in lower than 40% of the total yield (Munns and Tester 2008). Recently, it achieved salt-tolerant gene, *GmCHX1*, in *Glycine soja*, the wild progenitor of cultivated soybean [*Glycine max* (L.) Merr.] (Qi et al. 2014). The discovery of *GmCHX1* and another salt-tolerant gene, *GmSALT3*, reflected that during the domestication process these salt-tolerant alleles would have been lost in soybean.

Drought stress: The application of several drought-tolerant quantitative trait locus (QTL) or genes identified in many CWRs has not proven successful as we expected because they belong to polygenic quantitative trait controlled by many genes. Recently it developed the advance backcross introgression libraries in wild crop that give the alternative way to transfer the drought-tolerant genes in barley (*Hordeum spontaneum*) (Honsdorf et al. 2014; Naz et al. 2014). The presence of drought-tolerant QTL in introgression libraries of barley had demonstrated the drought tolerance level in field trials (Arbelaez et al. 2015). For exclusion of "linkage drag" genes during construction of introgression libraries, it needs more backcrossing and selection process that make this approach unsuitable for transfer of drought-tolerant genes in the crop.

7.5.2 Biotic Stress Tolerance

The influence of biotic stressors like fungi, viruses, bacteria, nematodes and insect pests also contribute in lowering the yield potential of the main crop. The prolonged use of resistant varieties has not make effective measure to control these diseases infected in crop as mutation process occur in pathogens and insects make them resistant. Subsequently, plant breeders have been utilizing explored CWR resources to develop biotic stress-resistant broad-spectrum varieties by engaging metabolomics and transcriptomics strategies to identify resistant genes in CWRs. The remarkable cases like corn leaf blight disease infested in maize reduced lower yield by 50% in the United States (FAO 2005). Later on, it was identified that the blight-resistant alleles in wild relative of Mexican maize (*Tripsacum dactyloides* L.) have been transferred into commercial corn lines (Maxted and Kell 2009). The rootworm, a devastating pest of cultivated corn in the United States, had suffered major losses to the yield. Prischmann et al. (2009) identified rootworm resistance in CWR of maize (*Tripsacum dactyloides* L.) and introduced it into cultivated corn.

The improvement of biotic stress tolerance like bacterial blight, blast, brown plant hopper attacks and sheath blight has been reported as the most successful result in cultivated rice by using explored 22 wild rice species (Jena 2010). It reported the significant success in the management of one of the most challenging and devastating rice disease, that is, bacterial leaf blight, caused by *Xanthomonas oryzae* pv. *oryzae* (*Xoo*). The identification of some bacterial blight-resistant genes (*Xa3*, *Xa*4, *Xa21* and *Xa23*) reported in wild rice served as a way to increase the level of resistance among susceptible rice cultivars (Zhou et al. 2011). In the rice breeding programs of bacterial leaf blight resistance, further achievement has been reported by introgression of a rice blast-resistant gene, *Pi33*, from wild rice, *Oryza rufipogon* (Ballini et al. 2007), into the most used rice blast resistance variety (IR64).

Similarly, the utilization of CWRs of tomato such as *Solanum chilense*, *Solanum habrochaites, Solanum peruvianum, Solanum pennellii* and *Solanum pimpinellifolium* for the enhancement of various biotic resistant traits in cultivated tomatoes (*Solanum lycopersicum*) (Prasanna et al. 2015; Seah et al. 2004; Chunwongse et al. 2002). The different wild tomatoes expressing varying degrees of resistance to *tomato yellow leaf curl virus* (TYLCV) had successfully introgressed into cultivated varieties made durable and broad resistance to *TYLCV* (Prasanna et al. 2015; Menda et al. 2014).

7.5.3 Improvement in Yield and Quality-Related Traits

The identified yield-enhancing QTL in wild relatives of different crops such as rice, wheat, maize, barley, soybeans, beans and capsicum etc. has been successfully involved in crop improvement programmes. For example, the utilization of a small-fruited tomato ancestor (*S. pimpinellifolium*) in cultivated tomato species (*Solanum hirsutum*) led to 20% increased yield and also improved fruit qualities (Tanksley et al. 1996). There was significant enhancement in the yield content of hybrid rice when it backcrossed with low-yielding wild ancestor (*O. rufipogon*) (Xie et al. 2008). Significant enhancement in the yield of domesticated soybeans was reported when it transferred QTL from *G. Soja* (Li et al. 2007). The contribution of CWRs for the development of nutrient-loaded variety has become increasingly in demand for the improvement of human health. As we know, the metabolite named glucosinolate plays an important role to prevent the risk of various types of cancer (Dinkova-Kostova and Kostov 2012). The generated cross-hybrid between wild and cultivated broccoli is shown three times more content of glucosinolate than conventional varieties (Sarikamis et al. 2006).

7.6 Biotechnological Approaches for the Use of Wild Relatives for Crop Improvement

The practice of advanced biotechnology techniques makes the path easier for the utilization of untapped resources of CWR in the crop improvement programmes worldwide.

7.6.1 Genomics Approaches

Genomics approaches are widely used for the identification of genes or genomic regions controlling complex traits linked with crop improvement. Seeing the recent decline in the cost factor of high-throughput next-generation sequencing technologies, scientists enable genotyping thousands of individuals by sequencing as well as resequencing to generate increased number of SNPs data that are associated with agronomic traits in CWRs (Li et al. 2014; Zhou et al. 2015). It enhances the knowledge of genetic basis of genomics regions associated with domesticated traits and further signal for interpretation that allelic variation in those regions in CWRs can benefit crop improvement. This approach is helpful in the genotyping of the segregating population like [F_2, BC_2, near-isogenic lines and recombinant inbred lines (RILs)] that constructs high-resolution linkage maps for narrow QTL regions. The genomics approach is successful only if phenomics data is available in hand that will fasten the mining of genes linked with superior traits in CWR.

7.6.2 Functional Omics Approaches

The global analysis of those complex regulatory genes, expressed proteins or metabolite candidates that express remarkable traits in CWRs come under transcriptomics, proteomics and metabolomics approaches. This omics approaches are helpful in dissection of the variation in complex traits such as drought tolerance, salt tolerance and pest resistance. By omic methodology, various stress tolerance genes present in the CWRs have been identified under different conditions/parameters. It reported that the presence of dehydrin genes expressed in CWRs of wild barley (*H. spontaneum*) as well as wild tomato species (*S. chilense* and *S. peruvianum*) has significant contribution to their involvement in drought tolerance capacity (Fischer et al. 2013; Suprunova et al. 2004). Earlier it was a herculean task to pinpoint the involvement of particular gene, protein or metabolite for the expression of traits of interest, but it is feasible with the help of "omic" approach as well as "linkage mapping." Suprunova et al. (2007) quantified the gene expression level pattern in the target QTL in wild barley (*H. spontaneum*) in which searched novel gene *Hsdr4* was responsible for water-stress tolerance trait. Similarly for drought tolerance in wild wheat, two candidate genes (*KNAT3* and *SERK1*) have been identified (Placido et al. 2013). The accomplishment of large-scale study of gene function at different levels has been fulfilled with the help of transcriptomics, proteomics and metabolomics profile combination with QTL tool. The consideration for application of metabolomics-assisted breeding (MeAB) in crop improvement programmes is limited up to some level due to high cost factor for extraction and quantification of metabolites in the crop and poor heritability of metabolites. Bleeker et al. (2011) extracted 7-epizingiberene metabolite from wild (*S. habrochaites*) and observed the repulsion of whiteflies on the sprayed susceptible cultivated tomatoes as well as resistance to spider mites (Bleeker et al. 2012).

7.6.3 Genome-Wide Association Studies (GWAS)

Genome-wide association studies (GWAS) of metabolomics is the perfect way to analyze profiling distribution pattern of thousands of metabolites typically produced in the explored CWRs as these wild crops showed higher level of diversity in their metabolic profiles due to evolutionary force effects than their cultivated descendants. The minimized linkage disequilibrium content in CWR typically needed higher density of genomic markers for metabolomics association studies. Due to the database of many markers publicly available as well as lower reduction cost factor for genotyping, it was reported that high-density SNP markers for wild soybean, wild tomatoes and wild rice have been successfully developed and that information can be shared with metabolomics mapping in CWRs. The main problem faced is the data generated in the metabolite profiling are poor in annotations and unknown.

7.6.4 Genetic Modification (GM) Technology

Genetic modification (GM) technology offers a convenient way to manipulate the crop cultivar by introducing desired gene traits. Genetic engineering techniques are answerable in those cases in which conventional breeding methods are unable to improve the traits in the target germplasm. This technology successfully developed the transgenic *Bacillus thuringiensis* (Bt) crops in commercial cotton crop and earned a good source of income in the international market. Recently, the literature revealed that the scope of GM technology for drought tolerance trait is not meeting the result of our expectations breeding methods including introgression (Honsdorf et al. 2014). Also, it is a matter of great concern regarding the safety level of foods developed by GM technology.

The approach of inducing mutations in existing genes is an alternative way to facilitate increase in the yield of crop rather than exogenous entry of new genes (Lusser et al. 2012). This approach includes cisgenesis, intragenesis, genome editing, RNA-dependent DNA methylation and oligo-directed mutagenesis techniques. Cisgenesis refers to the genetic modification of crop plants in which a gene is transferred from a native or cross-compatible donor species such as a CWR to the commercial cultivated crop for enhancing their overall performance in the field (Krens et al. 2015). For example, this technique reflected resistance disease to late blight in potatoes and scab resistance in apples (Vanblaere et al. 2011). It could be the alternative way for crop improvement as it requires less period of time for gene introduction and limited chances of any linkage drag problem in comparison with traditional breeding strategies.

An efficient way to utilize the CWRs deposited in gene banks for crop improvement programme is high-throughput genomics approach that enables breeders to effectively select the stored accessions of interest based on their genetic relationship. The phylogenetic tree generated from these genomic data gives us the collection of core germplasm representing the allelic richness of the gene bank. Therefore, it is a convenient way to execute high-throughput phenotyping of the representative core collection for the agronomical important traits than characterizing the entire large-scale phenotyping that will be time-consuming and more laborious exercise (Honsdorf et al. 2014)

7.7 Conclusion

There is widened scope of utilizing the potential of genetic diversity present in stored wild species bank with the help of advanced biotechnology to mitigate the problem of food security existing worldwide. In the recently published literature, the use of potential novel genes present in the wild resources involved in the crop improvement programme has been revealed, but much remains to be explored. Recently, the identification of SNPs in the region of QTL associated with agronomical important traits in the wild species was performed with the assistance of linkage mapping analyses and GWAS. Due to rapid progress in advanced biotechnologies such as diverse omics approaches, it is bridging genotype–phenotype gaps in CWRs for crop improvement. To prioritize the accountability of CWRs for effective CWR conservation, the collection should be intensively conducted in those geographic regions harboring the greatest richness of taxa (Castaneda-Alvarez et al. 2016). The involvement of global initiatives by regional institutions or local organizations includes increasing in situ or ex situ conservation efforts of CWR and build agreements for sharing of wild resources.

References

Abedinia M, Henry RJ, Blakeney AB et al (2000) Accessing genes in the tertiary gene pool of rice by direct introduction of total DNA from *Zizania palustris* (Wild rice). Plant Mol Biol Rep 18:133–138

Arbelaez JD, Moreno LT, Singh N et al (2015) Development and GBS-genotyping of introgression lines (ILs) using two wild species of rice, *O. meridionalis* and *O. rufipogon*, in a common recurrent parent, *O. sativa* cv. Curinga. Mol Breed 35: 81

Ballini E, Berruyer R, Morel JB et al (2007) Modern elite rice varieties of the 'Green Revolution' have retained a large introgression from wild rice around the *Pi33* rice blast resistance locus. New Phytol 175:340–350
Barazani O, Perevolotsky A, Hadas R (2008) A problem of the rich: prioritizing local plant genetic resources for ex situ conservation in Israel. Biol Conserv 141:596–600
Berlingeri C, Crespo MB (2012) Inventory of related wild species of priority crops in Venezuela. Genet Resour Crop Evol 59:655–681
Bleeker PM, Diergaarde PJ, Ament K et al (2011) Tomato-produced 7-epizingiberene and R-curcumene act as repellents to whiteflies. Phytochemistry 72:68–73
Bleeker PM, Mirabella R, Diergaarde PJ et al (2012) Improved herbivore resistance in cultivated tomato with the sesquiterpene biosynthetic pathway from a wild relative. Proc Natl Acad Sci 109:20124–20129
Buckler ES, Thornsberry JM, Kresovich S (2001) Molecular diversity, structure and domestication of grasses. Genet Res 77:213–218
Castaneda-Alvarez NP, de Haan S, Juarez H et al (2015) *Ex situ* conservation priorities for the wild relatives of potato (*Solanum* L. section petota). PLoS One 10 : e0122599
Castaneda-Alvarez NP, Khoury CK, Achicanoy HA et al (2016) Global conservation priorities for crop wild relatives. Nat Plants 2:16022
Chunwongse J, Chunwongse C, Black L et al (2002) Molecular mapping of the *Ph-3* gene for late blight resistance in tomato. J Hortic Sci Biotechnol 77:281–286
Clark RM, Linton E, Messing J et al (2004) Pattern of diversity in the genomic region near the maize domestication gene *tb1*. Proc Natl Acad Sci 101:700–707
Dempewolf H, Eastwood RJ, Guarino L et al (2014) Adapting agriculture to climate change: a global initiative to collect, conserve, and use crop wild relatives. Agroecol Sust Food 38:369–377
Dinkova-Kostova AT, Kostov RV (2012) Glucosinolates and isothiocyanates in health and disease. Trends Mol Med 18:337–347
Fetch T, Johnston PA, Pickering R (2009) Chromosomal location and inheritance of stem rust resistance transferred from *Hordeum bulbosum* into cultivated barley (*H. vulgare*). Phytopathology 99:339–343
Fischer I, Steige KA, Stephan W et al (2013) Sequence evolution and expression regulation of stress-responsive genes in natural populations of wild tomato. PLoS One 8:e78182
Foolad MR, Panthee DR (2012) Marker-assisted selection in tomato breeding. Crit Rev Plant Sci 31:93–123
Ford-Lloyd BV, Schmidt M, Armstrong SJ (2011) Crop wild relatives-undervalued, underutilized and under threat? BioSci 61:559–565
Godfray HCJ, Beddington JR, Crute IR et al (2010) Food security: the challenge of feeding 9 billion people. Science 327:812–818
Honsdorf N, March TJ, Berger B (2014) High-throughput phenotyping to detect drought tolerance QTL in wild barley introgression lines. PLoS One 9:e97047
Hua L, Wang DR, Tan LB et al (2015) *LABA1*, a domestication gene associated with long, barbed awns in wild rice. Plant Cell 27:1875–1888
Hyten DL, Song QJ, Zhu YL et al (2006) Impacts of genetic bottlenecks on soybean genome in a common recurrent parent, *O. sativa* cv. Curinga. Mol Breed 35: 81
Jena KK (2010) The species of the genus *Oryza* and transfer of useful genes from wild species into cultivated rice, *O. sativa*. Breed Sci 60:518–523
Kastner T, Rivas MJI, Koch W et al (2012) Global changes in diets and the consequences for land requirements for food. Proc Natl Acad Sci 109:6868–6872
Khoury C, Laliberte B, Guarino L (2010) Trends in *ex situ* conservation of plant genetic resources: a review of global crop and regional conservation strategies. Genet Resour Crop Evol 57:625–639
Khoury CK, Greene S, Wiersema J et al (2013) An inventory of crop wild relatives of the United States. Crop Sci 53:1496–1508
Khoury CK, Bjorkman AD, Dempewolf H et al (2014) Increasing homogeneity in global food supplies and the implications for food security. Proc Natl Acad Sci 111:4001–4006

Khoury CK, Castaneda-Alvarez NP, Achicanoy HA (2015) Crop wild relatives of pigeonpea [*Cajanus cajan* (L.) Millsp.]: distributions, *ex situ* conservation status, and potential genetic resources for abiotic stress tolerance. Biol Conserv 184:259–270
Konishi S, Lin SY, Ebana K (2006) A SNP caused the loss of seed shattering during rice domestication. Science 312:1392–1396
Krens FA, Schaart JG., van der Burgh AM et al (2015). Cisgenic apple trees; development, characterization, and performance. Front Plant Sci 6: 286
Landucci F, Panella L, Lucarini D et al (2014) A prioritized inventory of crop wild relatives and wild harvested plants of Italy. Crop Sci 54:1628–1644
Larson G, Piperno DR, Allaby RG et al (2014) Current perspectives and the future of domestication studies. Proc Natl Acad Sci 111:6139–6146
Li D, Pfeiffer TW, Cornelius PL (2007) Soybean QTL for yield and yield components associated with Glycine soja alleles. Crop Sci 48:571–581
Li YH, Zhou GY, Ma JX et al (2014) De novo assembly of soybean wild relatives for pan-genome analysis of diversity and agronomic traits. Nat Biotechnol 32:1045–1052
Lusser M, Parisi C, Plan D (2012) Deployment of new biotechnologies in plant breeding. Nat Biotechnol 30:231–239
Marais GF, Pretorius ZA, Marais AS et al (2003) Transfer of rust resistance genes from *Triticum* species to common wheat. South African J Plant Soil 20:193–198
Maxted N, Kell SP (2009) Establishment of a global network for the in situ conservation of crop wild relatives: status and needs. Food and Agriculture Organization of the United Nations Commission on Genetic Resources for Food and Agriculture, Rome
Maxted N, Ford-Lloyd BV, Jury S et al (2006) Towards a definition of a crop wild relative. Biodivers Conserv 15:2673–2685
Maxted N, Kell S, Ford-Lloyd B et al (2012) Toward the systematic conservation of global crop wild relative diversity. Crop Sci 52:774–785
McKey D, Elias M, Pujol B et al (2010) The evolutionary ecology of clonally propagated domesticated plants. New Phytol 186:318–332
Menda N, Strickler SR, Edwards JD et al (2014) Analysis of wild-species introgressions in tomato inbreds uncovers ancestral origins. BMC Plant Biol 14:287
Meyer RS, DuVal AE, Jensen HR (2012) Patterns and processes in crop domestication: an historical review and quantitative analysis of 203 global food crops. New Phytol 196:29–48
Miller AJ, Gross BL (2011) From forest to field: perennial fruit crop domestication. Am J Bot 98:1389–1414
Munns R, Tester M (2008) Mechanisms of salinity tolerance. Annu Rev Plant Biol 59:651–681
Munns R, James RA, Xu B et al (2012) Wheat grain yield on saline soils is improved by an ancestral Na+ transporter gene. Nat Biotechnol 30:360–364
Naz AA, Arifuzzaman M, Muzammil S et al (2014) Wild barley introgression lines revealed novel QTL alleles for root and related shoot traits in the cultivated barley (*Hordeum vulgare* L.). BMC Genet 15:107
Nevo E, Chen GX (2010) Drought and salt tolerances in wild relatives for wheat and barley improvement. Plant Cell Environ 33:670–685
Palaisa K, Morgante M, Tingey S et al (2004) Long-range patterns of diversity and linkage disequilibrium surrounding the maize *Y1* gene are indicative of an asymmetric selective sweep. Proc Natl Acad Sci 101:9885–9890
Placido DF, Campbell MT, Folsom JJ et al (2013) Introgression of novel traits from a wild wheat relative improves drought adaptation in wheat. Plant Physiol 161:1806–1819
Prasanna HC, Kashyap S, Krishna R et al (2015) Marker assisted selection of *Ty-2* and *Ty-3* carrying tomato lines and their implications in breeding tomato leaf curl disease resistant hybrids. Euphytica 204:407–418
Prischmann DA, Dashiell1 KE, Schneider DJ, Eubanks MW (2009) Evaluating Tripsacum-introgressed maize germplasm after infestation with western corn rootworms (Coleoptera: Chrysomelidae). J Appl Entomol 133:10–20

Qi XP, Li MW, Xie M et al (2014) Identification of a novel salt tolerance gene in wild soybean by whole-genome sequencing. Nat Commun 5:4340

Raizada P, Singh A, Raghubanshi AS (2009) Comparative response of seedlings of selected native dry tropical and alien invasive species to CO_2 enrichment. J Plant Ecol 2:69–75

Saintenac C, Zhang WJ, Salcedo A et al (2013) Identification of wheat gene *Sr35* that confers resistance to Ug99 stem rust race group. Science 341:783–786

Sarikamis G, Marquez J, MacCormack R et al (2006) High glucosinolate broccoli: a delivery system for sulforaphane. Mol Breed 18:219–228

Seah S, Yaghoobi J, Rossi M et al (2004) The nematode-resistance gene, Mi-1, is associated with an inverted chromosomal segment in susceptible compared to resistant tomato. Theor Appl Genet 108:1635–1642

Suprunova T, Krugman T, Fahima T et al (2004) Differential expression of dehydrin genes in wild barley, *Hordeum spontaneum*, associated with resistance to water deficit. Plant Cell Environ 27:1297–1308

Suprunova T, Krugman T, Distelfeld A et al (2007) Identification of a novel gene (*Hsdr4*) involved in water-stress tolerance in wild barley. Plant Mol Biol 64:17–34

Sweeney M, Mccouch S (2007) The complex history of the domestication of rice. Ann Bot 100:951–957

Tanksley SD, Grandillo S, Fulton TM et al (1996) Advanced backcross QTL analysis in a cross between an elite processing line of tomato and its wild relative L-pimpinellifolium. Theor Appl Genet 92:213–224

Tester M, Langridge P (2010) Breeding technologies to increase crop production in a changing world. Science 327:818–822

Vanblaere T, Szankowski I, Schaart J (2011) The development of a cisgenic apple plant. J Biotechnol 154:304–311

Vincent H, Wiersema J, Kell S et al (2013) A prioritized crop wild relative inventory to help underpin global food security. Biol Conserv 167:265–227

Warschefsky E, Penmetsa RV, Cook DR et al (2014) Back to the wilds: tapping evolutionary adaptations for resilient crops through systematic hybridization with crop wild relatives. Am J Bot 101:1791–1800

Xie XB, Jin FX, Song MH et al (2008) Fine mapping of a yield-enhancing QTL cluster associated with transgressive variation in an *Oryza sativa* × *O-rufipogon* cross. Theor Appl Genet 116:613–622

Xu X, Liu X, Ge S et al (2012) Resequencing 50 accessions of cultivated and wild rice yields markers for identifying agronomically important genes. Nat Biotechnol 30:105–111

Zhou YL, Uzokwe VNE, Zhang CH et al (2011) Improvement of bacterial blight resistance of hybrid rice in China using the *Xa23* gene derived from wild rice (*Oryza rufipogon*). Crop Prot 30:637–644

Zhou ZK, Jiang Y, Wang Z et al (2015) Resequencing 302 wild and cultivated accessions identifies genes related to domestication and improvement in soybean. Nat Biotechnol 33:408–414

Next-Generation Sequencing Technologies and Their Implications for Efficient Utilization of Genetic Resources

8

Mehak Gupta, Romesh Kumar Salgotra, and Bhagirath Singh Chauhan

Contents

Abstract

Plant genetic resources are the main component of the agricultural biodiversity. Agriculture plays a key role in feeding millions and protecting our natural resources and the environment with nurturing and utilizing plant diversity. Nature has devised an extraordinary mechanism by developing intraspecific genetic diversity in crop plants and their wild relatives. New high-throughput technologies like next-generation sequencing (NGS) facilitate cost-effective study of genetic diversity, identify desirable genes and alleles as well as make possible their transfer during crop improvement, thus reducing the time to deliver new varieties. This chapter focuses on how next-generation sequencing technologies can be involved for proper exploitation of PGR for crop improvement. The real

M. Gupta (✉) · R. K. Salgotra
School of Biotechnology, Sher-e-Kashmir University of Agricultural Sciences and Technology of Jammu, Jammu, Jammu and Kashmir, India

B. S. Chauhan
Queensland Alliance for Agriculture and Food Innovation (QAAFI), The University of Queensland, Gatton, QLD, Australia
e-mail: b.chauhan@uq.edu.au

R. K. Salgotra, S. M. Zargar (eds.), *Rediscovery of Genetic and Genomic Resources for Future Food Security*,
https://doi.org/10.1007/978-981-15-0156-2_8

and potential application of the current advances in genomic technologies can be used for efficient utilization of PGR for crop improvement to secure food security.

Acronyms

B	Billion
bp	Base pairs
CWRs	Crop wild relatives
FAO	Food and Agriculture Organization
Gb	Gigabytes
GRs	Genetic resources
HT	High throughput
M	Million
NA	Not available
NGS	Next-generation sequencing
PE	Paired end
PGR	Plant genetic resource
SE	Single end
SNPs	Single nucleotide polymorphisms

8.1 Introduction

Genetic resources (GRs) refer to genetic material containing functional units of heredity, which are of prospective value and upon which the world relies to improve the productivity and quality of crops as well as to maintain healthy populations of wild species. Plant genetic resource (PGR) is the only source of plant genetic diversity which provides valuable rare traits needed for meeting the challenges of adapting crop varieties in response to environmental stresses, plant disease and pests. In fact, they form the basis of all crop varieties that are bred to produce more, withstand stresses and yield quality output. The plant resources form an integral part of a huge interdependent system that encompasses the physical components and the biological community of life. These resources have enormous prospective benefits to humankind. Approximately 300,000 species of higher plants represent plant diversity on earth, and only about 7000 species have been cultivated by humans over the millennia for food, fodder and feed. Today, an individual genotype with seemingly useless set of characters may suddenly become essential tomorrow due to changing climatic conditions or outbreaks of disease. Therefore, it has been long realized that preservation and sustainable use of genetic resources for food and agriculture are at the core of food security and we should "conserve" all the diversity we have.

Plant genetic diversity is vulnerable to "genetic erosion". Over the years, ever-increasing population and climate changes have contributed to a large extent of stress on plant resources, thus leading to their decline or loss in nature. According to the Food and Agriculture Organization of the United Nations, replacement of local varieties by modern varieties resulting in reduction of the sheer number of cultivars is the main cause of genetic erosion. This is exaggerated by the emergence of new pests, weeds and diseases, environmental degradation, urbanization and land clearing. Hence, there is an immediate need to develop suitable conservation strategies for proper utilization and sustainability of these important resources. Traditional agriculture has a large number of diverse landraces in contrast to modern intensive agriculture which has a narrow genetic base. To preserve the diversity found within species of cultivated plants, many researchers take up an approach that comes together by storing diversity in genebanks (ex situ conservation) along with on-farm (in situ conservation).

Over the past decade, there have been significant advances in DNA sequencing technologies, which are driving many areas of plant science. Next-generation sequencing (NGS), also known as high-throughput sequencing, can be described as a number of different modern sequencing technologies which have revolutionized genomic research. NGS methods allow millions of bases to be sequenced in one round, at a fraction of the cost relative to traditional Sanger sequencing (Sanger et al. 1977). As the costs and capabilities of these technologies continue to improve, whole new fields of study are being opened, allowing us to analyse a variety of data sets and move towards questions which were not possible earlier. All NGS platforms perform sequencing of millions of small fragments of DNA in parallel which has led to dramatic changes in read length, sequencing chemistry, instrumentation, throughput and cost. Reference genome sequences for a large number of model and non-model species have been published, and others continue to be released at a highly unprecedented rate.

Prior to the advent of molecular characterization, accessions in germ plasm collections were mainly examined based on morphological characters and phenotypic traits (Boerner et al. 2006). The improvement of molecular techniques provides more accurate analysis of large collections. High-throughput (HT) technologies including DNA isolation, genotyping, phenotyping and next-generation sequencing provide new tools to add substantial value to genebank collections. The amalgamation of genomic data into genebank documentation systems with taxonomic, phenotypic and ecological data will lead in a new era for the valorization of PGR. Every aspect of genomics, from the determination of phenotypic traits to the application of NGS to whole genomes, will have a great impact on PGR conservation and their utilization in plant breeding (Van et al. 2011). Identification of genetic variation has become so efficient and precise that thousands of candidate genes can be trailed within large genebank collections (Varshney et al. 2009). Using NGS technologies, it is possible to resequence candidate genes, entire transcriptomes or entire plant genomes more efficiently and economically than ever before. Advances in sequencing technology will allow for whole-genome resequencing of hundreds of individuals. In this way, information on thousands of candidate genes and candidate regions can be harnessed for thousands of individuals to sample genetic diversity within and between germ plasm pools, to map quantitative trait loci (QTLs), to identify individual genes and to determine their functional diversity.

8.2 NGS: A New Era of Sequencing Technologies

The appearance of sequencing technologies has played an important role in the analysis of genomic sequences of organisms. Earlier, Sanger and Maxam-Gilbert sequencing technologies (1977) were the most common sequencing technologies used. But with the advent of 454 Life Sciences new sequencing machine GS20 in 2005 (Qiang-long et al. 2014), a new era of sequencing technologies was emerged, which opened new perspectives for genome exploration and analysis and, later, commercialized as technologies capable of producing sequences with very high throughput and at much lower cost than the first sequencing technologies. These new sequencing technologies are known as "next-generation sequencing (NGS) technologies" or "high-throughput sequencing technologies". NGS technologies produce a massively parallel analysis with a high throughput from multiple samples at much reduced cost (Mardis 2011). NGS technologies can be sequenced in parallel millions to billions of reads in a single run, and the time required to generate the gigabase-sized reads is only a few days or hours making it best than the first-generation sequencing such as Sanger sequencing. Unfortunately, NGS are incapable to read the complete DNA sequence of the genome; they are limited to sequence small DNA fragments and generate millions of reads. This limit remains a negative point especially for genome assembly projects because it requires high computing resources. NGS technologies continue to improve, and the number of sequencers increases these last years. However, the literature divided NGS technologies into three types: a) second-generation sequencing technologies which are characterized by the need to prepare amplified sequencing banks before starting the sequencing of amplified DNA clones (Schatz et al. 2010); b) third generation, also referred to as long-read sequencing technologies which are classified as single molecule sequencing technology (Vezzi 2012), in which a single molecule is capable of generating longer reads at much lower costs and in a shorter time; and c) fourth-generation sequencing technologies in which a single nanopore molecule is capable of producing very long contiguous reads—up to four orders of magnitude larger than a short read—and that too ultraportable connects via USB to a laptop. Several studies presented the sequencing technologies and detailed the chemical mechanisms of each sequencing platform (Michael 2010; Heather and Chain 2015; Reuter et al. 2015; Kulski 2016; Goodwin et al. 2016). In the following, we present a brief review of the newest generations of sequencing technologies (second, third and fourth) focusing on sequencing methods and platforms characterizing each generation of sequencing (Table 8.1).

Table 8.1 Newest generations of sequencing technologies

Platform	Instrument	Reads per run	Avg read length (bp)	Read type	Error type	Error rate (%)	Data generated per run (Gb)	Year
Second-generation sequencing technologies								
Roche/454	GS20	200	100	SE, PE	Indel	1	0.02	2005
Roche/454	GS FLX	400	250	SE, PE	Indel	1	0.1	2007
Roche/454	GS FLX Titanium	1 M	450	SE, PE	Indel	1	0.45	2009
Roche/454	GS Junior	100	400	SE, PE	Indel	1	0.04	2010
Roche/454	GS FLX Titanium+	1 M	700	SE, PE	Indel	1	0.7	2011
Illumina	MiSeq	25 M (maximum)	300	SE, PE	Mismatch	0.1	15 (maximum)	2011
SOLiD/ABI	5500 W	3B	75	SE	Mismatch	~0.1	160	2011
Ion Torrent	PGM 314 chip v2	400.000–550.000	400	SE	Indel	1	0.06 to 0.1	2011
Ion Torrent	PGM 316 chip v2	2–3 M	200	SE	Indel	1	0.6 to 1	2011
Illumina/Solexa/Solexa	HiSeq	5B (maximum)	150	SE, PE	Mismatch	0.1	1.5 Tb (maximum)	2012
Ion Torrent	Ion Proton	60–80 M	200	SE	Indel	1	10	2012
Illumina/Solexa	MiniSeq	25 M (maximum)	150	SE, PE	Mismatch	1	7.5 (maximum)	2013
SOLiD/ABI	5500xl W	6B	75	SE	Mismatch	~0.1	320	2013
Ion Torrent	PGM 318 chip v2	4–5.5 M	400	SE	Indel	1	1.2 to 2	2013
Roche/454	GS Junior+	100	700	SE, PE	Indel	1	0.07	2014
Illumina/Solexa	NextSeq	400 M (maximum)	150	SE, PE	Mismatch	1	120 (maximum)	2014
Illumina/Solexa	HiSeq X	6B (maximum)	150	SE, PE	Mismatch	0.1	1.8 Tb (maximum)	2014

(continued)

Table 8.1 (continued)

Platform	Instrument	Reads per run	Avg read length (bp)	Read type	Error type	Error rate (%)	Data generated per run (Gb)	Year
Ion Torrent	Ion S5/S5XL 520	3–5 M	400	SE	Indel	1	1.2 to 2	2015
Ion Torrent	Ion S5/S5XL 530	15 M–20 M	400	SE	Indel	1	3 to 5	2015
Ion Torrent	Ion S5/S5XL 540	60 M - 80 M	400	SE	Indel	1	NA	2015
Third-generation sequencing technologies								
PacBio	RS C1	432	1300	SE	Indel	15	0.54	2011
PacBio	RS C2	432	2500	SE	Indel	15	0.5 to 1	2012
PacBio	RS C2 XL	432	4300	SE	Indel	15	0.5 to 1	2012
PacBio	RS II C2 XL	564	4600	SE	Indel	15	0.5 to 1	2013
PacBio	RS II P5 C3	528	8500	SE	Indel	13	0.5 to 1	2014
PacBio	RS II P6 C4	660	13,500	SE	Indel	12	0.5 to 1	2014
PacBio	Sequel	350	10,000	SE	NA	NA	7	2016
Fourth-generation sequencing technologies								
Oxford Nanopore Technologies	MinION Mk	100	9545	1D, 2D	Indel/ mismatch	12	1.5	2015
Oxford Nanopore Technologies	PromethION	NA	9846	1D, 2D	NA	NA	2000 to 4000	2016
Oxford Nanopore Technologies	SmidgION							2018

NA not available, *SE* single end, *PE* paired end, *M* million, *B* billion, *Gb* gigabytes, *bp* base pairs

8.3 Implications of NGS Technologies for PGR Conservation

8.3.1 Genebanking

Advances in DNA sequencing technology and the development of high-throughput systems for multiparallel interrogation of thousands of polymorphic nucleotide now provide a suite of technological platforms facilitating the analysis of several hundred of gigabases per day using state-of-the-art sequencing technology. As NGS technologies enable the efficient sequencing of large numbers of samples and genebank collections generally consist of many poorly studied and underutilized accessions, it is not surprising that the impact of NGS technologies on the functioning of genebanks has also received attention. Therefore, genomic resources are nowadays conserved in genebanks. It has been suggested that the advances in sequencing may fundamentally change the functioning of genebanks like management, pereception and organisation. Increased collaboration between genebank managers and the user community is also recommended (Kilian and Graner 2012; McCouch et al. 2012). Regarding collection management, NGS technologies could be useful to basically support all management areas. For example, DNA sequence data of genebank accessions may be used to determine the genetic structure of collections and to improve the composition thereof by eliminating redundancies (Van Treuren et al. 2009). Ample sequence data of the existing collection allow genebank curators to take more informed decisions about acquisition by evaluating potentially interesting materials for their added value to the genetic diversity already present in the collection (Van Treuren et al. 2008). NGS data could also be used to monitor the regeneration of accessions in order to ensure the maintenance of genetic integrity thereof, for example, by comparing sequence data of samples before and after regeneration (Van Hintum et al. 2007). These high expectations, however, may not be valid for the genebank community at large. For example, at the research centres of the Consultative Group on International Agricultural Research (CGIAR), the focus is on a single or just a very limited number of staple crops, while ample research facilities and expertise are usually present. Consequently, the CGIAR genebanks are better positioned to profit from genomic research, in contrast to many institutional and national genebanks that manage a variety of crops and often have limited access to facilities and expertise. Even such genebanks may strongly differ in the ability to access modern technologies and to adapt to changing demands, as between national genebanks of developed and those of non-developed countries. Application of NGS technologies is therefore less straightforward for most genebanks when compared with genebanks in institutes such as the International Rice Research Institute and the International Maize and Wheat Improvement Center (CIMMYT). It is therefore not surprising that regarding sequencing applications, most progress is achieved for crops such as rice and maize (McCouch et al. 2012).

8.3.1.1 Status of PGR Conservation in Genebanks

As evidenced by the huge number of accessions that are conserved in genebanks for various species (Table 8.2), it is clear that enormous progress has been made in

Table 8.2 Top 10 crop species in the world's genebanks

Species	Common names	Number of accessions
Triticum aestivum	Wheat	856,168
Oryza sativa	Rice	773,948
Hordeum vulgare	Barley	466,531
Zea mays	Maize	327,932
Phaseolus vulgaris	Bean	261,963
Sorghum bicolor	Sorghum	235,688
Glycine max	Soybean	229,944
Avena sativa	Oat	130,653
Arachis hypogaea	Groundnut	128,435
Gossypium hirsutum	Cotton	104,780

conserving germ plasm in seed banks. These reference genome sequences provide perhaps the most important genome resource for promoting use of these species. However, PGR conservation is lagging behind in embracing advances in molecular biology especially in genome sequencing compared with other areas of plant science (FAO 2010). As observed by McCouch et al. (2012), the application of these approaches, which have been referred to as next-generation genebanking (Treuren and Hintum 2014), has the capacity to dramatically transform previously dormant genebanks into research centres with robust research activities. Currently, 7.4 million accessions of the world's PGR are conserved in about 1625 genebanks spread globally (FAO 2010). Wheat has the highest number of conserved accessions followed by rice, barley and maize in a decreasing order (Table 8.2). While great efforts have been put in the collection of major crops, thereby resulting in tremendous success in their conservation, minor crops, crop wild relatives (CWRs) as well as neglected and underutilized crops remain grossly underrepresented in genebanks (Wambugu et al. 2013; Bordoni and Hodgkin 2012). Plant identification, phylogenetic relationships, DNA barcoding and a clear biosystematics framework will be required for proper conservation and effective utilization of PGRs. Morphology-based plant identification, which is common in genebanks, increases chances of misidentification especially in case of morphologically similar and closely related species. Well-resolved phylogenetic relationships between cultivated species and their CWRs are vital in making germ plasm conservation management decisions. Additionally, they aid in gene discovery as well as defining strategies for gene transfer during crop improvement.

DNA barcoding is an effective species identification tool, but there is no universally agreed locus for plant barcoding (CBOL Plant Working Group 2009; Lahaye et al. 2008). Recently, the potential of whole chloroplast genome sequences as a universal barcode in plant identification as well as in resolving phylogenetic relationships has been demonstrated (Nock et al. 2011; Parks et al. 2009; Yang et al. 2013; Erickson et al. 2008; Waters et al. 2012). The ongoing advances in sequencing coupled with decreased sequencing costs as well as the high multiplexing capacity for chloroplast genomes will continue to make the whole plastid sequences a popular tool that may eventually replace Sanger-based DNA barcoding. Owing to the challenges of plastid enrichment (McPherson et al. 2013), sequencing of total DNA and then isolating chloroplast sequences are now the method of choice for most researchers. Chloroplast sequences can be assembled by reference-guided assembly where reads are mapped to a reference (Nock et al. 2011) or by de novo assembly followed by selection of chloroplast contigs through homology searches (McPherson et al. 2013). Recently, a robust approach using both of these approaches has been used to define genetic and evolutionary relationships between wild and cultivated species that constitute the primary gene pool for rice (Wambugu et al. 2015). The use of whole chloroplast sequences eliminates the need to have a priori information on a locus of choice, a difficulty that acts as a major hindrance to single or multilocus studies. With a much longer sequence than most commonly used DNA barcodes, it has more variation that can help discriminate closely related species. Whole plastid sequences have also been used to identify novel genetic resources (Brozynska et al. 2014). In addition to the use of chloroplast genomes, a set of well-selected informative single nucleotide polymorphisms (SNPs) has been used to detect cases of species misidentification in genebanks (Orjuela et al. 2014), which is one of the factors that hinder utility of conserved germ plasm. The capacity for rapid and inexpensive analysis of complete plastid genomes as well as analysis of large numbers of nuclear loci is offering unprecedented opportunities in the field of plant systematics. Previously intractable phylogenetic relationships are now easily being resolved using genomic-based approaches. The routine sequencing of complete nuclear genomes might in the future make whole-genome sequences a tool for use in plant systematics. These advances are expanding the types of questions that genebank managers can ask in the area of plant systematics, thereby potentially addressing challenges that have always limited germ plasm utilization.

Current improvements in genomic approaches have increased impetus towards studying the functional genetic variation of genebank collections. The increased availability of high-quality reference sequences has opened almost unlimited possibilities in deciphering the molecular and genetic basis of biologically and economically important traits. Resequencing of several genotypes through whole-genome shotgun sequencing followed by mapping to the reference is currently the most popular approach for genetic analysis and marker discovery (Li et al. 2012; Huang et al. 2009; Arai-Kichise et al. 2011; Xu et al. 2012; Wang et al. 2014; Li et al. 2010).

8.4 Challenges

A large proportion of accessions conserved in genebanks remain unidentified or identified up to genus level. Though it is difficult to accurately assess the extent of use of PGR, it is estimated that <1% of accessions conserved in various germ plasm repositories globally have been used in crop improvement (Sharma et al. 2013). Since 2006, the Food and Agriculture Organization (FAO) reported little change in the use of PGR in developing new varieties. The low use of PGR can be attributed to the lack of information on the potential value of conserved germ plasm, which is arguably one of the greatest challenges that face genebanks (Wambugu and Muthamia 2009; Khoury et al. 2010). The capacity to identify and transfer useful alleles to improved varieties has also been inadequate. However, this is usually constrained by the lack of taxonomic expertise (Chase and Fay 2009). The limited availability of accurate phenotypic data now presents a challenge in studying the value of genetic resources by linking genotypes and phenotypes. Owing to the large number of unstructured natural populations in genebanks, association mapping studies render themselves better suited for characterizing natural variation of genebank samples as opposed to QTL mapping.

8.5 Conclusion and Future Perspectives

PGRs form the natural variations that have supported human kind for several millennia. These resources are the basis for food security in addition to being sources of energy, animal feed, fibre as well as other ecosystem services. They are important in addressing the global challenges that are currently facing the human population, particularly the twin challenge of climate change and food scarcity. Owing to their great importance, effective conservation and sustainable utilization of these resources is critically important and has never been more urgent. Promoting the sustainable utilization of biodiversity is a key goal of various global and regional efforts and initiatives as well as international agreements and treaties governing genetic resources. In addition to these administrative, legal and political measures, which have been put in place, the use of scientific advances particularly genome sequencing has the potential to address some of the challenges that limit sustainable utilization of PGR.

References

Arai-Kichise Y, Shiwa Y, Nagasaki H et al (2011) Discovery of genome-wide DNA polymorphisms in a landrace cultivar of japonica rice by whole-genome sequencing. Plant Cell Physiol 52(2):274–282

Boerner A, Freytag U, Sperling U (2006) Analysis of wheat disease resistance data originating from screenings of Gatersleben genebank accessions during 1933 and 1992. Genet Resour Crop Evol 53:453–465

Bordoni P, Hodgkin T (2012) Climate change and the conservation of plant genetic resources. J Crop Improv 26(3):329–345

Brozynska M, Omar ES, Furtado A et al (2014) Chloroplast genome of novel rice germplasm identified in northern Australia. Trop Plant Biol 7(3–4):111–120

CBOL Plant Working Group (2009) A DNA barcode for land plants. Proc Natl Acad Sci U S A 106:12794–12797

Chase MW, Fay MF (2009) Barcoding of plants and fungi. Sci 325(5941):682–683

Erickson DL, Spouge J, Resch A et al (2008) DNA barcoding in land plants: developing standards to quantify and maximize success. Taxon 57(4):1304–1316

FAO (2010) Second report on the world's plant genetic resources for food and agriculture. Rome, Italy: FAO pp \

Goodwin S, McPherson JD, Richard McCombie W (2016) Coming of age: ten years of next-generation sequencing technologies. Nat Rev Genet 17:333–351
Heather JM, Chain B (2015) The sequence of sequencers: the history of sequencing DNA. Genomics 107:1–8
Huang X, Huang T, Dong G et al (2009) High-throughput genotyping by whole-genome resequencing. Genome Res 19(6):1068–1076
Khoury C, Laliberte´ B, Guarino L (2010) Trends in ex situ conservation of plant genetic resources: a review of global crop and regional conservation strategies. Genet Resour Crop Evol 57(4):625–639
Kilian B, Graner A (2012) NGS technologies for analyzing germplasm diversity in genebanks. Brief Funct Genom 11:38–50
Kulski JK (2016) Next-generation sequencing – an overview of the history, tools, and "Omic" applications, next generation sequencing-advances, applications and challenges. InTech
Lahaye R, van der Bank M, Bogarin D et al (2008) DNA barcoding the floras of biodiversity hotspots. Proc Natl Acad Sci U S A 105(8):2923–2928
Li J, He W, Yang G et al (2010) Resequencing of 31 wild and cultivated soybean genomes identifies patterns of genetic diversity and selection. Nat Genet 42(12):1053–1059
Li S, Xie K, Li W et al (2012) Re-sequencing and genetic variation identification of a rice line with ideal plant architecture. Rice 5(1):18
Mardis ER (2011) A decade's perspective on DNA sequencing technology. Nat 470:198–203
Maxam AM, Gilbert WA (1977) A new method for sequencing DNA. Proc Natl Acad Sci 74:560–564
McCouch SR, McNally KL, Wang W et al (2012) Genomics of gene banks: a case study in rice. Am J Bot 99(2):407–423
McPherson H, van der Merwe M, Delaney SK et al (2013) Capturing chloroplast variation for molecular ecology studies: a simple next generation sequencing approach applied to a rainforest tree. BMC Ecol 13(1):8
Michael LM (2010) Sequencing technologies – the next generation. Nat Rev Genet 11:31–46
Nock CJ, Waters DLE, Edwards MA et al (2011) Chloroplast genome sequences from total DNA for plant identification. Plant Biotechnol J 9(3):328–333
Orjuela J, Sabot F, Che'ron S et al (2014) An extensive analysis of the African rice genetic diversity through a global genotyping. Theor Appl Genet 127(10):2211–2223
Parks M, Cronn R, Liston A (2009) Increasing phylogenetic resolution at low taxonomic levels using massively parallel sequencing of chloroplast genomes. BMC Biol 7(1):84
Qiang-long Z, Shi L, Peng G, Fei-shi L (2014) High-throughput sequencing technology and its application. J Northeast Agricultural University 21:84–96
Reuter JA, Spacek DV, Snyder MP (2015) High-throughput sequencing technologies. Mol Cell 58:586–597
Sanger F, Nicklen S, Coulson AR (1977) DNA sequencing with chain terminating inhibitors. Proc Natl Acad Sci 74:5463–5467
Schatz MC, Delcher AL, Salzberg SL (2010) Assembly of large genomes using second generation sequencing. Genome Res 20:1165–1173
Sharma S, Upadhyaya HD, Varshney RK et al (2013) Pre-breeding for diversification of primary gene pool and genetic enhancement of grain legumes. Front Plant Sci 4:309
Treuren R, Hintum TJL (2014) Next-generation genebanking: plant genetic resources management and utilization in the sequencing era. Plant Genet Resour 12:298–307
Van Hintum TJL, van de Wiel CCM, Visser DL, van Treuren R, Vosman B (2007) The distribution of genetic variation in a Brassica oleracea genebank collection related to the effects on diversity of regeneration, as measured with AFLPs. Theor Appl Genet 114:777–786
Van K, Kim DH, Shin JH et al (2011) Genomics of plant genetic resources: past, present and future. Pl Genet Res 9:155–158
Van Treuren R, Engels JMM, Hoekstra R, van Hintum TJL (2009) Optimization of the composition of crop collections for ex situ conservation. Plant Genet Resour Characterization and Utilization 7:185–193

Van Treuren R, van Hintum TJL, van de Wiel CCM (2008) Marker-assisted optimization of an expert-based strategy for the acquisition of modern lettuce varieties to improve a genebank collection. Genet Res Crop Evol 55:319–330

Varshney RK, Nayak SN, May GD et al (2009) Next-generation sequencing technologies and their implications for crop genetics and breeding. Trend Biotechnol 27:522–530

Vezzi F (2012) Next generation sequencing revolution challenges: Search, assemble, and validate genomes. Ph.D, Universita degli Studi di Udine, Italy

Wambugu P, Furtado A, Waters D et al (2013) Conservation and utilization of African *Oryza* genetic resources. Rice 6(1):29

Wambugu PW, Brozynska M, Furtado A et al (2015) Relationships of wild and domesticated rice (Oryza AA genome species) based upon whole chloroplast genome sequences. Sci Rep 5(1):13957

Wambugu PW, Muthamia ZK (2009) Second report on the state of plant genetic resources for food and agriculture in Kenya. FAO, Rome

Wang MH, Yu Y, Haberer G et al (2014) The genome sequence of African rice (Oryza glaberrima) and evidence for independent domestication. Nat Genet 46(9):982–988

Waters DLE, Nock CJ, Ishikawa R et al (2012) Chloroplast genome sequence confirms distinctness of Australian and Asian wild rice. Ecol Evol 2(1):211–217

Xu X, Liu X, Ge S et al (2012) Resequencing 50 accessions of cultivated and wild rice yields markers for identifying agronomically important genes. Nat Biotechnol 30(1):105–111

Yang JB, Tang M, Li HT et al (2013) Complete chloroplast genome of the genus Cymbidium: lights into the species identification, phylogenetic implications and population genetic analyses. BMC Evol Biol 13(1):84

Comparative Genomics for Exploring New Genes and Traits for Crop Improvement

9

Gayatri Jamwal and Inderpal Singh Bijral

Contents

Abstract

Comparative genomics has brought a paradigm shift in our understanding of the evolutionary relationships among organisms and helped in elucidating the functions of newly discovered genes. With copious amount of information regarding plant genomic sequences being generated, it has become imperative to store such information in a manner where it is easy to retrieve and analyse. Aided with the tools of bioinformatics, comparative genomics approach is now being utilised for studying evolution of gene families and transposable elements in plants, identification of functional regions in their genomes, genetic basis of phenotypic variation and looking for DNA markers associated with desirable phenotype for breeding. This chapter will discuss the need of comparative genomics in crop improvement and the various online plant genomic databases available to do so.

G. Jamwal (✉) · I. S. Bijral
Sher-E-Kashmir University of Agricultural Sciences and Technology, Jammu, Jammu and Kashmir, India

R. K. Salgotra, S. M. Zargar (eds.), *Rediscovery of Genetic and Genomic Resources for Future Food Security*,
https://doi.org/10.1007/978-981-15-0156-2_9

Acronyms

BLAST	Basic local alignment and sequencing tool
EST	Expressed sequence tag
GAB	Genomics-assisted breeding
GBS	Genotyping by sequencing
GWAS	Association mapping and genome-wide association studies
PlantGenIE	The Plant Genome Integrative Explorer
PGDBj	Plant Genome Database Japan
PIECE	Plant Intron Exon Comparison and Evolution
POGs	Putative Orthologous Groups database
SALAD	Surveyed conserved motif Alignment diagram and the Associating Dendrogram
SNP	Single nucleotide polymorphism
TAIR	The Arabidopsis Information Resource
QTL	Quantitative trait loci

9.1 Introduction

Comparative genomics essentially talks about comparing the genomes of different organisms in order to have a better understanding of their evolutionary relationships and to divulge the functions of newly discovered genes. The information encompassed in the gene structure and organisation of a species is conserved to a greater extent in another species which shares a common parentage with it, or in other words if organisms from two different species show conservation in the way their genes are structured and organised, then they must have evolved from a common ancestor. The observation of homology among organisms in nature has formed the basis of comparative genomics. It is important here to mention the concepts of paralogs and orthologs. In genetics, homology is studied in the context of protein and DNA sequence similarity. When a speciation event happens, i.e. copies of a single gene get inherited by vertical descent in two different species originating from a common parental species, then these two genes are referred to as orthologs ("ortho" means exact). Such genes encode proteins with same function in different species. By studying orthologous genes, one can identify the function of newly sequenced genes. If, on the other hand, a gene is duplicated in a species and occupies a different position in its genome, then these two gene copies are called as paralogs ("para" means parallel). Paralogs are further divided into in-paralogs and out-paralogs. In-paralogs are paralogous genes that arose after speciation, whereas out-paralogs are those genes which arose before speciation. Paralogous genes encode for proteins which have similar but not identical functions as during duplication one copy of the gene receives a mutation which generates a new gene with a new function but related to the original one. In addition to these, another term used is xenologs which refers to genes that have arisen due to horizontal gene transfer.

Comparison of viral genomes in the early 1980s helped in paving a way for the field of comparative genomics. One simple way of comparing genomes of two species is to compare their genome size, gene number and chromosome number. However, with the advancement in genome sequencing technologies which resulted in reduction in the time and costs of sequencing, there has been a huge amount of sequence data generated which is being annotated and deposited in open-access online repositories. Such data confers information regarding synteny (order of genes in chromosomes) and homology among species. Using the tools of bioinformatics, this genomic information can now be used to derive the function of newly sequenced genes by comparing them with that of a reference sequence. It can also be used to align homologous segments of protein and DNA sequences of different species in order to assess the phylogenetic distance between them. As of now comparative genomics is being useful in understanding the evolutionary relationships among various plants, vertebrates, invertebrates, bacteria and other eukaryotes species.

9.2 Comparative Genomics in Plants

Plants form an integral part of human existence on this planet as we depend on them for primary sources of energy. Since the dawn of agriculture, it has been human endeavour to improve crop plants in terms of quality and quantity in order to meet the ever-growing nutritional demands of the population. With limited agricultural land, problems created by crop diseases and climatic changes, it has now become a challenge to develop plant varieties that can sustain themselves in the face of biotic and abiotic stress factors and meet the global food demand. Researchers across the globe are trying to identify various qualitative and quantitative traits that can be used for improving crop plants through classical and molecular breeding. First step towards it includes identification of desirable phenotypic traits in crops associated with improved yield, nutrition and or disease resistance followed by selection and then breeding to produce improved crop varieties. Plant phenotypes are a result of differences in the expression of genes and the effect of environmental conditions on them. Therefore, one needs to have an understanding of the plant genomic structure, organisation and the interaction of its genes with environment in order to exploit these for establishing improved crop varieties. Plant phenotyping includes measuring the structural traits of plant organs like leaves, roots, fruits, etc. at various stages in their life cycle. Previously phenotypic traits were studied and collected manually. However now image-based methods are being used. The images generated are analysed in a high-throughput, robust and accurate manner to get reliable results. The data generated has been helpful in identifying genes/quantitative trait loci through QTL mapping, association mapping and genome-wide association studies (GWAS). This information has been utilised for genomics-assisted breeding (GAB) for crop improvement. Since plants are complex systems, studying their phenotypic traits comes with its own challenges. There is a large population and many species in plants that need to be studied. There is difficulty in understanding the mechanisms involved in interaction between plant and its environment and lastly is the issue of phenotypic plasticity. In the current scenario, plant genotyping has emerged to complement phenotyping methods. Molecular marker assays are being developed for discovering and mapping genes for crop improvement. Till now 50 SNP arrays and 15 GBS (genotyping by sequencing) have been developed for more than 20 crop species. But more needs to be done for developing high-throughput and cost-effective genotyping platforms for applied breeding programmes. These technologies will help in targeting traits of economic importance and functional polymorphism. Efforts are underway to develop sequence- or array-based genotyping methods that can have a range of application wide enough to include crops with different genetic backgrounds. The goals of such platforms are to discover genes associated with important traits, provide a deep understanding of quantitative traits and help in identification of genes and pathways responsible for important breeding traits and, last but not the least, cost-effectiveness. Data generated by these genotyping methods is stored in online databases and can be accessed for comparing sequences of crop plants with that of a model plant in order to identify DNA markers associated with traits of agronomic importance. It is very important to characterise gene and genome structure-function relationships in model plants because this information is to be used for finding the same in plants with complex genomes or those whose gene functions have not been elucidated so far. In a nutshell, genotyping assisted with comparative genomics can be used for identifying phenotypes in plants (Kono et al. 2018).

9.3 Genomic Resources for Plant Biology

Till now more than 55 plant species have their genomes sequenced and annotated in plant genomic databases. These include plant species that belong to Arabidopsis, maize, wheat, rice, etc. The first plant to have its entire genome sequenced was *Arabidopsis thaliana* in 2000 followed quickly by *Oryza* in 2001. Formerly, the first database created to harbour the genomic sequence of a plant species was Resource, TAIR (http://www.arabidopsis.org/). Prior to that GenBank served the purpose of storing sequence data from all species. For a long time, *Arabidopsis thaliana* remained our single model sequenced diploid genome, and rice was the only sequenced monocot genome. But with the passage of time as the sequencing tech become cheaper and high-throughput along with advancements in bioinformatics, the pace of discovery has also increased. By 2017 a total of 250 plant species have fully sequenced genomes (Wang et al. 2017).

The 1000 plant genome project (one KP or 1 KP) initiative announced in 2008 aimed at sequencing the transcriptomes of plants as opposed to their genomes. This approach was adopted to focus on biochemical pathways, so that only those genes whose products are involved in these pathways are studied. Baring the 90% genomes which has noncoding regions, the rest of the genome is capable of generating data sufficient for constructing evolutionary relationships through sequence comparison. This approach is fundamentally different from EST as the entire sequence of each gene is acquired with high coverage rather than just a small portion of the gene sequence with an expressed sequence tag (Figs. 9.1 and 9.2).

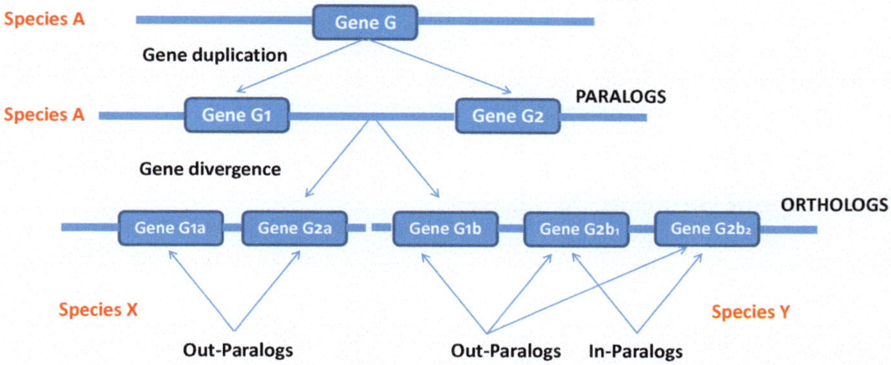

Fig. 9.1 Definition of orthologs, in-paralogs and out-paralogs

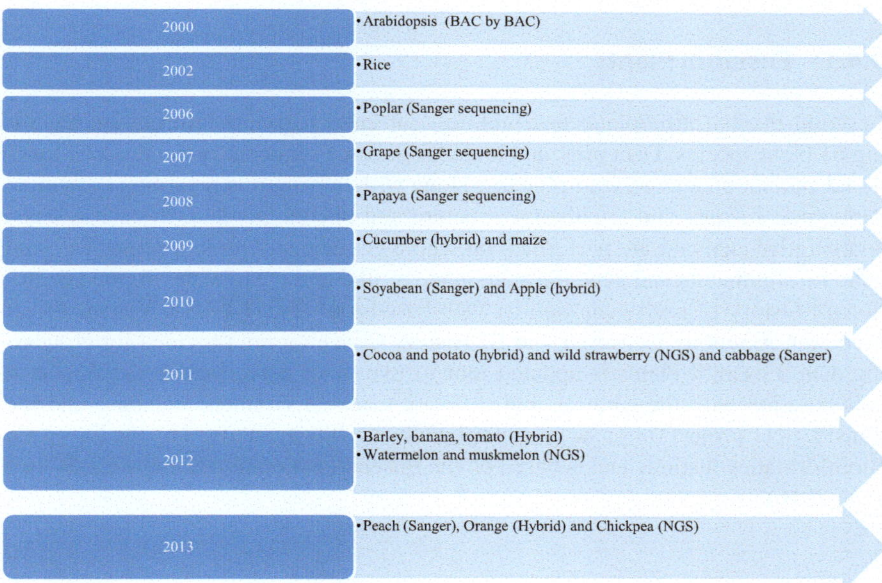

Fig. 9.2 Timeline of genome sequencing of various plant species

Such projects have uncloaked several species-specific details such as size of genomes, number of genes, sequence duplication pattern, syntenic relationships and presence of transposable elements. This information has been utilised during comparative genomic analysis to study the extents and rates at which changes occurred in the structure and size of genomes, large-scale genome duplications patterns (including polyploidization), how genes originate and become extinct, natural selection acting on large-scale genome structure and organisation variation. It has also revealed information regarding the factors that determine transposable element activity and what effects they have ultimately on the function and for epigenetic studies (Caicedo and Purugganan 2005). Currently, there are various bioinformatic tools available to carry out such comparisons among genomes. These tools are described in the following section.

9.4 Plant Genomic Databases

Plant genomic databases currently host sequencing information regarding approx. 145 plants. One simply needs to type in the species name and get the required information. Various algorithms are employed for comparing sequences and inferring results from them. Some dedicated genome browsers, such as GBrowse or JBrowse, serve as a graphical interface for users to browsing and analysing genomic sequences (Mochida and Shinozaki 2010). Below is the list of some of the comparative genomic databases.

9.4.1 Ensembl Plants

Ensembl Plants (http://plants.ensembl.org) currently harbours information regarding 61 plant species. Data provided includes genome sequence, gene models, functional annotation and polymorphic loci along with information regarding individual genotypes, linkage and population structure and phenotype data. In each release, comparative analyses are performed on whole genome and protein sequences, and genome alignments and gene trees are made available that show the implied evolutionary history of each gene family. Web-based tools like BLAST, Variant Effect Predictor, Assembly Converter and ID History Converter are provided for processing data. Ensembl Plants is updated four to five times a year and is developed in collaboration with the curation and software teams at Gramene database and the transPLANT project (http://www.transplantdb.eu). It is maintained by the European Bioinformatics Institute and is based on the Ensembl annotation framework (Bolser et al. 2016).

9.4.2 Gramene

Gramene (http://www.gramene.org/) is a curated, open-source, integrated data resource for comparative functional genomics in crops and model plant species. It is used for comparing plant genomes and biological pathways currently hosting 61 sequenced reference genomes. Gramene uses two Ensembl comparative analysis pipelines for (i) pairwise whole-genome alignments at the DNA level and (ii) phylogenetic gene trees with classification of ortholog and paralog gene relationships. Output from either of these methods may be further used to build synteny maps. Gramene currently incorporates SNP and/or structural variation datasets also. It uses the same web-based tools as Ensembl Plants (Monaco et al. 2014).

9.4.3 PlantGDB

PlantGDB (http://www.plantgdb.org/) provides sequence data for >70,000 plant species, custom EST assemblies (PUT) for over 150 species, web tools and plant genome browsers, as well as an outreach portal for plant genomics. Sequence data from GenBank and UniProt is downloaded every 4 months corresponding to every other GenBank release. Individual sequences are filtered to detect vector and repeat sequence. The spliced transcripts are aligned accurately with the genome for correct genome annotation. From July 2015 the PlantGDB's funding has ended, and the website is no longer being updated (Duvick et al. 2008).

9.4.4 PlantsDB

PlantsDB (http://pgsb.helmholtz-muenchen.de/plant/index.jsp) is a database that has been developed by the plant genomics group of the PGSB (Plant Genome and Systems Biology), formerly MIPS. The PGSB plant genomics group focuses on the analysis of plant genomes, using bioinformatics techniques. The PlantsDB provides a platform for integrative and comparative plant genome research. It has information regarding triticeae genome data, especially for barley, wheat and rye. PlantsDB uses web-based tools like Crows Nest, to visualise syntenic relationships between genomes. Data can be exchanged and cross-linked between PlantsDB and other plant genome database by the transPLANT project (http://transplantdb.eu/) (Spannagl et al. 2016).

9.4.5 Phytozome

Phytozome (http://phytozome.jgi.doe.gov/pz/portal.html), the Plant Comparative Genomics portal of the Department of Energy's Joint Genome Institute, is a hub for accessing, visualising and analysing JGI-sequenced plant genomes, as well as selected genomes and datasets that have been sequenced elsewhere. Phytozome v12.1.6 hosts 93 assembled and annotated genomes, from 82 Viridiplantae species. All gene sets in Phytozome have been annotated with KOG, KEGG, ENZYME, Pathway and the InterPro family of protein analysis tools. InParanoid pairwise orthology and paralogy groups have been calculated across all Phytozome proteomes. Families of related genes representing the modern descendants of putative ancestral genes are constructed at key phylogenetic nodes. These families provide additional insight into clade-specific orthology/paralogy relationships as well as clade-specific novelties and expansions. Query-based data access is provided by Phytozome's InterMine and BioMart instances, while bulk data sets can be accessed via the JGI's Genome Portal. JBrowse genome browsers are available for all genomes (Goodstein et al. 2012).

9.4.6 PLAZA

PLAZA (http://bioinformatics.psb.ugent.be/plaza) is an online resource for comparing plant genomes and studying their evolutionary and functional genomics. Each PLAZA entry contains structural and functional gene annotations, gene family data and phylogenetic trees and detailed gene colinearity information. The PLAZA 4.0 version consists of entries for Dicots 4.0 and Monocots 4.0. The present version of this database has information for 71 species as compared to the previous version that had only 37 species. The new PLAZA instances can also be accessed through a RESTful web service (Proost et al. 2015).

9.4.7 GreenPhylDB

GreenPhylDB (www.greenphyl.org/cgi-bin/index.cgi) is a web resource designed for comparative and functional genomics in plants jointly developed by Biodiversity International and the International Cooperation Center for Agricultural Research for Development (CIRAD). The database contains a catalogue of gene families based on gene predictions of genomes, covering a broad taxonomy of green plants. Result of our automatic clustering is manually annotated and analysed by a phylogenetic-based approach to predict homologous relationships. It supports evolution and functional studies to identify candidate gene affecting agronomic traits in crops (Rouard et al. 2011).

9.4.8 PlantOrDB

PlantOrDB (http://bioinfolab.miamioh.edu/plantordb) is a database for land plants and green algae. It classifies genes in families and identifies orthologous gene clusters (Li et al. 2015).

9.4.9 SALAD

SALAD (http://salad.dna.affrc.go.jp/salad/) is an acronym for Surveyed conserved motif ALignment diagram and the Associating Dendrogram. It is a comparative genomics database which uses proteome data derived from plant genome for comparison among species. This is because proteins with similar motifs share similar biochemical properties and have similar biological functions. These motifs are evolutionarily conserved; therefore they can be used to study evolutionarily relationships among species which is represented in the database by pre-computed dendrogram. In addition to this, linked to the same dendrogram, there is 'SALAD on ARRAYs' which is for viewing arbitrary microarray data sets of paralogous genes (Mihara et al. 2010).

9.4.10 PlantGenIE.org

PlantGenIE.org (The Plant Genome Integrative Explorer) is a collection of interoperable web resources for searching, visualising and analysing genomica and transcriptomics data from different plant species. Currently it includes dedicated web portals for enabling in-depth exploration of poplar, Norway spruce, and Arabidopsis. The PlantGenIE platform uses Chado databases and is based on the GenIE CMS (content management system) (Sundell et al. 2015).

9.4.11 POGs2

Putative Orthologous Groups database (http://pogs.uoregon.edu/) is a database designed to compare the information generated from proteomes of four plant species (*Arabidopsis thaliana*, *Zea mays*, *Oryza sativa* and *Populus trichocarpa*). It uses Gramene's ENSEMBL orthology prediction output. It uses proteome data from PPDB and gene annotations imported from TAIR, Rice Annotation Project, MaizeSequence.org and Phytozome.net (Tomcal et al. 2013).

9.4.12 Genomicus Plants

Genomicus (http://genomicus.biologie.ens.fr) is a database for flowering plants genomes. It enables users to navigate in genomes in several dimensions: linearly along chromosome axes, transversally across different species and chronologically along evolutionary time. Once a query gene has been entered, it is displayed in its genomic context in parallel to the genomic context of all its orthologous and paralogous copies in all the other sequenced angiosperm genomes. Moreover, Genomicus stores and displays the predicted ancestral genome structure in all the ancestral species within the phylogenetic range of interest. All the data on extant species displayed in this browser are from Ensembl Genome Plant, Phytozome, and specific servers (Louis et al. 2015).

9.4.13 Piece

Plant Intron Exon Comparison and Evolution database (https://wheat.pw.usda.gov/piece/) is a gene structure comparison and evolution database with information regarding 25 species. It compares exon-intron plant gene structures and their evolutionary and functional relationships. Annotated genes were extracted from the species and classified based on Pfam motif. Phylogenetic tree was reconstructed for each gene category integrates exon-intron and protein motif information (Wang et al. 2013).

9.4.14 PlantSEED

PlantSEED (http://modelseed.org) is a database for metabolic model reconstruction for plant genomes. It has information regarding primary metabolism of 39 plant species. It provides tools that help create metabolic reconstruction from transcriptome data by using algorithms that can identify metabolic enzymes with 97% accuracy. Users can upload their data, reconstruct and simulate metabolic activities of the proteins to understand and engineer metabolism in plants (Seaver et al. 2014).

9.4.15 PGDBj

PGDBj (http://pgdbj.jp) stands for Plant Genome Database Japan. It has an integrative approach for plant genome-related information from other databases. It has a database for orthologs, which comprises clusters of homologous sequences. PGDBj also provides information regarding DNA markers for breeding and QTL that can be used for crop improvement (Asamizu et al. 2014).

9.5 Conclusion

Complete genome sequences from model crops like rice, maize and wheat and partial genomic information for many other plants are helping us in understanding their genomic structure and organisation. Data generated from other sources like microarrays, high-throughput resequencing technology and EST has added to this knowledge. By comparing genomes of two or more plant species, one can infer the rate at which a particular gene or gene family evolved and study gene loss or retention due to duplications and chromosomal rearrangements which have contributed towards variation among species. In future with the availability of more information regarding sequenced genomes of other plant species along with the usage of bioinformatics tools, our understanding of the relationships among genes of different crop species, isolation of novel genes and determination of their functions will be further improved. As far as crop improvement is concerned, comparative genomics is helping in the identification of regions in the genome associated with traits of interest and DNA markers for marker-assisted breeding.

References

Asamizu E, Ichihara H, Nakaya A, Nakamura Y, Hirakawa H, Ishii T, Tamura T, Fukami-Kobayashi K, Nakajima Y, Tabata S (2014) Plant Genome DataBase Japan (PGDBj): a portal website for the integration of plant genome-related databases. Plant Cell Physiol 55(1):e8

Bolser D, Staines DM, Pritchard E, Kersey P (2016) Ensembl plants: integrating tools for visualizing, mining, and analyzing plant genomics data. Methods Mol Biol 1374:115–140

Caicedo AL, Purugganan MD (2005) Comparative plant genomics. Frontiers and prospects. Plant Physiol 138(2):545–547

Duvick J, Fu A, Muppirala U, Sabharwal M, Wilkerson MD, Lawrence CJ, Lushbough C, Brendel V (2008) PlantGDB: a resource for comparative plant genomics. Nucleic Acids Res 36:959–965
Goodstein DM, Shu S, Howson R, Neupane R, Hayes RD, Fazo J, Mitros T, Dirks W, Hellsten U, Putnam N, Rokhsar DS (2012) Phytozome: a comparative platform for green plant genomics. Nucleic Acids Res 40:1178–1186
Kono TJY, Lei L, Shih CH, Hoffman PJ, Morrell PL, Fay JC (2018) Comparative genomics approaches accurately predict deleterious variants in plants. G3 (Bethesda) 8(10):3321–3329
Li L, Ji G, Ye C, Shu C, Zhang J, Liang C (2015) PlantOrDB: a genome-wide ortholog database for land plants and green algae. BMC Plant Biol 15:161
Louis A, Murat F, Salse J, Crollius HR (2015) Genomicus plants: a web resource to study genome evolution in flowering plants. Plant Cell Physiol 56(1):e4
Mihara M, Itoh T, Izawa T (2010) SALAD database: a motif-based database of protein annotations for plant comparative genomics. Nucleic Acids Res 38:835–842
Mochida K, Shinozaki K (2010) Genomics and bioinformatics resources for crop improvement. Plant Cell Physiol 51(4):497–523
Monaco MK, Stein J, Naithani S, Wei S, Dharmawardhana P, Kumari S, Amarasinghe V, Youens-Clark K et al (2014) Gramene 2013: comparative plant genomics resources. Nucleic Acids Res 42:1193–1199
Proost S, Van Bel M, Vaneechoutte D, Van de Peer Y, Inzé D, Mueller-Roeber B, Vandepoele K (2015) PLAZA 3.0: an access point for plant comparative genomics. Nucleic Acids Res 43:974–981
Rouard M, Guignon V, Aluome C, Laporte MA, Droc G, Walde C, Zmasek CM, Périn C, Conte MG (2011) GreenPhylDB v2.0: comparative and functional genomics in plants. Nucleic Acids Res 39:1095–1102
Seaver SM, Gerdes S, Frelin O, Lerma-Ortiz C, Bradbury LM, Zallot R, Hasnain G, Niehaus TD, El Yacoubi B et al (2014) High throughput comparison, functional annotation, and metabolic modelling of plant genomes using the PlantSEED resource. Proc Natl Acad Sci U S A 111(26):9645–9650
Spannagl M, Nussbaumer T, Bader KC, Martis MM, Seidel M, Kugler KG, Gundlach H, Mayer KF (2016) PGSB PlantsDB: updates to the database framework for comparative plant genome research. Nucleic Acids Res 44:1141–1147
Sundell D, Mannapperuma C, Netotea S, Delhomme N, Lin YC, Sjödin A, Van dePeer Y, Jansson S, Hvidsten TR, Street NR (2015) The plant genome integrative explorer resource: PlantGenIE. Org. New Phytol 208(4):1149–1156
Tomcal M, Stiffler N, Barkan A (2013) POGs2: a web portal to facilitate cross-species inferences about protein architecture and function in plants. PLoS One 8(12):e82569
Wang W, Cao XH, Miclăuş M, Xu J, Xiong W (2017) The promise of agriculture genomics. Int J Genomics 5(9743749):2
Wang Y, You FM, Lazo GR, Luo MC, Thilmony R, Gordon S, Kianian SF, Gu YQ (2013) PIECE: a database for plant gene structure comparison and evolution. Nucleic Acids Res 41:1159–1166

Potential of Wild Species in the Scenario of Climate Change

10

Suzana Jordanovska, Zoran Jovovic,
and Violeta Andjelkovic

Contents

Abstract

Crop wild relatives (CWR) as an enormous source of genetic diversity for crop improvement are under great climate change pressure today, as well as from adverse effects caused by excessive exploitation, fragmentation, degradation, and loss of habitat. Ex situ conservation has been the main strategy for their protection until recently. Nowadays, more attention is paid to in situ conservation. Preservation of wild relatives in their natural environments allows the continuation

S. Jordanovska (✉)
Institute of Agriculture Skopje, University "Ss Cyrilus and Methodius" in Skopje, Skopje, Republic of Macedonia

Z. Jovovic
Biotechnical Faculty Podgorica, University of Montenegro, Podgorica, Montenegro

V. Andjelkovic
Maize Research Institute 'Zemun Polje', Belgrade, Serbia
e-mail: avioleta@mrizp.rs

R. K. Salgotra, S. M. Zargar (eds.), *Rediscovery of Genetic and Genomic Resources for Future Food Security*,
https://doi.org/10.1007/978-981-15-0156-2_10

of evolutionary processes and the creation of a new genetic variation capable of adapting to changing environmental conditions. This approach, besides increasing the interest for wild parents, also enables the creation of a much wider genetic basis needed to create varieties resistant to various stress situations.

Agricultural production based on current diversity faces numerous challenges caused by intense climate change. Without the use of germplasm of wild relatives and continuous genetic expansion, the creation of high-yielding varieties with tolerance to biotic and abiotic stresses will be increasingly uncertain in the future, while yields of agricultural crops will grow much slower or even decline. Therefore, the genetic variation contained in crop wild relatives will be of increasing importance for the improvement of crops, in particular for tolerance to biotic and abiotic stress.

Acronyms

ABA	Abscisic acid
BNF	Biological nitrogen fixation
Bt	*Bacillus thuringiensis*
CWR	Crop wild relatives
CO_2	Carbon dioxide
FACE	Free-Air Carbon dioxide Enrichment
HNVs	High natural values
HSPs	Heat shock proteins
IPCC	Intergovernmental Panel on Climate Change
IFAD	International Fund for Agricultural Development
PGR	Plant genetic resources
PROSEA	Plant Resources of Southeast Asia
PPP	Plant protection products
QTL	Quantitative trait loci
ROS	Reactive oxygen species
TCA	Tricarboxylic acid cycle
UNICEF	United Nations International Children's Emergency Fund
WEF	World Economic Forum
WEPs	Wild edible plants
WHO	World Health Organization

10.1 Introduction

All of the living beings on the planet have been under numerous and enormous environmental pressures during their long evolutionary history and, thanks to their adaptive qualities, managed to successfully respond to them (Reusch and Wood

2007). Since nowadays environment is changing faster than ever, it is not known if the living world has an adaptive capacity sufficient to adapt to such changes (Thomas et al. 2004).

Crop wild relatives (CWR) are a valuable source of genetic variation, and therefore their significance for modern agriculture is enormous. Wild relatives possess a wide range of potentially useful genes important for food safety and adaptation to climate change (Damiana 2008). The ever-present loss of CWR germplasm and habitat suggests that climate change and climate variability in the future will increasingly affect the viability of their survival. It is therefore necessary to approach their protection as soon as possible in order to preserve these resources while the diversity they possess is still available. Although there are conserved seeds of many CWR species (ex situ conservation) in world banks, this is still only a small fraction of the overall genetic variability that exists in the CWR. In addition, only a small proportion of conserved accessions have been characterized, which additionally restricts their use. Ex situ collections in gene banks are isolated from natural selection processes, thus preventing the development of new variations capable of gradually adapting to changing environmental conditions (Pressey et al. 2003). For this reason, in situ conservation is perceived as a conservation of evolutionary processes. In order to maintain the highest possible reservoir of genetic diversity, it is necessary to complement the in situ conservation with ex situ collections, since this enormous wealth of nature, regardless of whether it is a genetic material of actual or potential value, must be permanently preserved.

In the process of domestication, genetic uniformity increases, and cultural plants end up with less genetic variation than those of wild species. Greater homogenization of crops increases their vulnerability to biotic and abiotic stresses. The enormous genetic diversity provides CWR with a very high degree of adaptability within quite wide boundaries of ecological conditions (Nevo 2004). For this reason, CWR has been used for more than 100 years in formal breeding programs to improve tolerance to stressful abiotic conditions and to increase resistance to disease and pests (Sheehy et al. 2005). The extent of use of CWR in plant breeding significantly varies among crops. Crops with long history of breeding benefit most from wild relatives. Due to well-established pre-breeding programs and the use of advanced genomic tools, these activities are most advanced in wheat, rice, and tomatoes (Nemeth et al. 2015). On the other hand, in some crops such as maize, alfalfa, cassava, chickpea, cowpea, sweet potato, etc., a significant number of genes identified as potential sources of specific properties have been discovered, but for now there is not enough interest in their widespread application in breeding programs.

Crop genetic improvement for higher tolerance to biotic and abiotic stresses by the introgression of genes important for agriculture from CWR is one of the promising approaches to increase crop stability and productivity (Xie and Nevo 2008), but also to transfer genetic tolerance of wild relatives from experimental fields in primary agricultural production. Therefore, one of the most important goals of future plant breeding strategies should be the creation of tolerant varieties to various forms of biotic and abiotic stress, which will not be possible without the transfer of new genetic variation from CWR. Greater availability of CWR variability creates the conditions for more intense research of mechanisms of stress tolerance and opens up the possibility of discovering new superior alleles that are not present within the elite germplasm. There is a high likelihood that many of these properties can be found in the CWR.

Global climate change and an increase in average temperatures will undoubtedly lead to further increase in spatial distribution and inertness of existing pests, diseases, and weeds, as well as the emergence of new pests. The obvious need to find new sources of resistance directs researchers to wild relatives of crop species.

10.2 CWR as a Source of Novel Genetic Variation for Crop Responses to Biotic and Abiotic Stresses

Climate is one of the main factors on which CWR distribution depends. Relatively stable climatic conditions in the previous century had a positive effect on the distribution, abundance, phenology, and physiology of a large number of CWR, which led to the movement of their distribution toward the poles and higher altitudes. The global climate change has been significantly accelerated over the past 30 years (Osborn and Briffa 2005). Some modeling studies indicate that the mean temperature by 2025 could be increased by 0.4–1.1 °C, by 2050 0.8–2.6 °C, and by 2100 for 1.4–5.8 °C (IPCC 2007), which would certainly lead to the disappearance of wild plant species in vast proportions (Thomas et al. 2004). Therefore, there is no doubt that climate change in the future will increasingly affect the survival of wild species and require much greater mobility in their protection.

Today, there is not even the smallest corner on the planet where CWR are not at least to some extent exposed to some kind of stress or a combination of a number of stress factors. These populations are subjected to local environmental conditions and thus to global changes (climate change, land-use change, habitat destruction, etc.). Ecological factors determine living conditions in one habitat, so living beings are forced to constantly adapt to new conditions. Adaptation of CWR to habitat conditions is based on structural and functional modifications. CWR can defend against main abiotic stresses in different ways. Natural populations of plant species can follow the appropriate climate and thus move into areas where the climate conditions are suitable or adapt to the new conditions in the areas of their natural distribution. Different types of abiotic stress affect the plants individually, but often in combination, causing numerous morphological, physiological, biochemical, and molecular changes that negatively affect the growth, development, and productivity of plants.

Drought, salinity, and heat are major abiotic stresses that dramatically endanger global food supply. Drought in plant production occurs due to the lack of water in the root zone and regularly leads to a decrease in yield (Salekdeh et al. 2009). Nature during the evolution ensured that plants, including CWR, adapt in different ways to different environmental pressures. CWR tolerance to drought is their ability to maintain their functions at the level of a certain water potential in two ways – one is to avoid drought by not growing or growing slowly in the dry season, while the other is based on an active adaptation to the arid conditions (Munns et al. 2006). Water tolerance to drought is obtained through various physiological, morphological, and other adaptations that allow plants to establish a balance between the adoption and loss of water or by means of adaptations that ensure that the plant is tolerant to water deficit. The reaction of plants to drought is a very complex phenomenon due to unpredictable factors in the environments, but also interactions with other abiotic and biotic factors. Plants that have the ability to avoid drought factors (Reynolds et al. 2006) usually have a very well-developed root system, a weaker vegetative growth, and a specific structure of stomas that react to an increased evaporation intensity. On the second hand, the main mechanism of tolerance is osmotic adaptation. Tolerance on lack of water enables plants to maintain all important functions in the conditions of a reduced turgor. The adaptation of this type is characteristic for xerophytes (plants of dry habitats), but also for halophytes (plants of saline habitats).

The reaction of plants to different environmental conditions is conditioned by their genetic properties, as well as by interaction with the external environment (Jovovic et al. 2012). Gene expression is strongly influenced by environmental factors, so genes that contribute to yield in severe drought conditions cannot function under conditions of moderate water saturation, while under well-water conditions can even cause adverse effects. For this reason, drought tolerance testing should be performed in special targeted environments (Von Korff et al. 2008).

In contrast to drought tolerance, salt tolerance is based on the control of the salt absorption by the roots and its transport within the plant organism (Munns and Richards 2007). Due to the pronounced interaction of the genotype x environment, the level of drought tolerance is, unlike salt tolerance, very difficult to estimate. In general, CWR that shows drought tolerance also exhibits tolerance to increased salt concentration in soil (Farooq and Azam 2001; Farooq et al. 1989). *Triticum dicoccoides* and *Hordeum spontaneum*, progenitors of cultivated wheat and barley, have a very pronounced potential for drought tolerance and increased soil salinity, as confirmed by numerous experiments. The genes responsible for the drought and salt tolerance and quantitative trait loci (QTLs) identified in these progenitors have enormous significance in future wheat and barley breeding programs (Xie and Nevo 2008). Useful properties of CWR by using marker-assisted selection could be introduced into elite germplasm which would accelerate the use of the non-random genomic variation that these wild relatives possess (Waugh et al. 2009). As a proven worthwhile source of drought and salt tolerance, these wild species have been gaining importance recently. Due to differences in salinity, climate conditions, properties of irrigation water, agricultural practices, etc., a real salt-tolerant wheat genotype derived from CWR has still not been developed to be used in wider production practice. Therefore, in the coming period, more attention should be paid to varieties that demonstrate a high level of tolerance in specific local conditions. Although these resources may provide great advances in crop improvement, most of them are still largely neglected.

Global warming will generally have a negative impact on plant production due to the harmful effects of high temperatures on plant development. The increasing frequency of climatic extremes, including very high temperatures, can lead to an incomprehensible loss of productivity of crops (Jovovic et al. 2016). Stress caused by high temperatures has multiple effects on plants in terms of its morphology, physiology, biochemistry, and genetic regulation pathways. With a large number of crop species, the impact of high temperature stress is more pronounced on reproduction than on vegetative growth. An abrupt fall in yield in high temperature conditions is mainly associated with pollen infertility (Zinn et al. 2010). Although it is often increased by additional abiotic stresses such as drought and salt stress, heat stress exerts an independent way of action on the physiology and metabolism of plant cells. Although the effect of heat stress on the plant depends to a large extent on the stage of plant development, it has to some extent affected all vegetative and reproductive stages. In addition, the influence of high temperatures also depends on species and genotype, with abundant inter- and intraspecific variations.

During the evolution, the plants developed various mechanisms to combat high temperatures. They are divided into long-term phenological and morphological evolutionary adaptations and short-term stress avoidance and acclimation mechanisms. Stress proteins, ion transporters, osmo-protectants, free-radical scavengers, etc., have a key role in controlling stress effects (Wang et al. 2004). The ability to maintain leaf gas exchange in conditions of heat stress is directly related to heat tolerance. One of the important mechanisms of adaptation to extreme temperatures is the accumulation of osmo-protectants that directly participate in osmotic adaptation. Accumulation of proline, glycine betaine, and soluble sugars has a crucial role in the regulation of osmotic activities and the protection of cellular structures (Farooq et al. 2008). There are a number of options for improving the resistance of plants to heat stress, because the tolerance to high temperatures has a multigene character. It is therefore important to develop a set of markers of heat tolerance that will be used in further heat tolerance crop improvement. Identification of adaptive QTLs for resistance to high temperature plants is one of the ways to understand the mechanisms of tolerance (Roy et al. 2011). Various properties were used in mapping QTLs, such as thousand grain weight, grain filling duration, grain filling rate, canopy temperature depression, yield, etc. (Pinto et al. 2010). A realistic assessment of the impact of the genotype on the resistance of plants to heat stress is possible only through field research because laboratory testing does not reflect the conditions governing the field. The lack of a precise phenotyping protocol is a limiting factor in the genetic analysis of quantitative traits and an important prerequisite for identifying complex genetic networks associated with QTLs. In addition, the problem is also regulatory issues associated with GMOs, as well as the length of the process of breeding for tolerance. For the above reasons so far, very few tolerant crop varieties have been created.

An increased amount of water-soluble salts in the soil causes hyperionic and hyperosmotic stress in the active rhizosphere layer that slows down the growth of plants and significantly reduces yield (Shrivastava and Kumar 2015). Salinity stress is achieved through the inhibition of osmosis, the adoption of water through the root, or the direct effect of certain ions on the plant. Even at a relatively low salinity level, the concentration of salt in the soil is significantly higher than the level required for the normal functioning of the plant organism. Due to active leaf transpiration, salt accumulation occurs in the initial stages of growth, which causes the photosynthetic activity of leaves to decrease and, later, their premature death. Increasing the concentration of salt in the soil can be so severe that it causes serious problems in growing the plants.

The mechanisms of tolerance to salinity are directed to ionic imbalance in the cells or toward the minimization of osmotic stress. The ability to adopt the K+ ion and the high K+/Na+ ratio is one of the key elements of tolerance to salinity. Increased Na + ion concentration in the soil leads to their increased absorption by the plant. On the other hand, the adoption of K+ ions decreases, resulting in a cell functionality disorder (Genc et al. 2007).

To fight against salinity stress, plants are also struggling with the activities of certain genes. Bread wheat combats using slower transport of Na+ from the roots to the aboveground parts of the plant and using high ratio of K+/Na+ ions in leaves (Gorham 1990). These mechanisms are under the influence of three major genes:

- Kna1 – mapped to the long arm of chromosome 4 in genome D (Gorham et al. 1987)
- Nax1 – mapped to a longer arm of chromosome 2 in genome A (Lindsay et al. 2004)
- Nax2 – mapped at the distal end of the long arm of chromosome 5 in genome A (Nelson et al. 1995)

These three genes belong to the same gene family labeled HKT (high-affinity K+ transporter). Bread wheat inherited gene Kna1 from the wild species of *Aegilops squarrosa*, while *Triticum monococcum* was the donor of Nax1 and Nax2 genes (Gorham et al. 1997). Gene Nax1 influences the exclusion of Na+ from the root, while Nax2 is responsible for decreasing Na+ concentration in leaves (Byrt et al. 2007). The mechanism of action of Kna1 is similar to that of *Nax2* gene.

Living in community with pests and pathogens, encompassing bacteria, fungi, oomycetes, viruses, nematodes, and insects, plants have built up their own defense systems that, with more or less success, oppose the pathogen. The plant is healthy when it normally performs its physiological functions (cell division, adsorption of water and nutrients, photosynthesis, reproduction, etc.), while in the event of pathogen attack, there is a disorder in the functioning of the plant organism. Plant resistance to pathogens represents their ability to resist the penetration and spread of the parasite in plant tissues, as well as the resistance to the products of their metabolism. If all the plants within a single plant species exhibit complete resistance to infection, then it is an extremely effective resistance – immunity. The resistance and sensitivity of plants to pathogens are two extremes, among which there are a whole series of transitional types. A particular group in this sense is made up of tolerant plants. Tolerant plants in fungal and bacterial infections indicate those individuals that show the symptoms of the disease in volume and intensity as sensitive, while their yield is at the level of resistant plants. In the case of viral infection, tolerant plants are those plants that do not respond to infection with visible symptoms. Plant defense mechanisms against pathogens can be classified into two basic groups: passive (pre-infectious) and active (post-infectious).

Under the influence of various types of biotic stress in the world, more than 42% of the potential crop yield is lost annually – 15% can be attributed to insects, 13% to weeds, and 13% to other pathogens. For these reasons, finding mechanisms to reduce the resulting losses can be one of the more important options for increasing total production (Pimentel 1997), so improving crop for biotic stress tolerance appears as a long-term goal of many breeding programs. This goal can be achieved through:

- Improvement of plant material (use of tolerant/resistant plant material)
- Improving the health condition of the roots (field rotation, soilborne disease control, etc.)
- Improving irrigation practices (providing the optimal water quality and availability)
- Prevention of airborne hazards (foliar diseases, etc.) (Cook 2000)

The first resistant plants were created in traditional breeding systems more than 100 years ago (wheat disease resistance breeding programs), when the mechanism of action of resistant genes was unknown. During this period, new sources of resistance for crop improvement breeders were found by classical identification (Haverkort et al. 2008). Since 1993, genetic engineering has been used to introduce genes of interest to existing high-yielding varieties without undesired pleiotropic effects. So far, several transgenic R genes have been introduced into cultivated crops. Some of them originate from CWR germplasm (Dangl et al. 2013). Compared to classical breeding, this sort of selection method reduces screening time by 15 years. However, by deploying single R genes into the field, it has been shown that this can be a temporary solution and that the introduced resistance is defeated after a certain period of time (i.e., the Brassica Rlm1) (Sprague et al. 2006). There are several strategies to overcome this problem. One of them is the use of multiline varieties that differ in their gene-specific resistance. Multiline varieties reduce inoculums of disease compared to a susceptible monoculture. The second approach is based on combining several R genes. The deployment of multiple NLRs at once in a single cultivar ensures that in the event that the pathogen mutates to overcome one R gene, the other resistance sources take over the protection of crops. Such an approach has proven to be highly efficient in potato cultivars (Jo et al. 2014).

In recent decades, significant efforts have been made to create plants resistant to insects. For this purpose in classical selection, three methods are used:

- Morphological barrier method for insects
- Insect-repelling substance method
- Method of introducing toxic substances

In all three cases, it is a host nonspecific resistance that implies that certain plant species are not attractive to insects for which immunity is introduced. So far, with the help of conventional hybridization, a significant number of disease- and insect-resistant hybrids and cultivars have been created using a resistance gene from a variety of sources. Since this method of introducing resistance is slow and unpredictable, this increasingly turns to a direct way of introducing the resistance gene into the plant organism. So for this purpose from the *Bacillus thuringiensis* strain, it is isolated gene that determines the synthesis of proteins with insecticidal activity (Bt). Transfer Bt gene from the bacteria to the plant species significantly reduces the damage caused by an insect attack compared to the use of synthetic insecticides (Van Frankenhuyzen 2009). The advantage of this method compared to classical insecticides is also the fact that the cultivation of Bt plants provides protection of the whole plant, there are no residues of insecticides in the soil, and there is no harmful effect on useful insects, mammals, and humans. Some potential defects of Bt plants are the possibility that the bacterial gene is not synthesized in sufficient quantity but also that in time the insect resistance can occur (Agaisse and Lereclus 1995).

10.3 The Effect of Climate Changes on Crop Physiology

Global climate change is bringing unpredictable responses of plants to new meteorological and environmental conditions. Anthropogenic factors caused ecological changes that are faster than the ability of plants to respond and adapt to new conditions. For most of plants, existing diversity is appropriate source for adaptation, but the others are lacking sufficient genetic variability for new conditions. For example, environmental changes occurred vary rapidly compared to long lives of trees and their slower adaptation to new conditions, together with consequences of biotic and abiotic stresses (Lindner et al. 2010). Existing genetic variability is a very important factor that limits plant responses and adaptations to rapidly changing conditions.

Environmental factors have different effect on every organ and/or tissue, at the molecular and cellular levels or physiological and morphological processes through vegetative and reproductive phases. The capability of plants to respond to new environment is of high importance for plant production and agriculture worldwide. Primary climate change factors that affect and are going to influence crop production in the future are increasing atmospheric CO_2, rising temperatures, and changing precipitation amount and patterns (Lobell and Gourdji 2012). Abiotic stress factors impair plant growth and development, initiating a wide range of responses, at the whole plant level, and likewise growth and productivity or changes in gene expression and metabolic processes. Although individual effects of above factors are well studied, consequences of their interactions, as well as physiological implications of changing climate to crop development, are difficult to predict, especially in small regions.

Crop physiology comprehends fundamental processes such as photosynthesis, respiration, water, nutritional and hormonal status, and allocation of assimilates to different plant organs (Mohammed and Tarpley 2009). Photosynthesis is the most important process in plants, depending on leaf chlorophyll content, stomatal conductance, and enzymatic activities. Plant respiration is another important process, significantly affected by abiotic stress factors. Increase of respiration as a consequence of global warming implicates greater consumption of photo assimilates. Some estimations are that carbon lost through respiration is within the range of 30–70% obtained from photosynthesis, and rising of temperature provoked increase in respiration up to 40% (Loka and Oosterhuis 2010).

The atmospheric concentration of CO_2 is predicted to rise to 1000 ppm by 2100. As photosynthetic organisms, plants remove about 150–175 Pg of carbon per year (Welp et al. 2011) and release about half via respiration.

Generally, plant responses to different stress factors are tested under controlled laboratory conditions. However, Leakey et al. (2009) applied increased CO_2 in field conditions via Free-Air Carbon dioxide Enrichment (FACE) technique. It is widely confirmed that increased concentration of CO_2 increased photosynthetic assimilation, and on various species in FACE, elevated CO_2 (475–600 ppm) increased net photosynthesis by about 40%. Concentrations of CO_2 are important for stomata opening, e.g., opened stomata enable carbon dioxide influx into leaves for photoassimilation, and water diffusion out of plant leaves. Stomatal conductance, under increased CO_2, decreases water loss in whole plant by 5–20% (Ainsworth and Rogers 2007), which influences total plant water status and distribution of water through plant tissues and organs.

Besides effects on photosynthesis and water loss, elevated CO_2 influences secondary changes in plant physiology; greater assimilation induces faster plant growth and higher dry matter accumulation of 17% in aboveground part to about 30% in the soil (Ainsworth and Long 2005), as well as yield increase in wheat, rice, and soybean of 12–14% (Ainsworth 2008). Elevated CO_2 increased aboveground biomass of grassland by about 33%, depending on available water and nitrogen that were lower under drought conditions (Reich et al. 2014). Increased CO_2 influenced changes in chemical content of plants in FACE experiments: sugar and starch content raised by 30–40%, and nitrogen content per leaf unit mass decreased by 13% (Ainsworth and Long 2005). Possible explanations of nitrogen decrease are dilution of nitrogen, less uptake of water that reduces mineral uptake from the soil, and decrease of nitrate assimilation in organic molecules (Taub and Wang 2008; Bloom et al. 2010). Changes in nitrogen content are closely influencing protein content, namely, protein content in cereal grains and potato tubers decreased by 5–14% in increased CO_2 conditions, as well as mineral content (Taub and Wang 2008).

Another important determination of plants in increased CO_2 conditions is photosynthetic type: majority of plants (about 90%) are using C3 process, while the rest of species are using either C4 or CAM photosynthesis.

Most tropical and subtropical plants, including important cereals, such as maize, sorghum, sugarcane, and millet, are C4 plants. They accumulate CO_2 in the leaf by biochemical processes, and additional atmospheric CO_2 is not significantly influencing photosynthesis, but stomatal closure reduces water loss. In FACE experiment, stimulation in photosynthesis by increased CO_2 concentration was only 30% in C4 plants, compared to obtained assimilation in C3.

Contrary, C3 plants, like legumes, are increasing photosynthesis and growth in response to elevated CO_2 concentrations. Numerous studies showed that legumes had the largest increase in photosynthesis by elevated CO_2, compared to the other plants. Soybean showed more increase in photosynthesis and growth, compared to wheat and rice in FACE experiment, but not in final grain yield (Long et al. 2006). Soybean showed decrease in nitrogen in response to increased CO_2 concentration, compared to the other C3 species (Taub et al. 2008).

Besides photosynthesis, respiration is an essential process in plants that released solar energy stored in photoassimilates to produce useable cellular energy. Carbon dioxide and water are formed as by-products of respiration. Increased concentration of CO_2 initiated elevation of both photoassimilation and respiration. Respiration rate per leaf area increases with CO_2 due to the larger leaf mass per area. However, there are examples that respiration is reduced if plants are exposed to increased concentrations of CO_2, because of activation of genes encoding respiratory enzymes that affect chemical structure of plant tissues.

The effect of elevated CO_2 on photosynthesis and respiration could be summarized as follows:

1. Induces photosynthesis and net carbon gain.
2. Improves nitrogen use efficiency.
3. Decreases water use.
4. Stimulates respiration in dark phase.
5. Directly, it does not stimulate C4 photosynthesis but stimulates carbon gain and water storage.

In conditions of elevated CO_2, leaf size is increasing, as well as the result of cell expansion and/or a greater number of produced cells. Vegetative growth is also elevated, e.g., the number of nodes in soybean (Dermody et al. 2006) and the number of branches and tillers in wheat and rice.

Molecular mechanisms that are activated by increased CO_2 are related to changes in carbon metabolism in leaves or genes related to starch and sugar metabolism, glycolysis, the tricarboxylic acid (TCA) cycle, and mitochondrial electron transport in soybean (Leakey et al. 2009). In *Arabidopsis*, abundance of transcripts involved in respiration increased together with dark respiration rates (Markelz et al. 2014). Response to elevated CO_2 in rice is related to upregulation of genes, sucrose synthesis, glycolysis, and the TCA cycle (Fukayama et al. 2011). Elevated CO_2 influences plant growth and development by increasing photoassimilation and by activating different metabolic pathways in respiration.

Root biomass and root/shoot ratio are increasing in response to elevated CO_2. Increase in root biomass is a result of changes in root architecture, length, root branches, and their diameter, as well as changes at the cellular level (reviewed in Madhu and Hatfield 2013). Similar changes, e.g., increased length of primary and lateral roots and expansion and elongation induced by increased concentration of CO_2, were observed in model plant *Arabidopsis* (Crookshanks et al. 1998). Responding changes in root architecture improve plant ability for water and nutrient uptake (Gray et al. 2016). In some crops, like spring wheat, winter wheat, cotton, and sorghum, elevated CO_2 concentration induces expansion and branching of lateral roots, through horizontal layers of the soil.

Increased CO_2 contributed to increased grain yield in agricultural crops, but it causes different changes in nutritional traits of grain (Myers et al. 2014). Some crops and wild plants are without changes in biomass in response to increased concentration of CO_2, but they show differences in allocation of assimilates. Increase in fruit production in crops in response to elevated CO_2 is about 28%, but in wild plants it is only about 4% (Jablonski et al. 2002). Specificity of wild plants is greater genetic variability that contributed to greater variability in fruit and seed production in response to CO_2, compared to cultivated crops. Plants exposed to elevated CO_2 could prolong development and reproductive phase and photoassimilation, bringing higher yield. However, that condition could reduce protein and nitrogen content, particularly in non-legume crops (Myers et al. 2014).

There is evidence that in some plants increase of CO_2 increases assimilation rate in the short term but decreases growth in the long term, by suppression of some genes. It is important to find appropriate relation between photosynthesis and translocation/respiration of assimilates, particularly because elevated CO_2 is a part of climate change. Carbon dioxide content influenced concentration of plant hormones in apical zones that initiate cell division, elongation, and differentiation, as well as increased content of starch and sugars (Teng et al. 2006).

Generally, increased yield in crops in response to elevated CO_2 is not always obtained, and optimization of photoassimilation and translocation and respiration is an important task for stable crop production in the future.

Increased temperature affects crops in different ways, with the most important as follows (Lobell and Gourdji 2012):

- Accelerate crop development, but shorten vegetative season, and consequently reduce final yield.
- Decrease or increase photosynthesis, depending on ratio between optimum and current temperature; increase respiration, particularly with temperature increase during night.
- Increase of vapor pressure deficit that reduces water use efficiency and further reduces photosynthesis rate.
- Occurrence of temperature extremes, frost, and heat stress that damage plant cells and high temperatures during flowering and grain filling lead to sterility or very low yield.
- Higher concentrations of CO_2 together with rising temperature are changing distribution of pests and diseases and appearance of new strains.

The increased human activities in the last and twenty-first centuries are contributing to global warming: average temperature increased by 0.85 °C from 1880 to 2012 (Hartmann et al. 2013), and surface temperature is predicted to increase by 1–3.7 °C by 2100 (IPCC 2014). The 12 warmest years since 1880 occurred between 1990 and 2005, and predictions are that heat waves and extreme events are going to be more frequent and intense, with few days having temperatures 5 °C above average optimum (Meehl et al. 2007). Predicted temperatures will differ between eco-geographical regions; for example, projected temperature increase for Arctic surface will be faster than average global (IPCC 2014).

Plant response to elevated temperatures differs by tissue and growing stage. Usually, a reproductive phase is most sensitive to heat stress and most responsible for low yield under changed weather conditions. For example, the highest vegetative growth in rice is at 33 °C, but yield reduction occurred above 25 °C; optimum temperature for vegetative growth in sorghum is within a range of 26–34 °C and for reproductive phase about 25–28 °C (Maiti 1996). Plants respond to increased temperature by earlier occurrence of flowering and pollination and by slower development at temperature above optimum for that phase and species (Hatfield et al. 2011), since male gametophyte is particularly sensitive to heat stress (Zinn et al. 2010). For example, temperature about 33 °C reduces pollen viability in rice and reduces viability to zero at temperature about 40 °C (Parish et al. 2012); and in *Arabidopsis*, growth of pollen tube is differentiating high temperature sensitive from tolerant genotypes.

Possible solution to minimize the effect of increased temperature is to look for genotypes that pollinate earlier in the morning (Shah et al. 2011) or could extend grain filling period (Barlow et al. 2015). Elevated temperature has more impact on yield than vegetative growth, since minimum temperature is increasing, too. Predictions (Meehl et al. 2007) are that minimal temperature is going to increase faster than maximum temperatures, as well as occurrence of extreme weather events. Extreme events in summer (heat waves, drought) seriously affect plant production but in wheat frost could induce sterility and zero yield (Barlow et al. 2015). However, both effects, permanent temperature increase (Ruiz-Vera et al. 2013) and extreme event (Siebers et al. 2015), seriously reduce reproductive phase and final yield. As an example, increased temperature is predicted to reduce yield in soybean by 2.4% in Southern USA and increase by about 1.7% in Midwestern USA (Hatfield et al. 2011). Estimations of Lobell and Asner (2003) are that every degree of increased temperature decreases yield by more than 10% in maize and soybean.

Plant growth is determined by the photosynthesis/respiration ratio. Net assimilation is associated with night temperature, although the decrease induced by low night temperature is more expressed (Jing et al. 2016). Warmer nights also induced mineralization in soil nitrogen (Patil et al. 2010) and further increased photosynthesis and respiration but decreased starch and sugar and total carbohydrate content (Turnbull et al. 2002). Assimilates synthesized during the day are faster used up through respiration during warmer nights and thus initiate further photosynthesis (Lin et al. 2010). Increased temperature induced phenological and morphological changes in plants (Wahid et al. 2007), e.g., leaf expansion and increase of leaf numbers and leaf area index, and thus contributed to higher photosynthesis rate. Optimal temperature for photosynthesis differs between species growing in deserts in comparison to crops from temperate or cold regions (Sage et al. 2008). Photosynthetic activity varies among C4 and C3 plants, and the response to global warming and increased temperature is species specific. Recent research (Xia et al. 2014) showed different impacts of increased temperature during the day and/or night. Night temperature influenced plant physiology in direct and indirect way, with photosynthesis being affected the most (Liu et al. 2011). Changes in photosynthesis are related to leaf chlorophyll content, fluorescence parameters, nitrogen content, enzyme activities, and stomatal conductance.

Leaf development, e.g., initiation, expansion, and growth of new leaves, is strongly related to temperature increase, although optimum level is species specific (Hatfield et al. 2011), and molecular mechanisms are mostly unknown. Projected temperature increase is going to increase soil temperature and either directly or indirectly affects root development through changes in shoot development or by both factors. Increased temperature could induce root growth, depending on species-specific optimum, but further will affect different root functions. Changes in soil temperature depend on numerous soil characteristics, such as soil texture and structure, latitude, insolation, vegetation, etc.

Basically, plant responses to increased temperature occurred at molecular and cellular levels in different organs. Cells respond by changing the structure and stability of membranes and by synthesis of heat shock proteins, stress hormones, and chaperones. Abiotic and biotic stresses also induced enhanced production of ROS (McDonald and Vanlerberghe 2005) that mediate with both photosynthesis and respiration by disrupting cell membrane stability and all transport processes across membranes.

During photosynthesis, plants produce assimilates from CO_2 and light, and certain amount of water, depending on leaf morphology and weather conditions. Net assimilation is not in linear relations with water, but water is necessary to maintain the opened stomata and enable metabolism and plant growth.

Drought is the most important abiotic stress that seriously reduced plant development and yield, and its occurrence is predicted to increase with global climate change. On the other hand, plants are growing in certain areas and have to develop protective and adaptation mechanisms to tolerate and survive under drought. Responses to abiotic stresses are complex, initiated at transcriptional level that caused different physiological changes, including interactive mechanisms (Krishnan and Pereira 2008). Expression analysis in *Arabidopsis* under drought showed down-regulation of photosynthetic genes, but enzymes in starch degradation were induced. In maize, after exposure to dehydration stress after pollination, many water stress-inducible genes were upregulated by abscisic acid (ABA), via ABA-responsive elements; common induced genes for drought and salt stress had protective function (HSP and chaperonins, proline, and glycine-rich proteins, responsible for plant defense and detoxification) (Andjelkovic and Thompson 2006). Another expression analysis in maize, after exposure to water and heat stress, and a combination of both, induces 54 genes, mostly involved in photosynthesis, transport, stress response, and metabolism; 11 transcripts, in common for all three stresses, were mostly related to plant defense and protection from abiotic and biotic stresses (Andjelkovic and Ignjatovic-Micic 2011).

The first reaction to water deficit is growth reduction of expanding tissues, e.g., leaves, internodes, or maize silks and tubers (Tardieu et al. 2011), as a consequence of reduced turgor, cell division, and cell wall extension. Reduction in vegetative growth further affects photosynthesis and transpiration, reproductive and storage organs, and harvest index (Gambin and Borras 2007), inducing accumulation of photoassimilates in stems and roots, instead in storage organs. Sensitivity or tolerance to drought is genetically determined and, after detection in controlled conditions, usually expressed in the field. Quantitative trait loci (QTL) responsible for growth and morphological traits are translated or correlated with QTL for yield in maize (Nikolic et al. 2012, 2013). However, decreased growth under water stress reduces transpiration and water loss that is desirable under severe drought but, under mild water stress, influences reduction of accumulated biomass.

The second effect of water stress is induction of stomatal closure, by pathways related to increased level of stress hormone abscisic acid (ABA). Closed stomata contributed to increased water use efficiency and maintained water in the soil and leaves but, if last longer, led to reduced photoassimilation and crop yield.

The third effect is shortening of crop cycle, by accelerating maturation. That is a plant adaptive response aiming to complete growing cycle earlier and to translocate assimilates to reproductive organs before termination of drought occurred. One of the breeding strategies was to develop varieties with longer maintenance of green, photosynthetic active leaves ("stay green," Borrell et al. 2000). However, this approach is appropriate only for the soils that could store water, because prolonged "stay green" will increase water and biomass loss through transpiration (Hammer et al. 2006).

Leaf development is defined by cell division and cell expansion, and both processes are under significant influence of water stress in different plants (Clauw et al. 2015). For example, in *Arabidopsis thaliana*, leaf and epidermal area, total rosette area, and leaf numbers are reduced under mild water stress (Clauw et al. 2015). Reduction in leaf growth depends on growth stage when drought stress occurred. In some species, likewise *Ricinus communis*, reduction in leaf growth influenced by water stress could be compensated by rehydration only if leaves are more than 12 cm long; if leaves were shorter when drought occurred, rehydration treatment was not efficient. This indicates that existence of optimal conditions for cell division is of high importance in early leaf development, and if water stress occurred in this phase, plants cannot be rehydrated. If mild stress happens, genes that could be expressed are mostly related to abscisic acid signaling pathways and proline metabolism (Clauw et al. 2015). Leaf developmental phase is responsible for up- or down-expression of genes during water stress, but genes included in synthesis of cell wall components are dominantly expressed in older leaves (Clauw et al. 2015).

Contrary to shoot, root growth is maintained or even provoked by mild drought stress. Ratio of shoot to root growth is changed in different crops in response to water deficit and growth stage (Kravic et al. 2013). Numerous studies confirmed plant response to water stress by development of deep root that indirectly contributes to shoot growth under drought. In soybean, grown in the field under drought, water shortage stimulates root growth, particularly at the end of vegetative phase and beginning of reproductive stage, in soil deeper than 0.6 m. In recent nondestructive study of root growth in *Arabidopsis*, drought induced increase and downward orientation of lateral roots (Rellán-Álvarez et al. 2015).

Flowering and grain filling are the most sensitive reproductive phases on water deficit. There are two strategies for plant response to drought stress occurrence during flowering:

1. The escape strategy: comprehended plant adaptation in order to shorten crop cycle by early flowering and finishing of reproductive development before severe water stress occurred (Sherrard and Maherali 2006). In agricultural production, farmers chose varieties with shorter growing season, for particular area. Some desert plants have a very short life cycle after rain and complete it before drought. Shortening of growing cycle reduces total required amount of water but lowers a total assimilated biomass and final yield.
2. The avoidance strategy: represents increased water use efficiency under drought either by improving characteristic of root system (size, depth, architecture, conductance) or by avoiding dehydration by transpiration reduction (smaller leaf area or stomatal closure) (de Dorlodot et al. 2007).

Reproductive stage, e.g., development of floral organs and gametophytes, is very sensitive to drought that could cause flowering inhibition, losing of pollen viability, and reducing seed set and yield in cereals and model plants like *Arabidopsis* (Su et al. 2013).

Climate change is affecting physiological processes in plants, changing their development and life cycle. Predicted are changes in atmosphere, rising temperature, and more frequent occurrence of extreme events, like heat waves, droughts, and flooding (IPCC 2014). Plant adaptability to changing climate is very important for the existence of natural ecosystems and stable agricultural production in the future. At present, our knowledge about plant responses to global warming is based on prediction, simulation, and numerous experimental conditions. A more comprehended research on plant responses to abiotic and biotic stresses induced by climate changes requires molecular studies of all biological processes through life cycle. Application of molecular studies on tissue and cell levels and on different species and environmental conditions through different growth stages is necessary. Climate change will affect plants by a combination of different factors simultaneously: increased temperature and CO_2 concentration, together with changes in precipitation and extreme events, which makes plant response more complex to investigate. However, understanding influence of changed interacting climatic factors is of high importance for future food production and security.

10.4 The Potential of Wild Species for Utilization

Growing population in the world demands increased food production, whereas arable land and water resources are decreasing, and predicted climate changes are contributing to new environment. Manifestation of global changes, climatic, ecological, and technological, highlighted the importance for food quality and quantity improvement by sustainable usage of natural resources. Today, agricultural production is based on increased yield of a small number of crops, rather than on introduction of new plants from wide natural resources. Sustainable agriculture should envelope land and resource management, with the human and ecosystem interface (Shelef et al. 2018). They proposed cultivation of wild plants for local food production, as a part of sustainable agriculture. Wild plants are plant species growing spontaneously in natural ecosystems, able to self-maintain without human activity. They are opposite to "cultivated" or "domesticated" plants that are product of planned human activities through selection, breeding, and controlled multiplication and distribution.

Some anthropologists highlighted that terms "wild" and "domesticated" are specific and are with different meaning in different areas, for example, to European farmer and Kayapó Indian (Posey 1992). However, all plants are a member of global agroecosystem, and it is not easy to strictly separate them on "wild" or "cultivated" plants.

Many thousands of plants are exploited by humans as natural resources. Out of 300,000 high plant species, only few hundred are fully domesticated. According to the Food and Agriculture Organization (FAO) of the United Nations, 90% of food in human diet comes from only 30 cultivated plants, and out of 30,000 edible plants, only 150 are cultivated, which decreased genetic diversity by about 75% (Sethi 2015). However, thousands of wild plants grow locally, and small portion is partly domesticated. In practice, it is not easy to distinguish wild plants from domesticated, since there is wide range from completely wild to completely cultivated, depending on the degree of human activity.

Humans grew plants 10,000 years ago, influencing plant evolution and domestication by using local species with desirable traits for food (Zohary et al. 2012). Domestication was a long-lasting, slow, and complex process, and many species growing today are in various stages of domestication or cultivation. Numerous species, particularly trees, are planted but are genetically close to wild species. For example, many plants from Mayan gardens are native for Yucatan area, mostly trees that are left after deforestation, but could grow in home gardens. Domestication in cereals started with growing wheat in the Fertile Crescent and further spread to Europe (Zohary et al. 2012). After domestication, many species were spread to new environments, far from their center of origin. That happened with most important species today, such as rice (*Oryza sativa*), wheat (*Triticum aestivum*), soybeans (*Glycine max*), sugarcane (*Saccharum* spp.), tomato (*Solanum lycopersicum*), maize (*Zea mays*), and potato (*Solanum tuberosum*) (FAO 2015).

Contemporary agriculture is focused on a small number of crops that were bred to obtain high yields, under intensive cultivation in particular growing and climatic conditions. That is in contrast with extensive agricultural practice in locally grown mixture of plants in a small area. Utilization of wild plants in enhancement of agricultural production, by their introduction as new crops or as sources of quality traits for already cultivated crops, is recommended by FAO (http://www.fao.org/3/w8801e05.htm).

Wild plants envelope flowers, grasses, lichens, fungi, shrubs, and trees that grow without human activity. They are part of nature's biodiversity, growing anywhere, in wild areas such as fields and meadows, wetlands, hills and mountain, and coastal areas. Wild edible plants (WEPs) are part of natural richness, sources of nutrients with high importance in small, rural and suburban areas. Although widely underutilized, these WEPs are traditionally used in different communities, but are not significant to the human diet in developed areas. However, in developing part of the world, wild plants are important for local population, as food, fuel, medical plants, fibers, wood, etc. Use of wild species is part of tradition, native knowledge, and practice created and kept over generations (Slikkerveer 1994). Ethnobotany is examining utilization of wild plants through local and traditional knowledge, with possible application in modern society, particularly medicine and pharmaceutical industry.

Wild species are distributed in various geographical regions and natural ecosystems worldwide. Local population has learned through generations to use plants for different purposes. For example, in South Africa, millions of people in the Maputaland-Pondoland region depend on plants in forests, swamps, and grasslands, for food, feed, fuel, and health care. About 900 species for medical purposes have been found in the region and are important source for herbal trading in that part of the world. Besides, in Andes Mountain, numerous wild species with different usages are estimated, as well as in Himalayan region where local population uses numerous wild edible plants for food, medicinal, and cultural purposes.

The study of wild and noncultivated edible plants evaluated 99 species, belonging to 59 families, of which 96 were angiosperms, 1 gymnosperm, and 2 pteridophytes in the Kailash sacred region, shared by Nepal, India, and China (Aryal et al. 2018). Forty species were used for fruit and 31 for vegetables, most significantly contributed in daily food requirements. The highest use among vegetables was for *Dryopteris cochleata*. In Indian Uttar Pradesh hills, 480 wild species are consumed by the local population, while another ethnobotanical study was performed in India, in West Sikkim area, because of its richness in ethnic and plant diversity and proximity of Khangchendzonga National Park, an important world biosphere reserve (Mahendra et al. 2017). They recorded 124 native edible plant species, 44 herbs, 10 shrubs, 54 trees, and 16 climbers. Their availability is decreasing, due to destruction of natural habitats and overexploitation that need coordinated activity for their protection and conservation.

The study of Baydoun et al. (2017) discovered that 130 wild plant species supply local population in Lebanon with wooden and other products. Medicinal species are contributing to 33%, food and beverage (15%), fuel (12%), environmental uses (10%), and materials (8%), and 22% belong to bee plants, poisons, cosmetics, and social uses. The trees with the highest importance were *Ceratonia siliqua*, *Laurus nobilis*, *Prunus* sp., and *Rhus coriaria*, while *Cichorium intybus*, *Eryngium creticum*, *Gundelia tournefortii*, *Matricaria chamomilla*, *Melissa officinalis*, *Origanum* sp., *Salvia fruticosa*, and *Viola odorata* were the most represented in medical plants.

Majority of partially domesticated or wild plants are found in the tropics. Tropical forests are native area for great part of Earth's biodiversity, with numerous wild plants used for different purposes. Hundreds of millions of inhabitants in tropical forests use wild species and their products, differing between communities and regions. Over 50% of population in Amazonia is, for their survival, dependent on forestry activities. Another example is the Plant Resources of Southeast Asia (PROSEA) project with 6000 species, used by people in that region, and approximated to tropical area, it comes up to 25,000 plant species. Thousands of plant species are estimated for Mediterranean and the other temperate regions. Approximately 25,000 wild species are used in herbal medicine in China, Tropical Asia, India, Central and South America, and Africa and thousands of ornamental plants in natural and public parks and gardens.

There are domesticated and cultivated plant species, such as *Pistacia vera*, still collected as wild plants, similar with rocket (*Eruca sativa* and *Diplotaxis* spp.) that is collected as a wild plant for the market or some medical and aromatic species, such as oregano, sage, and African stinkwood (*Prunus africana*). Nowadays, wild plants could be used as a source of desirable traits, such as drought tolerance in bean (Corteś et al. 2013); for phytoremediation, e.g., removal of toxic metal from the soils (Čudić et al. 2016); or to enrich vegetable oils like sunflower oil, by natural antioxidants from wild edible plants such as *Rosa canina*, *Quercus ballota*, or leaves of *Sanguisorba minor* (Romajoro et al. 2013). The last study is a good example of innovative natural tool to increase thermal stability of sunflower, soybean, or corn oil and to escape application of synthetic antioxidants.

In the developing countries, about 500 million people are small farmers that cultivated plants for food (FAO 2016). The main potential of wild plants is contribution to local farming system and natural ecosystem, in particular habitat, such as soil stabilization, water supply, and climatic influence. Although introduction of wild plants in cultivation is long term and complex, today, it is of high importance to look for new crops for small-scale agricultural production, particularly on marginal lands and soils lacking in nutrients. Some of them could be energy crops, since reserves of coal and oil are decreasing, fiber crops, and some aromatic and medical plants. Besides, in industrial countries, the trends are urban farming and local food production and distribution that include wider usage of native plants and development of food natural resources and ecosystems.

Growing of wild plants and new crops has numerous benefits (Shelef et al. 2017):

1. Promote biodiversity and protect natural heritage. Contemporary agriculture is based on high-yielding crops, with narrow genetic base, which increases a risk of diseases and pests and reduces adaptation potential for climate change scenario. Native populations of wild species are foundation of natural ecosystems and represent a great source of genetic diversity for different traits and purposes (Palmgren et al. 2015).
2. Involvement of new crops in local small production of fresh and healthier new food, medicines, and products with commercial use. In Italy, for example, wild plants were better than modern cultivars, with higher yield of 15%, fruit color, and sugar content of 40%, compared to improvement of 1% obtained by conventional breeding. There are some suggestions about introduction and breeding of perennial crops (Cox et al. 2006). Today, perennial crops are not widely used, although they could store more carbon and need less resources, and some

breeding programs are initiated in wheat, sorghum, sunflower, and wheatgrass (Cox et al. 2006).

Wild plants are adapted to natural habitat, local climate, soil characteristic, and animals. They are in interrelationship, living together and depending upon each other in a common ecosystem. Wild plants provide food sources (nectar, pollen, seeds, leaves, and stems) and natural shelters for native butterflies, insects, birds, and other wild animals.

Wild plants alleviate soil degradation and erosion by plant-microorganism-soil interaction. Microbes in rhizosphere have important role for the development and yield of plants (Drinkwater and Snapp 2007). Preservation of plant diversity includes flora, fauna, and microbes in the rhizosphere, which all contribute to plant growth under changing climatic and environmental conditions (Pérez-Jaramillo et al. 2016). Comprehensive management of natural ecosystems is very important in the protection of biotic stressors. Wild crops through the history developed their own protective mechanisms against diseases and pests, reducing or even eliminating pesticide application. That further diminishes water, soil, and air pollution.

Wild plants are important to prevent spreading of invasive species, which cause serious damage. For instance, the European Council adopted regulation on preventing invasive species (PE-CONS 70/14, 13,266/14, ADD 1), since about 12,000 non-native species caused damage of 12 billion euros each year.

Native plants are significant as intercrops in agricultural practice, contributing to maintenance of soil quality, increased nitrogen uptake, reduced weeds, and reduced soil degradation and erosion. They reduce overall maintenance, since they need no fertilization and pesticides and little or no watering once they are adapted to local environment.

Growing native plant materials includes knowledge of the geographic origin and genetic diversity. It is recommended to choose plants from similar area, helping plants to adapt to local soil and climatic conditions and be more resistant to abiotic and biotic stressors (Dempewolf et al. 2014). The maintenance of genetic diversity enables plants to be less susceptible to pathogens and environmental stresses and more competitive with introduced invasive species.

10.5 Adaptation of Agricultural Ecosystems with Forcing Intercropping Cultivation

Climate change with predictions of 2–6 °C temperature increase by the end of the century (Collier et al. 2008) is taking place rapidly, and therefore the plant demands for fertile soil and water (IPCC 2014). This is why cultivation of previously unprocessed soil has started, mainly in the barrier vegetation belts where soil fertility, rich soil microflora, good soil structure, and supply of micro- and macro-elements are high and meet the needs for agricultural production (Wheeler and von Braun 2013).

Intercropping is defined as a multiple cropping system where multiple crops simultaneously grow on the same parcel in alternative rows. The basis for this type of crop cultivation is that those who are not competing among themselves or who do not have the same requirements for natural resources such as water, light, heat, nutrients, space, etc., are selected.

Therefore, with the intercropping system, cultivated crops produce an expected amount of biomass, and many of them build symbiotic relationships, i.e., benefit from each other. Such is, for example, the case of leguminous plants and any other plants that use fixed nitrogen and soluble forms of potassium and phosphorus. Even in conditions of existence of "reserves" of inaccessible potassium and phosphorus, the cultivation of leguminous plants translates them into soluble forms that are further exploited and obtained high yields (Li et al. 2011). Hence, intercropping is a particularly favorable form for the cultivation of many different plant species, the acquisition of biomass, and the utilization of natural resources. This multiple cropping system opens up new perspectives in dealing with climate change and the utilization of natural ecosystems in which climate change has occurred with a particular focus on CWR.

In conditions of multi-cropping simultaneously in the multiple cropping system, the benefit is multiple, and it also applies to the appearance of pests, diseases, and weeds that additionally make agricultural production difficult. The soil cover is large, evapotranspiration is reduced to a minimum, the leaching of nutrients with rains and fumes is almost minimal, the soil structure is strong, and the microbial activity of the soil microflora is high (Mayer et al. 2018). The listed characteristics minimize the appearance of weeds and the use of chemical plant protection products (PPPs) and thus the amount of residues in the soil and groundwater; some of the cultivated plants, especially if they use CWR, are home to certain predators, avoiding their attack on other plants in the intercropping system, the occurrence of diseases, and the need for the use of chemicals and contamination of the soil substrate and groundwater. The loss of soil moisture is reduced because the soil cover is large and thus conserves the underground moisture that the plants use for their own needs; the strong soil structure eliminates the possibility of withdrawal of nutrients by leaching caused by rains, torrents, and other flooded waters (Ulén and Johansson 2009). The root system gives them a significant structure of the subsoil layers and thus provides a favorable environment for the development of microorganisms that process soil, increase the organic mass, improve fertility of the soil, and develop a soil substrate that is prepared for the following plants that will be grown with minimal inputs from nutrients. All this leads to economic benefits and sustainable agricultural production in conditions of climate change that divert all natural ecosystems and those adapted to the species of cultivated plants (Malézieux 2012).

The growing of plants in intercropping system is a highly developed and, in many respects, a justified, very useful, and suitable system. From the ecological aspect considered in this way, ecologically justified agricultural practices are applied: the reduction of harmful secondary metabolites, i.e., residues that destroy the living world in the soil and underground water are reduced, leaving behind "sterile" soil and dying of the living organisms in the rivers, streams, and other flowing waters and decreasing the potential danger of poisoning humans and livestock.

The intercropping system enables the adaptation of ecosystems that have already caused damage to the vegetative composition and withdrawal of the vegetation belt higher up the hills and mountains and their utilization for agricultural production of soil substratum that is environmentally safe and agrochemical relatively well-supplied with nutrients. Where high natural values (HNVs) can be preserved in such adapted ecosystems, this is of great benefit, and often the natural boundaries of borders and shrubs represent the buffer zone (Kratovalieva et al. 2012). This approach is in line with measures taken to protect biodiversity and conserve plant and animal species from permanent loss by disappearing. In the event that the same ecosystems are left to the "concern of climate change," the damage would be unpredictable, and there would be enormous areas with bare surfaces (Campbell et al. 2014). This approach reduces the damages from climate change, but at the same time, adaptation and conversion of ecosystems to others also take place in times of need.

The development of "climate-adapted" solutions has been exacerbated to reduce greenhouse gas emissions, provide agricultural production, and adapt the surfaces so as not to remain naked and uncovered. These solutions represent cohesion of adapted mechanisms applied for the ultimate goal of sustainability and economic gain. Of course, this is not at all easy; on the contrary, it requires joint action, teamwork, collective solutions, and collaboration of science, practice, and policy makers.

In the intercropping system, two to three (sometimes with a specific schedule of even four) different plants, such as the case of growing flowers at the corners of the parcel, can be included. In addition to dealing with climate change, the purpose of this concept is to obtain the maximum possible yield and a large amount of biomass, which, after harvesting activities, enriches the organic mass in the soil (Jensen et al. 2015). The placement of the grown plants is different and depends on their biology, needs, resistance, and susceptibility, and most of them is in rows, squares, and mixed or at the corners of the plot. Various plant genetic resources (PGR) or CWR, locally adapted landraces, or local populations combined with commercial ones are selected. It is therefore necessary to take into account two basic categories that need to be synchronized, and these are the needs of plants for the natural resources available to them, such as water, nutrients, and light, and the possibilities for their rational utilization (exactly a certain amount of available nitrogen, phosphorus, potassium, and inaccessible forms of potassium and phosphorus).

In designing an intercropping system, it is necessary to develop "well-oriented methods" and "custom-developed approach" to existing climate change and plant relationships complementary or competitive and to what extent are they competitive in relation to available resources. Such an approach is necessary to achieve optimum functionality similar to that in natural ecosystems with as little as possible inputs such as PPPs for the prevention and management of pests and diseases, fertilizers, and tillage (Kontturi et al. 2011). At the same time, pollutants that contaminate the soil, the water, and the surrounding environment leave residues that are difficult to decompose, are long lasting, and are mainly extremely harmful for the health of people, livestock, and the rest of the living world in nature. This type of custom-developed approach provides a functional ecosystem in which participating individuals are well adapted. Hence, there is a need for close cooperation between experts in the field of ecology and agriculture on the one hand and locals like biodiversity guards on the other hand.

Experimental research shows that satisfactory results are obtained with the intercropping system and even the utilization of nitrogen, phosphorus, and potassium is greater than in monoculture systems of growing. Bacteria that colonize the root system, i.e., rhizosphere, are called rhizobacteria. The interaction between plants and bacteria takes place in the rhizosphere where plant growth-promoting rhizobacteria is actually stimulated by plant root exudates and attracted by root mucilage. The composition and quantity of root exudates depend on plant species, cultivar or landrace, plant growth stage, and abiotic factors; it facilitates the availability of bound soil nutrients and also plays a crucial role in soil phytoremediation contaminated by organic pollutants. Namely, with the help of biological nitrogen fixation (BNF), the "captured" elemental and useless nitrogen in the soil tiny interspaces is used by plants and incorporated into the organic mass. The insoluble "bound" forms of phosphorus and potassium in the soil, with the help of leguminous plants whose root ends contain bacteria in irregularly rounded lump formations called nodules during their symbiotic relationships, are excreted as exudates by which such forms of potassium and phosphorus become available to plants. BNF is the most efficient mechanism for the use of soil nitrogen source. Having in mind that BNF inputs annually range from 139 to 175 million tons of nitrogen added to the soil, the importance of this natural process is very clearly illustrated which, with smart solutions in cropping systems in meadows and pastures, replaces the need for application of 80–90 million tons of nitrogen as a fertilizer. In addition, the long-known way of natural soil fertilization through the process of green manure should be used, which improves the carbon sequestration and returns organic matter to the soil.

The potential of the intercropping system in facing the climate change is that the resulting biomass means a larger amount of carbon bound in organic matter, i.e., with carbon sequestration, the direct amount of greenhouse gas is reduced. In addition, soils with lower carbon content are enriched with carbon, which improves its chemical composition and quality while simultaneously performing climate mitigation. Experience shows that surfaces must not be left "naked" or not be planted with certain plants (Arbuckle and Roesch-McNally 2015). On the contrary, they should be covered with animal feed crops and biofuel crops, which increase plant production. The cultivation of several different plants simultaneously allows the use of water reserves in deeper layers, of course, with the right choice of deep-rooted plants, but in the search for water and other plants in the agricultural system, they develop deeper roots (Hauggaard-Nielsen et al. 2016). The intercropping system allows plants to grow in a community, taking into account that they are of varying heights, making some shade for others that are more sensitive to direct sunlight.

At the core of the intercropping system lies the different way of growth and development, nutrient requirements, root depth, plant height, and abiotic factors of susceptibility to pests and diseases (Ruosteenoja et al. 2016). However, the long-standing activities of science in the field of biological and agrobiological diversity working in gene banks and exploring the different ways of conserving biological diversity should be used to include the various CWR genotypes in the intercropping system, particularly deep-rooted or drought-resistant crop genotypes that can reduce intercrop competition in facing climate change. The GB contains preserved PGR with data of characterization and evaluation that can be included in mixtures to limit pathogens that were raised, and this is in fact the most important effects of diversification due to the range of various pathogens (Wheeler and von Braun 2013). In this mixed agroecosystem, the problem with weeds is reduced with good soil cover and crop density. The increased coverage leads to reduced tillage as potential sources of additional gas emissions.

It seems that legumes as intercrops – both annual grain legumes and perennials like clovers and lucerne – have multiple potentials offered to farmers and food producers and the worried parties concern climate-adaptive agricultures (FAO 2013). Sharing knowledge of science and consultant experiences in collaboration with farmers is particularly needed to allow technical development and optimization of economic production. These perceptions should encourage policy makers and market support throughout and should raise awareness of the benefits of intercropping for climate change mitigation and food production in adverse climatic conditions.

10.6 Challenges: Can We Find Similarities and Interdependence Between Landrace and Crop Wild Relatives?

Local food producers usually use old traditional varieties and landraces that are tolerant to major stress factors such as water regime, drought, frost, soil salinity, and susceptibility to diseases and harmful organisms that in some way explain the strategic commitment of Bioversity International and many other organizations that are concerned with the collection and conservation of PGRFA, CWR, and landraces (Veteläinen et al. 2009). Usually, the household level is driven by the maintenance and food production of landraces, and although they are less yielded than commercial, however, their involvement in local and larger systems is increasingly intensified.

Landraces are genetically different, bound to a particular region, showing specific resistance to biotic and abiotic stress factors in the environment in which they are grown, bind to traditional agricultural production, and have historical origin and unique identity (Negri 2007). They are associated with traditional food, customs, festivals, etc., and are part of the ethnobotanical data for a particular region and of traditional agrobiodiversity (Mijatović et al. 2013). As a result of their ability to develop adaptive mechanisms, they can be autochthonous that refers to the region of origin and allochthonous that relates to another region where they are introduced into the farming systems and have developed specific adaptation mechanisms in the new environment. Both autochthons and allochthonous are primary landraces (Kell et al. 2009). Unlike primary landraces that are not included in the official breeding programs, secondary landraces that differ from the original initial genetic material are included in developing formative breeding programs (Zeven 1998).

The wild relatives of cultivated plants (CWRs) that are more or less close to certain crops are an inexhaustible resource for solving problems related to the needs for sufficient and safe food production. They live in natural habitats and are not domesticated. Their diversity in the genetic structure that causes their plasticity and stability leads to the development of systematic strategies that are comprehensive for the characterization and use of CWR to improve the properties of commercial varieties. For this purpose, it is essential to ex situ, but also in situ, conservation of the CWR (Jansky et al. 2013). Apart from landraces, a unique opportunity that offers great possibilities to improve the "weaknesses" of varieties is the approach of breeders to the enormously rich gene pool in terms of genetic diversity. The synergistic approach with the introgression of CWR's vital genetic material in commercial varieties is the only way to create resistant varieties and varieties with improved properties that will be the basis for the development of new production systems in response to climate change. Until the occurrence of climate change, the main research work of breeders was focused on landraces; although there are data on established crop improvement dating more than 60 years (Hajjar and Hodgkin 2007), especially intensively and in a strong coupling of all stakeholders in the last 15 years, the focus is on the use of CWR germplasm. It should be noted here that for some crops such as potato, CWR play an important role in the creation of genomes with improved quality and resistance to diseases and pests (Maxted et al. 2008). Advanced characterization with the use of molecular tools and the development of genomics represents an integrated and systematic strategy for adaptation to climate change and the reduction of damage they cause each year.

With regard to the diversity of cultivated plants, landraces that are much better adapted to hot temperatures than commercial varieties and bring economic benefits can be considered an exceptional contribution, although it must not be greatly emphasized (Gilles et al. 2017). And while commercial production faces a long-term food deficit, local production of landraces helps to tackle challenges of warming temperatures that are associated with climate change as reflected on reduced production but not declassified as is the case with commercial varieties. The role of landraces in crop rotation and risk management strategies enables the development of more tolerant varieties with innovative techniques (Habiba et al. 2011).

Farmers' decisions to face climate change prevail in a combination of activities such as the involvement of landraces and CWRs in production and the intensive reduction in the use of artificial fertilizers and other chemicals with the main purpose of acting systemically to preserve soil fertility and the health of ecosystems. Patterns of farmers' selection concerning seed exchange influence the preservation of the landrace purity; if this is done in small regions, then the unique genotype is stored. Within the interregional collaboration, the genetic diversity of the genotype increases; therefore, the response to climate change is suitable in most cases (Bellon and Risopoulos 2001). Humanity is upright before challenges involving food and dealing with climate change that have multiple effects emerged, and therefore maintaining crop portfolios is extremely important for crop resilience (Meldrum et al.

2018). Creation and production management involving landraces and CWR or improved varieties with valuable adaptive properties found in local varieties and their wild relatives will provide food during periods of extreme climatic stress.

Traditional knowledge and ethnobotanical data complemented by scientific knowledge is the basis for finding local solutions to challenges involving a plan of resilience versus climate change (Kumar 2014). Primarily, consideration should be given to balancing between farmers' selection for certain properties and natural selection that occurs spontaneously in nature's wild relatives. From this approach, plants with improved qualities should be spread in the region of their primary habitat and around it. These plants are not so rarely spread out as seeds through trade routes and through meetings between people. From this perspective, the number of varieties, populations, and plants with much better properties and higher quality could be very large, with favorable, sought-after properties that need to be transferred further to obtain stress-tolerant and high-quality nutritive aspect genotypes (Dey and Sarkar 2011). Without mapping and existence of databases in which the place of origin, geographical coordinates, climatic belts, and vegetation zones would be entered, inadequate choices would be made, and the recombination process would be unsuccessful.

Gene banks around the world represent "gold reserves" for future use in pre-breeding and plant breeding (Melchinger et al. 2017). The conserved material (seeds, whole plants, parts of plants) carries the epithet "source of innovation" from which the most advanced biotechnological tools transfer genes and obtain improved varieties. This process is continuing and is enriching the genetic fund with the landraces and CWR genes. This is an added value, and today's modern commercial varieties have CWR genes that are tolerant to certain pests, diseases, drought, frost, salinity, etc., and precisely in both sources, landraces and CWR, we find the mutual similarities that increase the challenges. When it comes to interdependence between landraces and CWRs, the scientific circles debate whether primary landraces need to be improved, and with such an improved genome, they would further be grown as secondary ones and would be placed in breeding programs with the ultimate result to get resistant, with improved quality properties of the commercial variety (Chatzav et al. 2010), or in situ conservation of CWR directly by the method of introgression of genes would be introduced to improve the properties of commercial and susceptible varieties. Of course, the process in both cases is long lasting. Taking into account the adaptability of landraces, but also their greater resemblance to CWR, we conclude that for a shorter period of 3–5 years, new improved varieties will be produced ready to be put into commercial production. In the latter case, there is a significant difference between the genome of the commercial variety and the CWR, and the process of pre-breeding and breeding is longer and takes more than 5 years to obtain a variety for commercial production (Fig. 10.1).

However, new strategies are based on bioclimatic modeling of landraces and existing commercial varieties. All relevant stakeholders such as policy makers, farmers, and experts work together, leading to decisions about the future of food production that presents as a challenge for future partnerships with clearly defined coordination and a framework of cooperation.

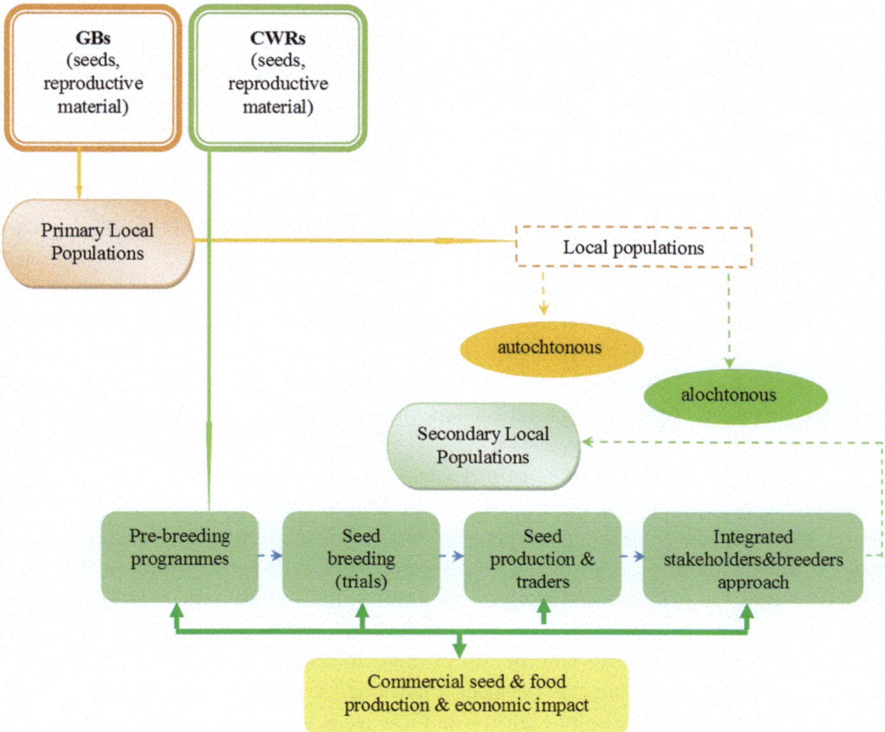

Fig. 10.1 A schematic illustration of the utilization of landrace and CWR in variety and plant trait seed breeding

10.7 Baseline Expectations Using Wild Species Before Head of Climate Changes

During the evolutionary process of subdivision of wild plant species in which properties have been identified to have a positive and beneficial effect on humans and cattle from a different aspect such as food and nutrition, alleviating symptoms in certain diseases and improving lactation, endurance, immunity, and their properties were further refined in the selection process.

The original forms of today's plants and crops have gone through a long way of improvement but always by obtaining better quality properties and more perfect adaptation to the microclimate conditions of the external environment. Climate conditions play a key role in the determination of the region, microclimate of location, composition of the local plant and animal biocoenosis, and functioning of natural ecosystems. Hence, from the point of view of climate change that occurred and prematurely contended with the envisaged scenarios of scientists, great hope is expected in the wild relatives from the aspect of the survival of species that will give an appropriate response (Ehrlén and Morris 2015).

While climate change seems to have irretrievably damaged the appearance of various ecosystems and their composition, we are all aware of its impact, and we should equally worry about our essential needs, the most obvious being the biodiversity of food and quality of the diet. And while the world is facing hunger, at the same time, climate change has evolved which strongly influences the wild and, as abovementioned, leads to resistant forms of impacts in terms of high and low temperatures, ultraviolet range (UVB, 290–320 nm that damages the plants and animals), changed humidity, increased CO_2 emissions in the atmosphere, pedological composition, and soil microflora.

And so far, if the result of human influence were global ecosystem changes, somehow progressed in some transitional phases for which we had time and various forums and formed advisory bodies to speak openly and look at them in the eyes. Today, facing them, we seem to be in panic and every scientific gathering ends with more terrible predictions and scenarios that unfortunately happen. We were and we are witnessing them.

Facing the irreversible losses of old varieties and traditional populations that survived through time and were well locally adapted through developed mechanisms of environmental adaptations, we are rectified before a disaster called rapid rural and change appraisal (RRA) that will affect food production in the local and small communities living in rural and underdeveloped areas (Pritchard et al. 1999).

On the other hand, the high temperatures have affected habitat conditions and the changing ecological boundaries between phytocoenological zones and belts (Holt and Keitt 2005). Thus, the belt of shrubs and low trees withdrew higher, pushing the forest belt to higher elevations, and in the composition of the meadows and pastures, there were changes in the plant communities which in turn reflected the appearance of the habitat and resulted in a change in the landscape.

Primarily, the following changes that arise and will occur in natural biocoenosis should be taken into account, where the key factor for the appearance, composition, and life in them is determined by the most common species that are determinant and whose alteration in the population numbers causes irreversible changes. They relate to several moments like physiological response to plants, changes in the phenological development of plants, the appearance of visible phenotypic changes, species distribution and abundance, and the changing ecosystem structure.

The simple conclusion is that by increasing the temperature on a global scale, photosynthetic activity is reduced and yields are declassified and produced much less food in the world. Here is the question: How can the adaptive potential and mechanisms of the wild relatives of the spontaneous flora be utilized in the wake of the climate changes that have arisen and that will be even more pronounced in the future?

CWR are natural source of genetic structure that carries genes with certain endurance to various unfavorable conditions on the outside environment through the development of environmentally adaptive mechanisms. But it should be noted that these adaptations have led to certain modifications that are somewhat phenotypically visible and perceptible. Collecting CWR has been intensified in the last

40–50 years, and with the use of modern biotechnological tools, specific resistant genomes have been successfully created. Contemporary trends and strategies created by FAO (1997) charted developmental directions with an interdependence in the use of CWR to the level of sustainable agricultural production and rural development. Climate change affected not only the cultivated crops but also the replacement of some plants with others that originate from nature. Mankind faced with rapidly emerging climate changes has accepted "with both hands" these "new" plants that were used only in underdeveloped regions and rural areas and now served as food for the table and as feed for livestock.

It has long been clear that agricultural systems must be revised and that agricultural plants and crops grown should be replaced by others with improved properties. Small farmers were the first to start with replacement by introducing suitable cropping plants. This is more easily applicable to smaller areas where plant properties adapt to microclimate conditions (increased temperature and reduced rainfalls). In contrast, large production facilities that extend on larger surfaces suffered the greatest damage, and replacement with other more resilient crops did not go as fast and easy. Namely, as the most affected by climate change, large areas have greater needs in terms of adaptation, because there is a need of plants that carry genes for resistance to drought, diseases, frost, etc. That is why PGRs originate from the spontaneous flora which contains germplasm that is well adapted to adverse conditions in nature almost "overnight," which has become a "hot spot" for scientists. The search for improved germplasm was supported by the characterization and evaluation of CWR in gene banks ex situ and in situ, and this was the basic response to improving the properties of cultivated plants. In breeding programs, modern biotechnological tools were used, and as a result, biotech plants emerged as the response to the climate change. It is also one of the possible biotechnological solutions that can be used in certain unfavorable conditions, as far as possible, but not unique. At the same time, there is a smooth production of food for people and animals. In this field, United States is the most advanced, where some regions are almost completely oriented to these biotech plants.

Areas and individual plants are generally protected, but the number of CWRs that can be used or can be conserved in situ is still unknown. The members of the traditional communities in the area are involved in their protection and rarely where this process takes place with the predicted dynamics. In this direction, it must be noted that farmers who are aware of climate change resort to land-use change and they are trying to achieve mitigation of impact, but not anywhere in any crops. The CWR germplasm can be used almost in any plant, but mainly in ex situ conditions, it is used in cereal, leguminous, industrial, and fruit plant species. Of course, their response in native conditions is also different depending on their biological and physiological mechanisms of adaptation as well as their utilization of the increased amount of carbon (Battisti and Naylor 2009). Hence, although scientists and researchers work intensively, it takes time to create drought- and pest-resistant plants that would alleviate damage from higher temperatures, would have dramatic impacts on agricultural production, and would not increase the level of hunger in the world.

It is difficult to predict the degree of drought that appears, as well as the area where it will appear, hence the greater number of declassified incomes and the strategic crops in most countries. Drought certainly reflects negatively and reduces potential irrigation, so creating drought-tolerant crops is necessary. As a result of climate change and the need for "new" suitable plants, crop shift has also occurred in order for farmers to have income, but this has been reflected in the strategic production of food in the world and on the quality of production. CWR shows the characteristics that are targeted and are taken into the breeding programs (Nevo 2004). Tolerance developed by CWR to unfavorable conditions in the environment where they grow gives the opportunity to transfer their properties/genes into the genome of cultivated plants.

Thus, in rice whose production is reduced by 10% for every 1 °C temperature increase, in response to heat stress and drought during the growing season (Peng et al. 2004), researchers have come to exploit the rich wild rice gene pool of the wild relative *Oryza officinalis* which blooms in the early morning hours of the day, using this feature by incorporating it into the cultivated rice varieties. Concerning rice drought tolerance, *Oryza longistaminata* is involved in breeding programs (Dempewolf et al. 2014). Wild emmer *Triticum dicoccoides* is a wild progenitor of durum wheat and possesses a rich gene pool for resistance to drought, so it is used to create varieties of wheat that will be adapted to dry conditions (Peleg et al. 2009). *Hordeum spontaneum* as progenitors of barley is valuable with developed adaptive mechanisms to tolerate drought and salt habitats. This species originated from water-limited habitats and thus developed deep- and narrow-spreading roots in order to reach the deep soil layer (Bengough et al. 2004). The wild ancestor of barley *Hordeum bulbosum* is also used in breeding programs because there are genes for resistance to drought. It therefore has an essential meaning for the barley breeding programs.

The original wild form of maize is known as the teosinte. Numerous genetic studies using molecular methods and tools as well as testing for positive selection suggest that *Zea mays* subsp. *parviglumis* is the species that has the closest genetic similarity to the domesticated maize and originates from the teosinte (Lukens and Doebley 2001). Teosinte has a wide ecological valence and is grown in conditions of hot and humid climate to moderately dry and dry regions as well as on different soil substrates and vegetation belts. Hence, the developed subspecies influenced by the diversification have different tolerance and resistance to flooding and drought. So, *Zea mays* ssp. *mexicana* has a short vegetation season and drought-resistant genes, while *Zea nicaraguensis* has unique resistance to frequent precipitation and flooding (Sánchez González 2018).

In order to utilize the unique potential of CWR as progenitors of cultural plants which have shown susceptibility, nonresistance, and poor adaptive ability to the occurrence of climate changes, there is necessity for development of new molecular marker technologies that would accelerate introgression breeding programs (Brar 2005). Only in this way the potential of CWR which they carry in themselves can be used appropriately, and at the same time it is a unique way of science to participate in the management of agricultural production in the conditions of today's climate change (Dempewolf et al. 2014). In the case of garden plants, a good example is the progenitor *Lycopersicon pimpinellifolium* that has a resistance gene to *Fusarium oxysporum* and *Lycopersicon hirsutum* that have shown resistance to insect attacks that have caused enormous damage. For potatoes, for example, the damages by as much as 30% have been reduced by using CWR in the late blight case, so the necessity of the wild relatives of the cultural plants, which still needs to be done today, has been fostered in the past with many examples.

The fact is that the public knows little about the importance of CWR; very few are aware of their potential for producing plants that will have better properties and will be resistant to certain pests and diseases that reduce food production associated with hungry people in the world, especially on the African continent (FAO, IFAD, UNICEF, WFP and WHO 2018), and even less people know that one of the strongest tools besides the rest is improving the genetic constitution of cultural plants by transferring genes from their wild relatives. One of the earliest examples of this claim dates from the late 1800s, when grape phylloxera (*Daktulosphaira vitifoliae*) appeared in vineyards in Europe and North America, destroying millions of hectares of vineyards (Prance 1987). The knowledge possessed by the professionals and scientific community that North American wild grape species *Vitis rupestris*, *Vitis berlandieri*, and *Vitis riparia* have roots resistant to phylloxera was a permanent solution to the destroyed vineyards. Since then, many scientific breeding studies have created grapevine hybrids and many root stocks (Eibach and Töpfer 2015).

CWRs contain genes that carry resistant properties to adverse external conditions, diseases, and pests which are useful and provide a source of solutions to many of the problems that agricultural production is facing today in terms of climate change. Their adaptability enables the use of rural areas that have not been cultivated till then but are a solid basis (soil fertility has been conserved for many years) for increasing the area under cultivated plants and increasing in this way the profitability in food production. All the wealth of conserved in situ and ex situ CWR in gene banks and their long-term storage in the world's global seed vault in Svalbard gives hope that humanity will be able to rationally deal with climate change and climate catastrophes. It should not be disregarded that today's cultural plants involved in commercial production have been originating from CWR over many years of breeding processes, and the least we can do is go back to the study of the CWR genetic structure and use their useful properties by using contemporary biotechnological tools toward transferring genes with the potential for increasing yields and avoiding climate disasters (Knight 2003). Increasing the resistance of agricultural plants leads to productive agricultural production, and this is the only tool for mitigating damage from climate change.

10.8 Conclusion

Climate change and extreme weather variability strongly influence global food safety, altering the basics of life on Earth. They are becoming an increasingly serious source of biodiversity endangerment and an increasing challenge for agriculture. The distribution and survival of wild plant species, including CWR, depends on the intensity of climate change. The CWR populations in the natural distribution area show significant genetic differences between them. At present, there is a lot of uncollected and non-conserved genetic diversity in nature that could be of importance for further crop improvement. Bearing in mind the importance of CWR in creating new varieties with improved adaptations to biotic and abiotic stresses, it is extremely important that crop wild relatives are adequately conserved. Diversity of wild plants is not intensively studied, and their identification and characterization are not well organized. Since wild species are partially domesticated in small areas and important agronomic characteristics are kept over a long time among farmers, it is necessary to record all information for future needs and utilization. Their preservation and use to broaden the genetic base of modern crops are vital to adopt agriculture on the impacts and consequences of climate change. Therefore, knowledge of the potential impact of climate change on CWR and finding acceptable systems for their protection are key activities to sustain agricultural production and world food security.

References

Agaisse H, Lereclus D (1995) How does *Bacillus thuringiensis* produce so much insecticidal crystal protein? J Bacteriol 177:6027–6032

Ainsworth EA (2008) Rice production in a changing climate: a meta-analysis of responses to elevated carbon dioxide and elevated ozone concentration. Glob Chang Biol 14:1642–1650

Ainsworth EA, Long SP (2005) What have we learned from 15 years of free-air CO_2 enrichment (FACE)? A meta-analytic review of the responses of photosynthesis, canopy properties and plant production to rising CO_2. New Phytol 165:351–372

Ainsworth EA, Rogers A (2007) The response of photosynthesis and stomatal conductance to rising (CO_2): mechanisms and environmental interactions. Plant Cell Environ 30:258–270

Andjelkovic V, Ignjatovic-Micic D (2011) ESTs analysis in maize developing kernels exposed to single and combined water and heat stress. Sci Agric 68(3):353–360

Andjelkovic V, Thompson R (2006) Changes in gene expression in maize kernel in response to water and salt stress. Plant Cell Rep 25:71–79

Arbuckle JG Jr, Roesch-McNally G (2015) Cover crop adoption in Iowa: the role of perceived practice characteristics. J Soil Water Conserv 70:418–429

Aryal KP, Poudel S, Chaudhary RP, Chettri N, Chaudhary P, Ning W, Kotru R (2018) Diversity and use of wild and non-cultivated edible plants in the Western Himalaya. J Ethnobiol Ethnomed 14:10–28

Barlow KM, Christy BP, O'Leary GJ, Riffkin PA, Nuttall JG (2015) Simulating the impact of extreme heat and frost events on wheat crop production: are- view. Field Crops Res 171:109–119

Battisti D, Naylor R (2009) Historical warnings of food insecurity with unprecedented seasonal heat. Science 323(5911):240–244

Baydoun SA, Kanj D, Raafat K, Aboul Ela M, Chalak L, Arnold-Apostolides N (2017) Ethnobotanical and economic importance of wild plant species of Jabal Moussa Bioreserve, Lebanon. J Ecosyst Ecograph 7:3

Bellon MR, Risopoulos J (2001) Small-scale farmers expand the benefits of improved maize germplasm: a case study from Chiapas, Mexico. World Dev 29:799–811

Bengough AG, Gordon DC, Al-Menaie H, Ellis RP, Keith R, Thomas WTB, Forster BP (2004) Gel observation chamber for rapid screening of root traits in cereal seedlings. Plant Soil 262:63–70

Bloom AJ, Burger M et al (2010) Carbon dioxide inhibits nitrate assimilation in wheat and *Arabidopsis*. Science 328:899–903

Borrell AK, Hammer GL, Douglas ACL (2000) Does maintaining green leaf area in sorghum improve yield under drought? I. Leaf growth and senescence. Crop Sci 40:1026–1037

Brar D (2005) Broadening the gene pool and exploiting heterosis in cultivated rice. In: Toriyama K, Heong KL, Hardy B (eds) Rice is life: scientific perspectives for the 21st century. Proceedings of the World Rice Research Conference November 2004, Tokyo/Tsukuba, pp 4–7

Byrt C, Platten D, Spielmeyer W, James RA, Lagudah ES, Dennis E, Tester M, Munns R (2007) HKT1;5-like cation transporters linked to Na + exclusion loci in wheat, Nax2 and Knal. Plant Physiol 143:1918–1928

Campbell BM, Thornton P, Zougmoré R, van Asten P, Lipper L (2014) Sustainable intensification: what is its role in climate smart agriculture? Curr Opin Environ Sustain 8:39–43

Chatzav M, Peleg Z, Ozturk L, Yazici A, Fahima T, Cakmak I, Saranga Y (2010) Genetic diversity for grain nutrients in wild emmer wheat: potential for wheat improvement. Ann Bot 105(7):1211–1220

Clauw P, Coppens F, De Beuf K, Dhondt S, Van Daele T, Maleux K, Storme V, Clement L, Gonzalez N, Inzé D (2015) Leaf responses to mild drought stress in natural variants of Arabidopsis. Plant Physiol 167:800–816

Collier P, Conway G, Venables T (2008) Climate change and Africa. Oxfors Rev Econ Policy 24-2(1):337–353

Cook JR (2000) Advances in plant health management in the twentieth century. Annu Rev Phytopathol 38:95–116

Corteś AJ, Monserrate FA, Ramírez-Villegas J, Madriñaǹ S, Blair MW (2013) Drought tolerance in wild plant populations: the case of common beans (*Phaseolus vulgaris* L.). PLoS One 8(4):e62898

Cox TS, Glover JD, Van Tassel DL, Cox CM, DeHaan LR (2006) Prospects for developing perennial grain crops. Bioscience 56:649–659

Crookshanks M, Taylor G, Dolan L (1998) A model system to study the effects of elevated CO2 on the developmental physiology of roots: the use of *Arabidopsis thaliana*. J Exp Bot 49:593–599

Čudić V, Stojiljković D, Jovović A (2016) Phytoremediation potential of wild plants growing on soil contaminated with heavy metals. Arh Hig Rada Toksikol 67:229–239

Damiana A (2008) History, achievements and current status of genetic resources conservation. J Agron 100(1):9–21

Dangl JL, Horvath DM, Staskawicz BJ (2013) Pivoting the plant immune system from dissection to deployment. Science 341:746–751

de Dorlodot S, Forster B, Pages L, Price A, Tuberosa R, Draye X (2007) Root system architecture: opportunities and constraints for genetic improvement of crops. Trends Plant Sci 12:474–481

Dempewolf H, Eastwood RJ, Guarino L, Khoury CK, Müller JV, Toll J (2014) Adapting agriculture to climate change: a global initiative to collect, conserve, and use crop wild relatives. Agroecol Sustain Food Syst 38(4):369–377

Dermody O, Long SP, DeLucia EH (2006) How does elevated CO2 or ozone affect the leaf-area index of soybean when applied independently? New Phytol 169:145–155

Dey P, Sarkar AK (2011) Revisiting indigenous farming knowledge of Jharkhand (India) for conservation of natural resources and combating climate change. Indian J Trad Knowl 10(1):71–79

Drinkwater L, Snapp S (2007) Understanding and managing the rhizosphere in agroecosystems. Elsevier, Burlington. https://doi.org/10.1016/B978-012088775-0/50008-2

Ehrlén J, Morris WF (2015) Predicting changes in the distribution and abundance of species under environmental change. Ecol Lett 18:303–314

Eibach R, Töpfer R (2015) 1 – traditional grapevine breeding techniques. Grapevine Breeding Programs for the Wine Industry. Traditional and Mol Techniques, pp 3–22

Food and Agriculture Organization (FAO) (1997) The state of the world's plant genetic resources for food and agriculture. FAO, Rome. online: www.fao.org/ag/AGP/AGPS/Pgrfa/pdf/swrfull.pdf

Food and Agriculture Organization of the United Nations (2013) Climate-Smart Agriculture Sourcebook. p 570. http://www.fao.org/docrep/018/i3325e/i3325e.pdf

FAO (2015) The state of food insecurity in the world. Meeting the 2015 international hunger targets: taking stock of uneven progress. FAO Rome

FAO (2016) The state of food and agriculture 2016.Climate change, agriculture and food security. Rome. (also available at www.fao.org/3/a-i6030e.pdf)

FAO, IFAD, UNICEF, WFP and WHO (2018) The State of Food Security and Nutrition in the World 2018: Building climate resilience for food security and nutrition. FAO, Rome. http://www.fao.org/state-of-food-securi ty-nutrition/

Farooq S, Azam F (2001) Co-existence of salt and drought tolerance in Triticeae. Hereditas 135(2–3):205–210

Farooq S, Niazi MLK, Iqbal N, Shah TM (1989) Salt tolerance potential of wild resources of the tribe Triticeae. II. Screening of species of the genus Aegilops. Plant Soil 119:255–260

Farooq M, Basra S, Wahid A, Cheema Z, Cheema M, Khaliq A (2008) Physiological role of exogenously applied glycinebetaine to improve drought tolerance in fine grain aromatic rice (*Oryza sativa* L.). J Agron Crop Sci 194:325–333

Fukayama H, Sugino M, Fukuda T, Masumoto C, Taniguchi Y, Okada M, Sameshima R, Hatanaka T, Misoo S, Hasegawa T, Miyao M (2011) Gene expression profiling of rice grown in free air CO2 enrichment (FACE) and elevated soil temperature. Field Crops Res 121:195–199

Gambin BL, Borras L (2007) Plasticity of sorghum kernel weight to increase assimilates availability. Field Crops Res 100:272–284

Genc Y, McDonald GT, Tester M (2007) Reassessment of tissue Na+ concentration as a criterion for salinity tolerance in bread wheat. Plant Cell Environ 30(11):1486–1498

Gilles LM, Khaled A, Laffaire J-B, Chaignon S, Gendrot G et al (2017) Loss of pollen-specific phospholipase NOT LIKE DAD triggers gynogenesis in maize. EMBO J 36:707–717

Gorham J (1990) Salt tolerance in the Triticeae: K/Na discrimination in synthetic hexaploid wheats. J Exp Bot 41(5):623–627

Gorham J, Hardy C, Wyn Jones RG, Joppa LR, Law CN (1987) Chromosomal location of a K/Na discrimination character in the D genome of wheat. Theor Appl Genet 74:584–588

Gorham J, Bridges J, Dubcovsky J, Dvora J, Hollington PA, Luo MC, Khan JA (1997) Genetic analysis and physiology of a trait for enhanced K+ /Na+ discrimination in wheat. New Phytol 137:109–116

Gray SB, Siebers M, Locke AM, Rosenthal DR, Strellner RS, Paul RE, Klein SP, Ruiz-Vera UM, McGrath J, Dermody O, Ainsworth EA, Bernacchi CJ, Long SP, Ort DR, Leakey ADB (2016) Intensifying drought eliminates the expected benefits of elevated [CO2] for soybean. Nat Plants 2(9):16132. https://doi.org/10.1038/nplants.2016.132

Habiba U, Shaw R, Takeuchi Y (2011) Chapter 2 socioeconomic impact of droughts in Bangladesh. Commun Environ Disaster Risk Manag Droughts Asian Monsoon Reg 8:25–48

Hajjar R, Hodgkin T (2007) The use of wild relatives in crop improvement: a survey of developments over the last 20 years. Euphytica 156(1):1–13

Hammer G, Cooper M, Tardieu F, Welch S, Walsh B, van Eeuwijk F et al (2006) Models for navigating biological complexity in breeding improved crop plants. Trends Plant Sci 11:587–593

Hartmann DL, Klein Tank AMG, Rusticucci M, Alexander LV, Brönnimann S, Charabi Y, Dentener FJ, Dlugokencky EJ, Easterling DR, Kaplan A, Soden BJ, Thorne PW, Wild M, Zhai PM (2013) Observations: atmosphere and surface. In: Stocker TF, Qin D, Plattner G-K, Tignor M, Allen SK, Boschung J, Nauels A, Xia Y, Bex V, Midgley PM (eds) Climate change 2013: the physical science basis. Contribution of Working Group I to the Fifth Assessment Report of the Intergovernmental Panel on Climate Change. Cambridge University Press, Cambridge/New York

Hatfield JL, Boote KJ, Kimball BA, Ziska LH, Izaurralde RC, Ort D, Thomson AM, Wolfe DW (2011) Climate impacts on agriculture: implications for crop production. Agron J 103:351–370

Hauggaard-Nielsen H, Lachouani P, Trydeman Knudsen M, Ambus P, Boelt B, Gislum R (2016) Productivity and carbon footprint of perennial grass–forage legume intercropping strategies with high or low nitrogen fertilizer input. Sci Total Environ 541:1339–1347

Haverkort AJ, Boonekamp PM, Hutten R, Jacobsen E, Lotz LAP, Kessel GJT et al (2008) Societal costs of late blight in potato and prospects of durable resistance through cisgenic modification. Potato Res 51:47–57
Holt RD, Keitt TH (2005) Species' borders: a unifying theme in ecology. Oikos 108:3–6
IPCC (2007) Climate change 2007: synthesis report. summary for policymakers. http://www.ipcc.ch/ipccreports/ar4-syr.htm
IPCC (2014) Climate change 2014: synthesis report. Contribution of Working Groups I, II and III to the Fifth Assessment Report of the Intergovernmental Panel on Climate Change. Pachauri RK, Meyer LA. IPCC, Geneva, p 151
Jablonski LM, Wang X et al (2002) Plant reproduction under elevated CO_2 conditions: a meta-analysis of reports on 79 crop and wild species. New Phytol 156:9–26
Jansky SH, Dempewolf H, Camadro EL, Simon R, Zimnoch-Guzowska E, Bisognin DA et al (2013) A case for crop wild relative preservation and use in potato. Crop Sci 53:746–754
Jensen S, Bedoussac EL, Carlsson G, Journet EP, Justes E, Hauggaard-Nielsen H (2015) Enhancing yields in organic crop production by eco-functional intensification. Sustain Agric Res 4:42–50
Jing P, Wang D, Zhu C, Chen J (2016) Plant physiological, morphological and yield-related responses to night temperature changes across different species and plant functional types. Front Plant Sci 7:1774
Jo KR, Kim CJ, Kim SJ, Kim TY, Bergervoet M, Jongsma MA et al (2014) Development of late blight resistant potatoes by cisgene stacking. BMC Biotechnol:14–50. https://doi.org/10.1186/1472-6750-14-50
Jovovic Z, Dolijanovic Z, Kovacevic D, Velimirović A, Biberdzic M (2012) The productive traits of different potato genotypes in mountainous region of Montenegro. Genetika 44(2):389–397
Jovovic Z, Micev B, Velimirovic A (2016) Impact of climate change on potato production in Montenegro and options to mitigate the adverse effects. Acad J Environ Sci 4(3):047–054
Kell SP, Maxted N, Allender C, Astley D, Ford-Lloyd BV et al (2009) Vegetable landrace inventory of England and Wales. The University of Birmingham, Birmingham, p 117
Knight J (2003) A dying breed. Nature 421:568–570
Kontturi M, Laine A, Niskanen M, Hurme T, Hyövelä M, Peltonen-Sainio P (2011) Pea-oat intercrops to sustain lodging resistance and yield formation in northern European conditions. Acta Agric Scand Sect B Soil Plant Sci 61:612–621
Kratovalieva S, Popsimonova G, de Vlieger A, Selamovska A, Mihovski T, Vlahova M, Carlier L (2012) HNV farming systems in Macedonia. J Mountain Agric Balkans 15(6):1415–1429
Kravić N, Marković K, Anđelković V, Hadži-Tašković Šukalović V, Babić V, Vuletić M (2013) Growth, proline accumulation and peroxidase activity in maize seedlings under osmotic stress. Acta Phys Plant 35(1):233–239
Krishnan A, Pereira A (2008) Integrative approaches for mining transcriptional regulatory programs in Arabidopsis. Brief Funct Genom Proteomics 7:264–274
Kumar V (2014) Role of indigenous knowledge in climate change adaptation strategies: a study with special reference to North-Western India. J Geogr Nat Disast 4:131. https://doi.org/10.4172/2167-0587.1000131
Leakey ADB, Ainsworth EA, Bernacchi CJ, Rogers A, Long SP, Ort DR (2009) Elevated CO2 effects on plant carbon, nitrogen, and water relations: six important lessons from FACE. J Exp Bot 60:2859–2876
Li QZ, Sun JH, Wei XJ, Christie P, Zhang FS et al (2011) Overyielding and interspecific interactions mediated by nitrogen fertilization in strip intercropping of maize with faba bean, wheat and barley. Plant Soil 339:147–161
Lin DL, Xia JY, Wan SQ (2010) Climate warming and biomass accumulation of terrestrial plants: a meta-analysis. New Phytol 188:187–198
Lindner M, Maroschek M, Netherer S, Kremer A, Barbati A, Garcia-Gonzalo J, Seidl R, Delzon S, Corona P, Kolstro M, Lexer MJ, Marchetti M (2010) Climate change impacts, adaptive capacity, and vulnerability of European forest ecosystems. Forest Ecol Manag 259:698–709
Lindsay MP, Lagudah ES, Hare RA, Munns R (2004) A locus for sodium exclusion (Nax1), a trait for salt tolerance, mapped in durum wheat. Funct Plant Biol 31:1105–1114

Liu YF, Li TL, Xu T, Qi MF, Xu CQ, Qi HY (2011) Effect of low night temperature treatment and recovery on photosynthesis and the allocation of absorbed light energy in tomato (*Lycopersicon esculentum* Mill.) leaves. J Hortic Sci Biotechnol 86:91–96

Lobell DB, Asner GP (2003) Climate and management contributions to recent trends in U.S. agricultural yields. Sci 299:1032

Lobell DB, Gourdji SM (2012) The influence of climate change on global crop productivity. Plant Physiol 160(4):1686–1697

Loka DA, Oosterhuis DM (2010) Effect of high night temperatures on cotton respiration. ATP levels, and carbohydrate content. Environ Exp Bot 68:258–263

Long SP, Ainsworth EA et al (2006) Food for thought: lower-than-expected crop yield stimulation with rising CO_2 concentrations. Sci 312:1918–1921

Lukens L, Doebley J (2001) Molecular evolution of the teosinte branched gene among maize and related grasses. Mol Biol Evol 18(4):627–663

Madhu M, Hatfield JL (2013) Dynamics of plant root growth under increased atmospheric carbon dioxide. Agron J 105:657–669

Mahendra T, Krishan P, Rai SK (2017) Traditional use of wild plants for food in West Sikkim, India. Int J Life Sci 5(4):730–741

Maiti RK (1996) Sorghum science. Science Publishers, Lebanon

Malézieux E (2012) Designing cropping systems from nature. Agron Sustain Dev 32:15–29

Markelz RJC, Lai LX, Vosseler LN, Leakey ADB (2014) Transcriptional reprogramming and stimulation of leaf respiration by elevated CO2 concentration is diminished, but not eliminated, under limiting nitrogen supply. Plant Cell Environ 37:886–898

Maxted N, Ford-Lloyd BV, Kell SP, Iriondo JM, Dulloo ME, Turok J (eds) (2008) Crop wild relative conservation and use. CABI, Wallingford

Mayer A, Hausfather Z, Jones AD, Silver WL (2018) The potential of agricultural land management to contribute to lower global surface temperatures. Sci Adv 4(8):eaaq0932. https://doi.org/10.1126/sciadv.aaq0932

McDonald AE, Vanlerberghe GC (2005) Alternative oxidase and plastoquinol terminal oxidase in marine prokaryotes of the Sargasso Sea. Gene 349:15–24

Meehl GA, Stocker TF, Collins WD, Gaye AJ, Gregory JM, Kitoh A, Knutti R, Murphy JM, Noda A, Raper SCB, Watterson JG, Weaver AJ, Zhao Z (2007) In: Solomon S, Qin D, Manning M, Chen Z, Marquis M, Averyt KB, Tignor M, Miller HL (eds) Global climate projections. Cambridge University Press, Cambridge/New York

Melchinger AE, Schopp P, Müller D, Schrag TA, Bauer E, UnterseerS HL, Schipprack W, Schön C-C (2017) Safeguarding our genetic resources with libraries of doubled-haploid lines. Genetics 206(3):1611–1619

Meldrum G, Mijatović D, Rojas W, Flores J, Pinto M, Mamani G, Condori E, Hilaquita D, Gruberg H, Padulosi S (2018) Environ Dev Sustain 20:703. https://doi.org/10.1007/s10668-016-9906-4

Mijatović D, Van Oudenhoven F, Eyzaguirre P, Hodgkin T (2013) The role of agricultural biodiversity in strengthening resilience to climate change: towards an analytical framework. Int J Agric Sustain 11(2):95–107

Mohammed AR, Tarpley L (2009) Impact of high nighttime temperature on respiration, membrane stability, antioxidant capacity, and yield of rice plants. Crop Sci 49:313–322

Munns R, Richards R (2007) Recent advances in breeding wheat for drought and salt stresses. In: Advances in molecular breeding toward drought and salt tolerant crops. https://doi.org/10.1007/978-1-4020-5578-2_22

Munns R, James RA, Läuchli A (2006) Approaches to increasing the salt tolerance of wheat and other cereals. J Exp Bot 57:1025–1043

Myers SS, Zanobetti A, Kloog I, Huybers P, Leakey AD, Bloom AJ et al (2014) Increasing CO2 threatens human nutrition. Nature 510:139–142

Negri V (2007) Towards a more comprehensive definition of 'landrace' than currently published. In: Del Greco A, Negri V, Maxted N (compilers) Report of a task force on on-farm conservation and management, second meeting, 19–20 June 2006, Stegelitz, Germany, Bioversity International, Rome, p 20

Nelson JC, Sorrells ME, Van-Deynze AE, Lu YH, Atkinson M, Bernard M, Leroy P, Faris JD, Anderson JA (1995) Molecular mapping of wheat: major genes and rearrangements in homoeologous groups 4, 5 and 7. Genet 141:721–731
Nemeth C, Yang C, Kasprzak P, Hubbart S, Scholefield D, Mehra S et al (2015) Generation of amphidiploids from hybrids of wheat and related species from the genera Aegilops, Secale, Thinopyrum, and Triticum as a source of genetic variation for wheat improvement. Genome 58:71–79
Nevo E (2004) Genomic diversity in nature and domestication. In: Henry R (ed) Diversity and evolution of plants. Genotypic and phenotypic variation in higher plants. CAB Publications, Wallingford, pp 287–316
Nikolic A, Ignjatovic-Micic D, Dodig D, Andjelkovic V, Lazic-Jancic V (2012) Identification of QTLs for yield and drought-related traits in maize: assessment of their causal relationship. Biotech Biotech Equi 26(3):2952–2960
Nikolic A, Andjelkovic V, Dodig D, Mladenovic Drinić S, Kravic N, Ignjatovic-Micic D (2013) Identification of QTL-s for drought tolerance in maize, II : yield and yield components. Genetika 45(2):341–350
Osborn TJ, Briffa KR (2005) The spatial extent of 20th-century warmth in the context of the past 1200 years. Science 311:841–844
Palmgren MG, Edenbrandt AK, Vedel SE, Andersen MM, Landes X, Østerberg JT et al (2015) Are we ready for back-to-nature crop breeding? Trends Plant Sci 20:155–164
Parish RS, Phan HA, Iacuone S, Li SF (2012) Tapetal development and abiotic stress: a centre of vulnerability. Funct Plant Biol 39:553–559
Patil RH, Laegdsmand M, Olesen JE, Porter JR (2010) Effect of soil warming and rain fall patterns on soil N cycling in Northern Europe. Agric Ecosyst Environ 139:195–205
Peleg Z, Fahima T, Krugman T, Abbo S, Yakir D, Korol AB, Saranga Y (2009) Genomic dissection of drought resistance in durum wheat x wild emmer wheat recombinant inbreed line population. Plant Cell Environ 32:758–779
Peng S, Huang J, Sheehy EJ, Laza CR, Visperas MR, Zhong X, Centeno SG, Khush SG, Gassman GK (2004) Rice yields decline with higher night temperature from global warming. PNAS 101(27):9971–9975
Pérez-Jaramillo JE, Mendes R, Raaijmakers JM (2016) Impact of plant domestication on rhizosphere microbiome assembly and functions. Plant Mol Biol 90:635–644
Pimentel D (1997) Techniques for reducing pesticide use. Wiley, Hoboken
Pinto RS, Reynolds MP, Mathews KL, McIntyre CL, Olivares-Villegas JJ, Chapman SC (2010) Heat and drought adaptive QTL in a wheat population designed to minimize confounding agronomic effects. Theor Appl Genet 121:1001–1021
Posey DA (1992) Interpreting and applying the "reality" of indigenous concepts: what is necessary to learn from the natives? In: Redford KH, Padoch C (eds) Conservation of neotropical forests. Working from traditional resource use. Columbia University Press, New York, pp 21–34
Prance G (1987) The first resource: Wild species in the north American economy. By C. and R. Prescott-Allen. Brittonia 39(4):427. https://doi.org/10.2307/2807319
Pressey RL, Cowling RM, Rouget M (2003) Formulating conservation targets for biodiversity pattern and process in the Cape Floristic Region, South Africa. Biol Conserv 112:99–127
Pritchard SG, Rogers HH, Prior SA, Peterson CM (1999) Elevated CO2 and plant structure: a review. Glob Chang Biol 5:807–837
Reich PB, Hobbie SE, Lee TD (2014) Plant growth enhancement by elevated CO2 eliminated by joint water and nitrogen limitation. Nat Geosci 7:920–924
Rellán-Álvarez R, Lobet G, Lindner H, Pradier PL, Sebastian J, Yee MC, Geng Y, Trontin C, Larue T, Schrager-Lavelle A, Haney CH, Nieu R, Maloof J, Vogel JP, Dinneny JR (2015) GLO-roots: an imaging platform enabling multidimensional characterization of soil-grown root systems. eLife 4:1–26
Reusch TB, Wood TE (2007) Molecular ecology of global change. Mol Ecol 16:3973–3992

Reynolds MP, Rebetzke G, Pellegrinesci A, Trethowan R (2006) Chapter 11, Drought adaptation in wheat. In: Ribaut JM (ed) Drought tolerance in cereals. Haworth's Food Products Press, New York, pp 402–436

Romojaro A, Sanchez-Bel P, Serrano M, Pretel MT (2013) Wild edible plants as potential antioxidants in vegetables oils. J Chem Article ID 457902, 4

Roy SJ, Tucker EJ, Tester M (2011) Genetic analysis of abiotic stress tolerance in crops. Curr Opin Plant Biol 14:232–239

Ruiz-Vera U, Siebers M, Gray SB, Rosenthal DR, Kimball BA, Ort DR, Ber-Nacchi CJ (2013) Global warming can negate the expected CO2 stimulation in photosynthesis and productivity for soybean grown in the Midwest United States. Plant Physiol 162(1):410–423

Ruosteenoja K, Räisänen J, Venäläinen A, Kämäräinen M (2016) Projections for the duration and degree days of the thermal growing season in Europe derived from CMIP5 model output. Int J Climatol 36:3039–3055

Sage RF, Way DA, Kubien DS (2008) Rubisco, Rubisco activase, and global climate change. J Exp Bot 59:1581–1595

Salekdeh GH, Reynolds M, Bennett J, Boyer J (2009) Conceptual framework for drought phenotyping during molecular breeding. Trends Plant Sci 14:488–496

Sánchez González de J, Ruiz Corral JA, García GM, Ojeda GR, LDIC L, Holland JB, Medrano RM, GEG R (2018) Ecogeography of teosinte. PLoS One 13(2):e0192676. https://doi.org/10.1371/journal.pone.0192676

Sethi S (2015) Bread, wine, chocolate: the slow loss of foods we love. Harper Collins, New York

Shah F, Huang J, Cui K, Nie L, Shah T, Chen C, Wang K (2011) Impact of high- temperature stress on rice plant and its traits related to tolerance. J Agric Sci 149:545–556

Sheehy JE, Elmido AE, Centeno HGS, Pablico PP (2005) Searching for new plants for climate change. J Agric Meteorol 60:463–468

Shelef O, Weisberg PJ, Provenza FD (2017) The value of native plants and local production in era of global agriculture. Front Plant Sci 8:2069

Shelef O, Fernández-Bayo J, Sher Y, Ancona V, Slinn H, Achmon Y (2018) Elucidating local food production to identify principles and challenges for sustainable agriculture. In: Galanakis CM (ed) Sustainable food systems from agriculture to industry: improving production and processing. Elsevier-Academic Press, New York, p 416

Sherrard ME, Maherali H (2006) The adaptive significance of drought escape in *Avena barbata*, an annual grass. Evolution 60:2478–2489

Shrivastava P, Kumar R (2015) Soil salinity: a serious environmental issue and plant growth promoting bacteria as one of the tools for its alleviation. Saudi J Biol Sci 22:123–131

Siebers MH, Yendrek CR, Drag D, Locke AM, Rios Acosta L, Leakey ADB, Ainsworth EA, Bernacchi CJ, Ort DR (2015) Heat waves imposed during early pod development in soybean (*Glycine max*) cause significant yield loss despite a rapid recovery from oxidative stress. Glob Chang Biol 8:3114–3125

Slikkerveer L (1994) Indigenous agricultural knowledge systems in developing countries: a bibliography. Indigenous Knowledge Systems Research and Development Studies No. 1. Special Issue: INDAKS Project Report 1 in collaboration with the European Commission DG XII. Leiden, the Netherlands, Leiden Ethnosystems and Development Programme (LEAD)

Sprague SJ, Balesdent MH, Brun H, Hayden HL, Marcroft SJ, Pinochet X et al (2006) Major gene resistance in Brassica napus (oilseed rape) is overcome by changes in virulence of populations of Leptosphaeria maculans in France and Australia. Eur J Plant Pathol 114:33–40

Su Z, Ma X, Guo H, Sukiran NL, Guo B, Assmann SM, Ma H (2013) Flower development under drought stress: morphological and transcriptomic analyses reveal a cute responses and long-term acclimation in Arabidopsis. Plant Cell 25:3785–3807

Tardieu F, Granier C, Muller B (2011) Water deficit and growth. Coordinating processes without an orchestrator? Curr Opin Plant Biol 14:283–289

Taub DR, Wang XZ (2008) Why are nitrogen concentrations in plant tissues lower under elevated CO_2? A critical examination of the hypotheses. J Integr Plant Biol 50:1365–1374

Taub D, Miller B et al (2008) Effects of elevated CO_2 on the protein concentration of food crops: a meta-analysis. Glob Chang Biol 14:565–575
Teng N, Wang J, Chen T, Wu X, Wang Y, Lin J (2006) Elevated CO2 induces physiological, biochemical and structural changes in leaves of *Arabidopsis thaliana*. New Phytol 172:92–103
Thomas CD, Cameron A, Green RE et al (2004) Extinction risk from climate change. Nature 427(6970):145–148. https://doi.org/10.1038/nature02121
Turnbull MH, Murthy R, Griffin KL (2002) The relative impacts of day time and night-time warming on photosynthetic capacity in *Populus deltoides*. Plant Cell Environ 25:1729–1737
Ulén B, Johansson G (2009) Long-term nutrient leaching from a Swedish arable field with intensified crop production against a background of climate change. Acta Agric Scand Sect B Soil Plant Sci 59(2):157–169
Van Frankenhuyzen K (2009) Insecticidal activity of Bacillus thuringiensis crystal proteins. J Invertebr Pathol 101:1–16
Veteläinen M, Negri V, Maxted N (2009) European landraces: on-farm conservation, management and use. Bioversity International, Technical Bulletin No. 15, Roma, Italy
von Korff M, Grando S, Del Greco A, This D, Baum M, Ceccarelli S (2008) Quantitative trait loci associated with adaptation to Mediterranean dryland conditions in barley. Theor Appl Genet 117:653–669
Wahid A, Gelani S, Ashraf M, Foolad MR (2007) Heat tolerance in plants: a novel review. Environ Exp Bot 61:199–223
Wang X, Liu L, Liu S, Sun X, Deng Z, Pi Y et al (2004) Isolation and molecular characterization of a new CRT binding factor gene from Capsella bursa-pastoris. J Biochem Mol Biol 37:538–545
Waugh R, Jannink JL, Muehlbauer GJ, Ramsay L (2009) The emergence of whole genome association scans in barley. Curr Opin Plant Biol 12:218–222
Welp LR, Keeling RF, Meijer HAJ, Bollenbacher AF, Piper SC, Yoshimura K et al (2011) Interannual variability in the oxygen isotopes of atmospheric CO_2 driven by El Niño. Nature 477:579–582
Wheeler T, von Braun J (2013) Climate change impacts on global food security. Science 341(6145):508–513. https://doi.org/10.1126/science.1239402
Xia J, Chen J, Piao S, Ciais P, Luo Y, Wan S (2014) Terrestrial carbon cycle affected by non-uniform climate warming. Nat Geosci 7:173–118
Xie W, Nevo E (2008) Wild emmer: genetic resources, gene mapping and transfer for wheat improvement. Euphytica 164:603–614
Zeven AC (1998) Landraces: a review of definitions and classifications. Euphytica 104:127–139
Zinn KE, Tunc-Ozdemir M, Harper JF (2010) Temperature stress and plant sexual reproduction: uncovering the weakest links. J Exp Bot 61:1959–1968
Zohary D, Hopf M, Weiss E (2012) Domestication of plants in the Old World: the origin and spread of domesticated plants in Southwest Asia, Europe, and the Mediterranean Basin. Oxford University Press on Demand, Oxford

Role of Wild Relatives for Development of Climate-Resilient Varieties

11

Manmohan Sharma, Punya, and Bharat Bhushan Gupta

Contents

Abstract

Climate change is a hot topic nowadays, and its impact on agriculture and related fields makes the scientific community to work toward innovating new technologies which proves resilient during fluctuations in climate. Climate resilience can be generally defined as the capacity for a socio-ecological system to absorb stresses and maintain function in the face of external stresses imposed upon it by climate change and to adapt, reorganize, and evolve into more desirable configurations that improve the sustainability of the system, leaving it better prepared for future climate change impacts. Climate changes possess a severe effect on plant genetic resources and wild plant species. These wild species are the rich source of novel alleles for biotic and abiotic stress resistance which can be used to develop varieties with superior traits. Thus, understanding of anomalies in

M. Sharma (✉) · Punya
School of Biotechnology, Sher-e-Kashmir University of Agricultural Sciences and Technology of Jammu, Jammu, Jammu and Kashmir, India

B. B. Gupta
Division of Plant Breeding and Genetics, Sher-e-Kashmir University of Agricultural Sciences and Technology of Jammu, Jammu, Jammu and Kashmir, India

R. K. Salgotra, S. M. Zargar (eds.), *Rediscovery of Genetic and Genomic Resources for Future Food Security*,
https://doi.org/10.1007/978-981-15-0156-2_11

climatic variables is essential to make the agriculture sector climate resilient. Thus, future crop species will need to be able to thrive in a drier, warmer, and more variable and extreme climatic conditions. To meet these challenges, plant breeders need to exploit genetic diversity available in the form of germplasm, landraces, and wild or weedy forms. Some of the genetic diversity may be found in landraces, traditional/farmer's varieties that are still being cultivated by farmers around the world. However, a much wider spectrum of biodiversity is found in wild plant species that are closely related to domesticated crops. They are of key importance to breeding crops for adaptation to climate changes.

Acronyms

CWR	Crop wild relatives
CO_2	Carbon dioxide
GHGs	Greenhouse gases
GPS	Global positioning system
GWAS	Genome-wide association studies
MAS	Marker-assisted selection
IPCC	Intergovernmental Panel on Climate Change
PGR	Plant genetic resources
QTL	Quantitative trait loci
SNP	Single nucleotide sequence

11.1 Introduction

Climate change is a phenomenon of transition of earth's climate system due to emission of greenhouse gases from natural and anthropogenic sources. The level of absorption, scattering, and emission of radiation within the atmosphere, ocean, and at the earth surface is highly affected by the amount of concentration of atmospheric greenhouse gases (GHGs), aerosols, soil type and moisture, vegetation and land cover, solar radiations, etc. Globally, climate change influences many areas of socioeconomic activities, such as agriculture, horticulture, forestry, etc., and is a major threat for biodiversity and ecosystem function (Lepetz et al. 2009). Vulnerability to climate change depends not only on physical and biological responses but also on socioeconomic characteristics. Low-income population especially those who cultivate crops under rain-fed and nonirrigated agriculture systems in dry land, arid, and semiarid areas are highly affected by severe hardship due to climate change (Grasty 1999). Climatic conditions play an important role in the development, productivity, and distribution of plants and affect the balance between different dominating species and abiotic factors. Agriculture is one of the sectors that is sensitive to global warming (e.g., through atmospheric temperature, precipitation, soil moisture, sea level, and humidity) which contributes to climate change. Increased global temperature along with associated carbon dioxide (CO_2) increase, altered pattern of rainfall and salinity, and emergence of new pest strains and diseases are the indicators of climate change (Tester and Langridge 2010). Temperature change will have variable impacts on vegetation and ecosystem productivity, structure, and composition depending on the actual temperature range at the location (Morison and Lawlor 1999).

11.2 Impact of Climate Change on Agriculture

Due to impact of climate change, agricultural productivity is directly affected in different geographical regions of the world (Alexandrov and Hoogenboom 2000), thus posing a great threat to food security. Agriculture is highly affected by changes in climatic parameters such as CO_2 concentration, rainfall, temperature, drought, etc. Food production can be negatively or positively affected following variation in weather patterns (short winter, long summer, earlier spring) and other extreme weather events such as drought (change in the amount and timing of precipitation), flooding, etc. The predicted climatic changes are expected to have fairly widespread impacts on agriculture. For example, rice flowers show increased sterility at high temperatures, maize is very sensitive to drought at the time of flowering, wheat senescence starts earlier and faster under warmer conditions, etc. (Lobell et al. 2012). The overall impacts of higher temperatures on crop responses at the plot level may be variable, without considering changes in the frequency of extreme events. For instance, moderate warming may benefit in crop and pasture productivity in temperate regions, while it may reduce productivity in tropical and semiarid regions. Modeling studies indicate small beneficial effect of temperate variation with increase in the range from 1 to 3 °C with associated increase in CO_2 levels and rainfall changes. The elevated CO_2 levels are predicted to reduce the nutritional quality of many crops, while some crops may become toxic due to changes in the chemical composition of their tissues (Dwivedi et al. 2013). Climate change will affect food supply unless actions are taken to increase the resilience of crops. Reports have shown a drastic decrease in the production of major cereals by 2020, including 9% for maize, 11% for rice, and 14% for wheat (Hisas 2011). In many cropping regions, the crop environment will tend to be warmer with more irregular rainfall, and spikes in stress levels will be more severe. The challenge is not only to raise agricultural production for an expanding population but to achieve this under more adverse environmental conditions in a sustainable manner.

In particular, the incidences of extreme conditions are expected to exceed those currently experienced in the temperate-tropical cropping regions (David et al. 2008). This poses a threat to world food security, especially for production of the annual staple food crops of the world: wheat, rice, maize, sorghum, and potato. A range of other crops from grain legumes to clonally propagated tubers are also expected to suffer yield declines, especially when extreme conditions occur in the reproductive period of grain crops or in the tuberization period for root crops (Schafleitner et al.

2011). Anthesis and pollination in many crops are adversely affected by temperatures above 35 °C, and the critical temperature may be as low as 32 °C (Singh et al. 2011). High night temperatures above 32 °C can impede gametogenesis and photosynthesis in rice (Lafarge et al. 2011). In potatoes, the initiation of tubers requires temperatures below 18 °C. Severe moisture stress in the pre-flowering and the flowering periods reduces formation of florets as well as affects fertilization and seed set (Vadez et al. 2011). Grain size is also reduced by terminal drought in the ripening period. Where rainfall is lower due to climate change, the reduced cloud cover may lead to increased frequency of frost in the reproductive period with further loss of yield potential. The rise in global world temperature is now expected to exceed 2 °C by around 2050 in the worst case scenario and certainly before 2100. In the temperate tropical latitudes, this will result in shorter crop growth periods and reduced yield potential (Lotze-Campden 2011). There are some positive aspects to the predicted climate changes. Cropping will become more feasible at high altitudes and at high latitudes, with longer periods above freezing.

11.3 Role of Crop Wild Relatives

Out of about 300,000 species of higher plants on earth, about 7000 species have been domesticated and cultivated by humans over the millennia for food, fodder, and feed, representing a relatively small and seemingly insignificant proportion. However, intraspecific genetic diversity in crop plants and their wild relatives has been maintained under natural conditions. These plant resources are, in general, designated as plant genetic resources (PGR). They form the basis of all modern-day high-yielding crop varieties that are bred to produce more, withstand stresses, and yield quality output. Crop wild relatives (CWRs), an important component of precious plant genetic resources, are potential source of plant genetic diversity and globally constitute a vital resource to secure food supply for mankind through improved agricultural practices. The key to successful crop improvement is a continued supply of genetic variability and beneficial traits contained in this diversity (Dwivedi et al. 2008); and wild relatives of modern crops are the source of much of this novel diversity, providing genes with improved nutritional quality, with resistance to pests and diseases, as well as with traits adapted to drought and extreme temperatures. The systematic assessment of CWR genetic resources under specific conditions of abiotic stress will allow better targeting of their use for breeding populations, lines, and clones and their further use in cultivar development. Association genetics along with population genetics, bioinformatics, and reverse genetic approaches will help to identify genes or quantitative trait locus (QTLs) from CWR that can enhance crop adaptation under abiotic stress-prone environments (Baute et al. 2015). The adaptation of crops to gradual change in climatic conditions will require screening of existing cultivars and breeding of new ones for adaptation to drought, temperature stresses, sustained productivity, disease resistance, and other factors, highlighting the importance of maintaining the pools of genetic variation in CWRs. Unfortunately, CWRs are facing threat of extinction due to climate change.

Under elevated CO_2 levels, CWR produce relatively less fruit and seed than domesticated crops (Jablonski et al. 2002), increasing their risk of extinction. They have remained at relatively low priority for breeders as well. The looming threat of climate change has put to risk the natural habitats and farming systems, thus directly or indirectly promoting genetic erosion of crop wild relatives and landraces. Thus, there is an urgent need to conserve CWRs both in the wild (in situ) and in gene banks (ex situ) to ensure that genetic diversity remains available for future generations.

11.4 Mechanisms of Adaptation of Crop Wild Relatives

Plant species distribution is limited not only by their absolute ability of survival but also through competition within species depending on which species acclimate and grow better in a given climate. In the context of population extinction, it is important to consider the effects during climate events. Temporal variability in environment is commonly believed to increase the probability of population extinction, particularly if environmental variability increases due to climate change. Climate is a potent source of selection in natural populations, yet the importance of adaptation in the response of wild plant species to climate change has remained questionable. Climate plays a major role in increasing or reducing yield levels in global perspective from temperate to tropics. Many experiments show that CO_2 is a limiting factor, in which higher concentration of CO_2 enhances photosynthesis and crop growth, modifying water and nutrient cycles (Tubiello et al. 2008); these responses are found to hold even for plants grown under different stressful conditions. The increased CO_2 concentration induces and makes an increase in the grain weight, and according to one of the observations, it was greater under average phosphorus treatment compared to higher phosphorus level. This influence of CO_2 and phosphorus supply was attributed to increase in the number of cells within endosperm, which is the result of enhanced rate of cell division during grain development or by greater amount of grain filling during ripening phase. However, it has also been shown that elevated CO_2 concentrations may have negative effects on the grain quality in wheat in terms of protein content (Pleijel and Uddling 2011), as it alters wheat grain lipids and doubled the number of mitochondria in wheat leaves, lowers seed nitrogen concentration, and decreases grain and flower protein (Qaderi and Reid 2009).

Several studies have shown that soil warming can affect availability of nutrient and increase soil N mineralization and nitrate leaching and organic matter decomposition, and a slight temperature increase can produce a significant enhancement of activities. An increase in N mineralization in soil can be predicted under favorable moisture conditions and substrate availability, mainly in those ecosystems where temperature is a limiting factor, which leads to increase NPP (net primary production), to increase N demand, and ultimately to decrease N availability in the soil. An increasing temperature will also speed up the release of nutrients locked up in organic soil fraction and minerals, while decreasing soil moisture may limit this process. A higher rate of weathering of nutrient-rich rocks generally leads to higher base saturation of the soil and maintains higher soil pH; both characteristics are favorable to plant growth. However, elevated CO_2 has not been thought to have a direct effect on weathering (Lukac et al. 2010).

There are many processes in plant growth which are affected by interaction of enhanced temperature and carbon dioxide; the processes that determine carbon balance in the shorter term and from the long time scales of development and growth lead to accumulation of biomass and yield. The two main reasons to expect progressively increasing CO_2 responsiveness of plant carbon balance at higher temperatures are (1) the decreased ratio of photosynthesis to photorespiration and (2) the decreased ratio of gross photosynthesis to dark respiration in warmer conditions (Morison and Lawlor 1999). The effect of elevated CO_2 on photosynthetic reactions is more pronounced at high temperature. Some reports indicate that future increase in temperature may increase root mortality more in N-rich soils in temperate forests than in N poor soils in boreal forest areas with important implications for the cycling between plant and soil (Lukac et al. 2010). Some studies found that changes in activation state and catalytic constant occur due to both CO_2 and temperature, and their interaction, which affected the photosynthetic rate demonstrating the underlying complexity of the photosynthetic regulation mechanisms (Morison and Lawlor 1999).

Environmental change has an impact on growth rate of individual trees and has a cumulative effect on different interactions and processes inside the forest, as well as ability to change the amount of living materials in the forest ecosystem as a whole (Lukac et al. 2010). Temperature is one of the decisive factors affecting growth and productivity by accelerating bud burst (BB), flowering and stem elongation during spring, extending duration of the growing season, and controlling species distribution. For instance, the predicted warming of 2–6 °C by 2100 in north temperate forest regions will have substantial impacts on growth and species composition (Gunderson et al. 2012). Environmental shift affects the extent of plant diseases, insect pests and weeds, and their occurrence and infestation. Following these changes, preventive actions are needed to reduce the effects on human health and ecosystems (Roos et al. 2010). Different chemical, biological, and physical processes in earth systems need various temperature ranges. Usually moderate and optimal temperature is essential for normal activities within the systems; a certain rise or fall from moderate temperature will affect many activities within the processes.

11.5 Strategies to Tackle the Impact of Climate Changes

In the last century, scientifically managed plant breeding programs have enabled gains in productivity in many crops, through using a very high selection intensity for a small number of key traits (increased harvest index, improved plant architecture), very large breeding populations, the pyramiding of complementary pest and disease resistances, and high crop input responsiveness, for various target environments (Lafarge et al. 2011). Under various environmental stresses, plant breeding shows dynamic techniques in crop development and betterment. It gives a way to potentially guarantee food security and safety under harsh weather variations and help plants escape from various stresses through a crucial phase of plant growth by developing stress-resistant cultivars (Blum 2018). Genetic divergence analysis is used for assessment of genetic distances among genotypes arising as a result of inbreeding, outbreeding, assortment, and recombination events. Genetic divergence analysis is considered a very important method for the development of new cultivars based on genetic distance and similarities. Some of the important strategies which can be devised to tackle impact of climate change for achieving sustainable food production include:

11.5.1 Breeding and Biotechnological Interventions for Utilization of CWRs in the Scenario of Climate Change

11.5.1.1 Identifying the Useful CWRs for Breeding Programs

Crop wild relatives are an important source of genetic diversity for crop improvement. A complementary option is to seek genetic variation for tolerance to abiotic stresses in wild relatives, for each of the important plant food species. Landrace populations or traditional local varieties with genetic variation have been selected through natural and manual selection for population complexes with specific adaptation to the range of seasonal crop environments in each locality (Bennett 1970). It is now possible to systematically explore the genetic variation in historic local landraces and CWRS by using GPS locators and world climate maps to describe the natural selection for local adaptation and to identify candidate germplasm for tolerances to extreme stresses. Novel genetic variation will be needed to extend the range of tolerances to high-temperature stresses and to severe droughts in the crop growing period (IPCC 2007). For genetic studies, landrace is a significant source and is a valuable basis for stress resistance as it contains cultivars adjustable to diverse environmental stress. Pre-breeding, molecular breeding, and integrated plant breeding are useful to develop biotic and abiotic stress-tolerant cultivars using genomics approaches like marker-assisted selection (MAS), genomic selection, and genome-wide-association studies (GWAS).

11.5.1.2 Marker-Assisted Selections for Precision Breeding

Marker-assisted selection (MAS) can increase the efficiency of incorporating desirable traits present in wild germplasm into domesticated, or elite, cultivars. MAS relies on genetic markers that are either causal for, or strongly linked to, a phenotype. The primary benefit of MAS is the ability to select individuals possessing a trait of interest at the seed or seedling stage using genetic markers. MAS allows the breeder to eliminate plants that do not possess the desired trait and may otherwise require a decade of cultivation to assess phenotypically. Instead, resources and space can be dedicated only to individuals with the desired characteristic. Plants with the desired trait can then be backcrossed to elite germplasm to maintain the wild trait of interest while preserving important commercial traits. Backcrossing to elite germplasm is crucial to ensuring traits of agricultural importance are maintained when breeding with wild relatives: the goal is to retain all desirable characteristics of the elite cultivars while introducing only the small number of desirable loci from the wild. In addition to saving time, MAS can decrease the cost of perennial breeding using wild relatives. When compared to traditional fruit breeding, MAS was estimated to save up to 43 percent of operational costs over the first 6–8 years of an apple breeding program (Edge-Garza et al. 2015). MAS eliminates the need to phenotype and therefore offers the greatest cost and time savings for traits that may be difficult or expensive to measure, such as disease resistance, as well as traits expressed late in development, such as fruit quality (Töpfer et al. 2011).

11.5.1.3 Genomic Approaches

Biotechnological approaches provide beneficial resources to elucidate biological functions of any genetic information for crop upgrading and development. Different molecular markers are studied in population genomics across the environment in many individuals to find out novel variation patterns and help to find if the genes have functions in significant ecological traits (Keurentjes et al. 2008). In many crops, the breeding programs are coupled with genomic approaches to achieve great heights in molecular breeding and to screen elite germplasms with multi-trait assembly (Bevan and Waugh 2007). Genomic approaches also enable investigation of the molecular mechanisms underlying the abiotic stress resistance. These approaches aid in the development of climate smart crops for better yield and production under different climate changes (Roy et al. 2011).

11.5.1.4 Genome-Wide Association Studies (GWAS)

Genome-wide association studies (GWAS) is a powerful tool for understanding the complete set of genetic variants in different crop cultivars to recognize allelic variants linked with any specific trait (Manolio 2010). GWAS generally highlights linkage among SNPs and traits and identifies the major genes/QTLs associated (Bush and Moore 2012). In plants, GWAS has widespread applications related to biotic and abiotic stresses. GWAS have been applied to abiotic stresses such as salt tolerance (Wan et al. 2017) and heat tolerance (Lafarge et al. 2017). In *Arabidopsis thaliana*, GWAS study was carried out by Verslues et al. (2014) aided by reverse genetic approaches to elucidate unique genes that accumulate proline under drought stress. This research gave insights for proline accumulation under drought stress conditions (Verslues et al. 2014). *Aegilops tauschii* is reported to have many resistance genes regulating the abiotic stresses (Ashraf 2009). A significant knowledge is required for the breeders to understand the genetic architecture of *Aegilops tauschii* to improve drought resilience. Qin et al. (2016) investigated 373 different varieties of *A. tauschii* to examine 13 traits controlling drought stress. For GWAS, 7185 SNPs were designated to study the phenotypic behavior to find the association between SNPs with phenotypic traits (Qin et al. 2016). Kumar et al. (2015) reported various genes regulating the salinity tolerance in rice by using high-throughput SNPs arrays. Six thousand genotype-based SNPs were detected for genes related to stress, and linkage among SNPs and phenotypic data was interpreted. A novel QTL present on chromosome number 1 was reported and was called "Saltol" which is associated with salt tolerance at seedling stage. Lafarge et al. (2017) performed GWAS for genotyping 167 rice varieties for spikelet sterility (SPKST) and panicle micronutrient and observed significant association between SPKST, secondary traits, and 14 loci. These loci were investigated for functions related to heat shock proteins, controlling plant responses, development of gametophyte, cell division, and detecting abiotic stresses. Chopra et al. (2017) reported various stress-tolerant genes in *Sorghum bicolor* associated with heat and cold stresses. Thirty SNPs were identified for genes related to anthocyanin expression and carbohydrate metabolism, which are powerfully associated with cold stress at the seedling growth phase of sorghum. Similarly, 12 SNPs were discovered for heat stress at the seedling stage and controlled by the genes having functions in ion transport mechanism and sugar metabolism. In another study, Chen et al. (2017) examined *Sorghum bicolor* for heat-tolerant traits such as leaf firing (LF) and leaf blotching (LB) at the vegetative phase of growth. To identify the association among SNPs with genotype and heat tolerance, GWAS was performed. Nine SNPs were observed to be closely linked with LF, and five SNPs were identified for LB traits. Furthermore, 14 genes associated with SNPs were discovered that have stress-responsive expression to abiotic stresses.

11.5.2 Situation-Specific Selection of Crops

Switching to more stress-tolerant crops may be an option for particular crop regions. For example, pearl millet is more drought tolerant than sorghum (Diakité et al. 2008); triticale and rye are more drought tolerant than wheat with different respective strategies for maintaining photosynthesis or for recovering from drought (Uprety and Sirohi 1987).

11.5.3 Agronomic Management

In order to tackle abiotic stress factors, farmers have adopted useful approaches such as altering planting and harvesting time, selection of crops with short lifecycles, crop rotation, irrigation techniques, and variation in cropping schemes. Under climatic stress conditions, all of these approaches are very beneficial for crop adaptability (Duku et al. 2018).

Another plant adaptability approach is by means of crop management techniques that have the ability to enhance crop development under various environmental stresses. The choice of sowing time, planting density, and optimum irrigation practices are crucial techniques to tackle weather stresses (Battisti et al. 2018). Fertilizers are also very vital to reduce the effect of global warming and supporting the plant for better adaptability. They provide substantial energy to plants and are beneficial to maintain the fertility of the soil and increased productivity. Hence, the importance of fertilizer in nourishing the world is undeniable (Henderson et al. 2018).

Modification in sowing time, use of drought-resistant cultivars, and the cultivation of new crops are some important strategies to lessen the climatic variability danger and provide better adaptability to crop plants for assuring food safety and security (Ali and Erenstein 2017).

11.6 Conclusion

There is a growing concern about the use of genetic resources in the scenario of climate change to secure food security. The genetic resources can be efficiently used to cope the adverse effect of climate change through genomic resources. The future crop species will need to be able to thrive in a drier, warmer, and more variable and extreme climatic conditions which can be achieved through efficient utilization of genetic resources in crop improvement. To meet these challenges, plant breeders need to exploit genetic diversity available in the form of germplasm, landraces, and wild or weedy forms by using various genomic resources such MAS, GWAS, high-throughput techniques, etc. However, a much wider spectrum of biodiversity is found in wild plant species that are closely related to domesticated crops. They are of key importance to breeding crops for adaptation to climate changes. Thus, there is a need for understanding of anomalies in climatic variables which is essential to make the agriculture sector climate resilient and to develop climate-resilient varieties.

References

Alexandrov VA, Hoogenboom G (2000) Vulnerability and adaptation assessments of agricultural crops under climate change in the southeastern USA. Theor Appl Climatol 67:45–63

Ali A, Erenstein O (2017) Assessing farmer use of climate change adaptation practices and impacts on food security and poverty in Pakistan. Clim Risk Manag 16:183–194

Ashraf M (2009) Biotechnological approach of improving plant salt tolerance using antioxidants as markers. Biotechnol Adv 27:84–93

Battisti R, Sentelhas PC, Parker PS, Nendel C, Gil MDS, Farias JR, Basso CJ (2018) Assessment of crop management strategies to improve soybean resilience to climate change in Southern Brazil. Crop Pasture Sci 69:154–162

Baute GJ, Dempewolf H, Rieseberg LH (2015) Using genomic approaches to unlock the potential of CWR for crop adaptation to climate change. In: Redden R et al (eds) Crop wild relatives and climate change, pp 268–280. https://doi.org/10.1002/9781118854396.ch15

Bennett E (1970) Adaptation in wild and cultivated plant populations. In: Frankel OH, Bennett E (eds) Genetic resources in plants—their exploration and cultivation; IBP handbook no 11. Blackwell Scientific Publications, Oxford, pp 115–129

Bevan M, Waugh R (2007) Applying plant genomics to crop improvement. BioMed Central, London, p 2007

Blum A (2018) Plant breeding for stress environments: 0. CRC, Boca Raton

Bush WS, Moore JH (2012) Genome-wide association studies. PLoS Comput Biol 8:e1002822

Chen J, Chopra R, Hayes C, Morris G, Marla S, Burke J, Xin Z, Burow G (2017) Genome-wide association study of developing leaves' heat tolerance during vegetative growth stages in a sorghum association panel. Plant Genome 10:1–15

Chopra R, Burow G, Burke JJ, Gladman N, Xin Z (2017) Genome-wide association analysis of seedling traits in diverse Sorghum germplasm under thermal stress. BMC Plant Biol 17:12

David B, Lobell DB, Burke MB, Tebaldi C, Mastrandrea MD, Falcon WP, Naylor RL (2008) Prioritizing climate change adaptation needs for food security in 2030. Science 319(5863):607–610

Diakité L, Sidibé A, Smale M, Grum M (2008) Seed value chains for Sorghum and Millet in Mali. A state-based system in transition; IFPRI Discussion Paper 00749; International Food Policy Research Institute: Washington, DC, USA

Duku C, Zwart SJ, Hein L (2018) Impacts of climate change on cropping patterns in a tropical, sub-humid watershed. PLoS One 13:e0192642

Dwivedi SL, Upadhyaya HD, Thomas Stalker H et al (2008) Enhancing crop gene pools with beneficial traits using wild relatives. Plant Breed Rev 30:180–230

Dwivedi SL, Sahrawat KL, Upadhyaya HD, Ortiz R (2013) Food nutrition and agrobiodiversity under global climate change. Adv Agron 120:1–128

Edge-Garza DA, Luby JJ, Peace C (2015) Decision support for cost-efficient and logistically feasible marker-assisted seedling selection in fruit breeding. Mol Breed 35:223

Grasty S (1999) Agriculture and climate change. TDRI Quarterly Review (eds Auger P, and Suwanraks R) 14(2):12–16

Gunderson CA, Edwards NT, Walker AV, O'Hara KH, Campion CM, Hanson PJ (2012) Forest phenology and a warmer climate - growing season extension in relation to climate provenance. Global Change Biol 18:2008–2025

Henderson B, Cacho O, Thornton P, van Wijk M, Herrero M (2018) The economic potential of residue management and fertilizer use to address climate change impacts on mixed smallholder farmers in Burkina Faso. Agric Syst 167:195–205

Hisas S (2011) The food gap. The impacts of climate change in food production: a 2020 perspective. Univ Ecol Fund, Alexandria

IPCC (2007) Summary for policymakers. In: Metz B, Davidson OR, Bosch PR, Dave R, Meyer LA (eds) Climate change 2007: mitigation. Contribution of working group III to the fourth assessment report of the intergovernmental panel on climate change. Cambridge University Press, Cambridge/New York

Jablonski LM, Wang X, Curtis PS (2002) Plant reproduction under elevated CO_2 conditions: a meta-analysis of reports on 79 crop and wild species. New Phytol 156:9–26

Keurentjes JJ, Koornneef M, Vreugdenhil D (2008) Quantitative genetics in the age of omics. Curr Opin Plant Biol 11:123–128

Kumar V, Singh A, Mithra SA, Krishnamurthy S, Parida SK, Jain S, Tiwari KK, Kumar P, Rao AR, Sharma S (2015) Genome-wide association mapping of salinity tolerance in rice (*Oryza sativa*). DNA Res 22:133–145

Lafarge T, Peng S, Hasegawa T, William P, Quick SV, Jagadish K, Wassmann R (2011) In: Yadav SS, Redden RJ, Hatfield JL, Lotze-Campen H, Hall AE (eds) Genetic adjustment to changing climates: Rice. In crop adaptation to climate change. Wiley-Blackwell, Chichester,. Chapter 12, pp 298–313

Lafarge T, Bueno C, Frouin J, Jacquin L, Courtois B, Ahmadi N (2017) Genome-wide association analysis for heat tolerance at flowering detected a large set of genes involved in adaptation to thermal and other stresses. PLoS One 12:e0171254

Lepetz V, Massot M, Schmeller DS, Clobert J (2009) Biodiversity monitoring: some proposals to adequately study species' responses to climate change. Biodivers Conserv 18:3185–3203

Lobell DB, Sibley A, Ivan Ortiz-Monasterio J (2012) Extreme heat effects on wheat senescence in India. Nat ClimChang 2:186–189

Lotze-Campden H (2011) Climate change, population growth, and crop production: an overview. In: Yadav SS, Redden RJ, Hatfield JL, Lotze-Campen H, Hall AE (eds) Crop adaptation to climate change. Wiley-Blackwell, Chichester,. Chapter 1, pp 1–11

Lukac M, Calfapietra C, Lagomarsino A, Loreto F (2010) Global climate change and tree nutrition: effects of elevated CO2 and temperature: Tree physiology [0829-318X] Lukac yr. 30, 1209–1220

Manolio TA (2010) Genomewide association studies and assessment of the risk of disease. N Engl J Med 363:166–176
Morison JIL, Lawlor DW (1999) Interaction between increase CO2 concentration and temperature on plant growth. Plant Cell Environ 22:659–682
Pleijel H, Uddling J (2011) Yields vs quality trade-offs for wheat in response to carbon dioxide and ozone. Glob Chang Biol 18:596–605
Qaderi MM, Reid DM (2009) Crop responses to elevated carbon dioxide and temperature (chp1). In: Singh SN (ed) Climate change and crops, environmental science and engineering. Springer, Berlin. https://doi.org/10.1007/978-3-540-88246-61
Qin P, Lin Y, Hu Y, Liu K, Mao S, Li Z, Wang J, Liu Y, Wei Y, Zheng Y (2016) Genome-wide association study of drought-related resistance traits in *Aegilops tauschii*. Genet Mol Biol 39:398–407
Roos J, Hopkins R, Kvarnheden A, Dixelius C (2010) The impact of global warming on plant diseases and vectors in Sweden. Eur J Plant Pathol 129:9–19
Roy SJ, Tucker EJ, Tester M (2011) Genetic analysis of abiotic stress tolerance in crops. Curr Opin Plant Biol 14:232–239
Schafleitner R, Ramirez J, Jarvis A, Evers D, Gutierrez R, Scurrah M (2011) Adaptation of the potato crop to changing climates. In: Yadav SS, Redden RJ, Hatfield JL, Lotze-Campen H, Hall AE (eds) Crop adaptation to climate change. Wiley-Blackwell, Chichester,. Chapter 11, pp 287–297
Singh RP, Vara Prasad PV, Sharma AK, Raja Reddy K (2011) Impacts of high-temperature stress and potential opportunities for breeding. In: Yadav SS, Redden RJ, Hatfield JL, Lotze-Campen H, Hall AE (eds) Crop adaptation to climate change. Wiley-Blackwell, Chichester,. Chapter 5.1, pp 166–185
Tester M, Langridge P (2010) Breeding technologies to increase crop production in a changing world. Science 327:818–822
Töpfer R, Hausmann L, Eibach R (2011) Molecular breeding. In: Adam-Blondon A-F, Martínez-Zapater JM, Kole C (eds) Genetics, genomics and breeding of grapes. CRC, Boca Raton, pp 160–185. https://doi.org/10.1201/b10948-8
Tubiello F, Schmidhuber J, Howden M, Neofotis P G, Park S, Fernandes E, Thapa D (2008) Climate change response strategies for agriculture: challenges and opportunities for the 21st Century, Agriculture and Rural Development Discussion paper 42 The World Bank
Uprety DC, Sirohi GS (1987) Comparative study on the effect of water stress on the photosynthesis and water relations of triticale, rye and wheat. J Agron Crop Sci 159:349–355
Vadez V, Kholova J, Choudhary S, Zindy P, Terrier M, Krishnamurthy L, Ratna Kumar P, Turner NC (2011) Responses to increased moisture stress and extremes: whole plant response to drought under climate change. In: Yadav SS, Redden RJ, Hatfield JL, Lotze-Campen H, Hall AE (eds) Crop adaptation to climate change. Wiley-Blackwell, Chichester,. Chapter 5.2, pp 186–197
Verslues PE, Lasky JR, Juenger TE, Liu TW, Kumar MN (2014) Genome-wide association mapping combined with reverse genetics identifies new effectors of low water potential-induced proline accumulation in Arabidopsis. Plant Physiol 164:144–159
Wan H, Chen L, Guo J, Li Q, Wen J, Yi B, Ma C, Tu J, Fu T, Shen J (2017) Genome-wide association study reveals the genetic architecture underlying salt tolerance-related traits in rapeseed (*Brassica napus* L.). Front Plant Sci 8:593

Strategies for Conservation of Genetic Resources

12

Jahangir Ahmad Dar, Masrat Kareem, Sajad Majeed Zargar, Aijaz A. Wani, Sheezan Rasool, and Kaisar A. Bhat

Contents

Abstract

Genetic resources are of prime historical and practical significance for survival of human beings, to generate revenue and support a balanced ecosystem. Several aspects of these genetic resources further depend on the conservation strategies for balancing the ecosystem. Conserving resources is of prime importance to safeguard mankind against growing demand of food and stabilizing the ecosystem. Plant genetic resources (PGR) refer to the heritable material contained within and among plant species of present and potential value. In recent past, the diversity in plant genetic resources found in wild and weedy cultivars, and landraces have been reported to save animal and plant population from diseases, pests, and environmental changes. However, the loss of genetic resources at

J. A. Dar (✉) · S. M. Zargar · S. Rasool · K. A. Bhat
Proteomics Laboratory, Division of Plant Biotechnology, SKUAST-K, Srinagar, Jammu and Kashmir, India

M. Kareem · A. A. Wani
Department of Botany, University of Kashmir, Srinagar, Jammu and Kashmir, India

R. K. Salgotra, S. M. Zargar (eds.), *Rediscovery of Genetic and Genomic Resources for Future Food Security*,
https://doi.org/10.1007/978-981-15-0156-2_12

alarming rates due to anthropogenic pressure, climate change, pollution explosion, genetic erosion, gross mismanagement of resources, and population growth is the immediate topic of debate. Hence, there is a need for conservation and sustainable utilization of genetic resources. Against this backdrop, it is important to generate the database to get deep insight into the approaches for conservation purpose. The important conservation methods employed for genetic resources include in situ and ex situ conservation. In-depth survey, collection, and documentation are of principal importance to have a realistic view on genetic resource diversity. International, national, and individual appreciation value of these resources would facilitate their sustainable utilization. Dire need of creating avenues for genetic resources mandates ease of excess, and effective policies may be enacted for their protection particularly in biodiversity hotspots and in regions of high endemism. Present chapter highlights various approaches for conservation of genetic resources as well as the lacunae that are desirable to combat the future generation against high demand of eco-stability and food insecurity.

Acronyms

BGCI	Botanic Gardens Conservation International
CO_2	Carbon dioxide
DNA	Deoxyribonucleic acid
IBPGR	International Board for Plant Genetic Resources
ICC	International Co-ordinating Council
LN	Liquid nitrogen
LOS	Low-oxygen storage
LPS	Low-pressure storage
MSBP	Millennium Seed Bank Project
MoEFCC	Ministry of Environment, Forest and Climate Change
MPCA	Medicinal Plant Protection Area
PGR	Plant genetic resources
PGRFA	Plant Genetic Resource for Food and Agriculture
UNFAO	United Nations Food and Agriculture Organization
UNESCO	United Nations Education, Scientific and Cultural Organization

12.1 Introduction

Plant genetic resources (PGR) are stakeholders for balancing the global food security and agriculture especially with expanding global population. PGR refers to the heritable materials contained within and among plant species of present and potential, economic, scientific, or societal value. They include materials considered of systematic importance and applicable in cytogenetics, phylogenetics, evolutionary biology, and physiological, biochemical, pathological, and ecological research and breeding, encompassing all cultivated crops and those of little to no agricultural value as well as their weedy and wild relatives (Ulukan 2011). The diversity of genetic resources for food and agriculture (i.e. plants/crops, animals, aquatic resources, forests, micro-organisms, and invertebrates) plays a crucial role in meeting basic human food and nutritional needs. It is essential for maintaining and enhancing the efficiency and the resilience of production systems, as well as contributing to sustainable diets and to the delivery of ecosystem services.

World food demand is expected to double or triple by 2050. This increase can further elevate against the rise in world population from 6 billion to 8–10 billion people as well as an increase in per capita consumption (Green et al. 2005). To address this serious issue, the development of new elite cultivars and improvement of the agricultural practices (e.g. fertilization, pesticide application, planting time and density, and irrigation) are prerequisite (Fehr 1984). It is the crop genetic diversity which acts as the raw material for development of the elite cultivars and is an indicator of sustainability. PGR is the only source of plant genetic diversity which provides valuable traits needed for meeting the challenges of adopting crop varieties. An individual genotype with a seemingly useless set of characters today may suddenly become essential tomorrow due to changing climatic conditions or outbreak of a disease. Crop genetic diversity provides an assurance of future genetic progress and an insurance against unforeseen threats to agricultural production such as disease epidemics or climate changes.

A species or a population sample of a particular part of its genetic variation can be maintained through in situ or ex situ conservation. In situ conservation is the preservation of species and populations of living organisms in their natural habitat. This method preserves both the population and the evolutionary processes that enable the population to adapt by managing organisms in their natural state or within their normal range. For example, large ecosystems may be left intact as protected reserved areas with minimal intrusion or alteration by humans. Ex situ conservation is the preservation and propagation of species and populations, their germ cell lines, or somatic cell lines outside the natural habitat. This method maintains the genetic diversity extant in the population in a manner that makes samples of the preserved material readily available. It includes botanical gardens, greenhouses, and the preservation of seeds or other plant materials in germplasm banks under appropriate conditions for long-term storage. This chapter discusses the role as well as methods of various conservation strategies of genetic resources.

12.2 Factors Influencing the Loss of Genetic Resources

The erosion of the genetic resources poses a severe threat to the world's food security in the long term. The degradation and destruction of habitats is a major cause of the loss of plant genetic resources. Overexploitation, invasive species, pollution, anthropogenic pressure, and climate change are among the other causes for loss of genetic resources. Genetic erosion or the reduction in genetic diversity in crop plants is the reduction in the number of different crop species being grown and decrease in genetic diversity within crop species. Today, due to the limitations of modern large-scale, mechanized farming, only 150 plant species are under extensive cultivation. The majority of humans live on only 15 plant species, which account for over 90% of human energy needs. Within crop species, landraces have been displaced by bred cultivars. The Green Revolution is widely blamed for affecting most crops including rice and wheat. Rubenstein et al. (2005) reported that only 15% of the rice area was planted to landraces (in irrigated lowlands). For wheat, landraces occupy 23% for the durum wheat area, 12% of the winter bread wheat area, and 3% of spring bread wheat area. In contrast, 60% of the maize area is planted to landraces in the developing world. Of the nearly 8000 varieties of apple that grew in the United States, more than 95% no longer exist. In Mexico, only 20% of the corn types recorded in 1930 can now be found. Only 10% of the 10,000 wheat varieties grown in China in 1949 remain in use (UNFAO Report 1996). The modern intensive agriculture calls for uniformity and consequently has a narrow genetic base. In contrast, traditional agriculture has large number of diverse landraces. Other causes for loss of genetic diversity include changes in agricultural production systems, overgrazing, excessive harvesting, deforestation, land clearance, introduction of new pests, and diseases.

12.3 Strategies for Conserving Plant Genetic Resources

Plant genetic resources conservation has become increasingly important as more plants have become threatened or rare. The reason behind collecting germplasm includes danger of genetic erosion or extinction; users at national and international level have expressed a clear need for the germplasm. The genetic diversity is missing or insufficiently represented in existing ex situ germplasm conservation, and that more needs to be explored (Engels et al. 1995). After the Second World War, efforts to conserve plant genetic resources came mainly from breeders' organizations in the United States and Europe, which led to crop-specific collections, primarily located in developed countries. In the 1960s and 1970s, more focus was put on the collection and conservation of plant genetic resources in face of genetic erosion by organizations such as the Rockefeller Foundation and the European Society of Breeding Research (Pistorius 1997). A key event in the conservation of plant genetic resources was the establishment of the International Board for Plant Genetic Resources (IBPGR) (now Bioversity International) in 1974, whose mandate was to promote and assist in the worldwide effort to collect and conserve the plant genetic resource needed for future research and production. IBPGR mobilized scientists to create a global network of gene banks, thus marking the international recognition of the importance of plant genetic resources (Pistorius 1997). Broadly there are two basic approaches for genetic resource conservation – in situ and ex situ conservation (Fig. 12.1). Plant genetic resources conserved by any of these methods are often referred to as germplasm, which is a shorthand term meaning "any genetic materials".

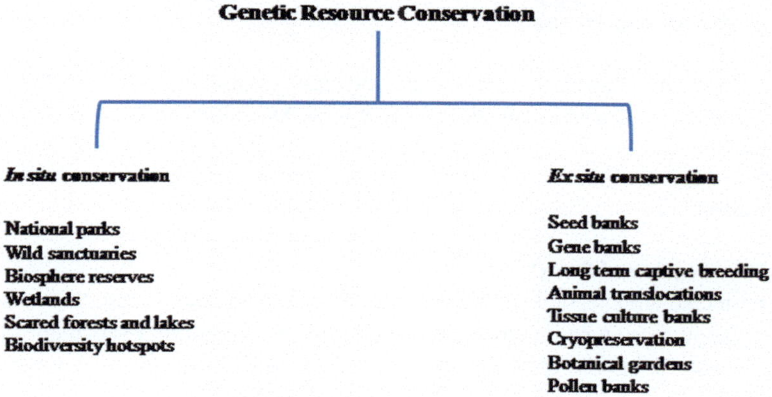

Fig. 12.1 Main strategies of for conserving plant genetic resources

12.4 In Situ Conservation

In situ conservation refers to conservation of genetic resources in natural populations. This type of conservation can take place in farmers' fields for domesticated materials or in natural environments for wild relatives of crop plants or wild species. For some species, such as tropical trees, this is the only form of conservation currently possible. Most of in situ conservation was applied to wild crop relatives, an important source of genetic variation to crop breeding programmes (Thayer and Alan 2005).

Also, most medicinal plants are endemic species, and their medicinal properties are mainly because of the presence of secondary metabolites that respond to stimuli in natural environments and that may not be expressed under culture conditions (Coley et al. 2003; Figueredo and Grelle 2009). In situ conservation of whole communities allows us to protect indigenous plants and maintain natural communities, along with their intricate network of relationships (Gepts 2006). Additionally, in situ conservation increases the amount of diversity that can be conserved (Forest et al. 2007) and strengthens the link between resource conservation and sustainable use (Long et al. 2003). In situ conservation efforts worldwide have focused on establishing protected areas and taking an approach that is ecosystem-oriented, rather than species-oriented (Ma et al. 2012). This approach was promoted by environmentalists and conservationists, who directed their efforts primarily toward the conservation of ecosystems and species diversity. In situ conservation is promoted because landraces are an essential component of indigenous cultures, it supposedly allows evolution to proceed, and it is the primary form of conservation for wild crop relatives. The evolutionary potential of in situ conservation is limited by the rapidity and magnitude of anthropogenic impacts on our planet, as outlined earlier. However, farmers can manage their genetic resources not only to maintain high levels of diversity in their fields but also for the selection of essential traits in their natural habitat, often by combining diversity from landraces and cultivars (Brush 1992, 1995; Birnbaum et al. 2003; Perales et al. 2003; Zizumbo-Villarreal et al. 2005).

12.4.1 Protected Areas

Protected areas include all the areas where the entire natural ecosystem of life forms can be conserved. Establishment of national parks, wild sanctuaries, biosphere reserves, forest area reserves, mangrove conservation, biodiversity hotspots, Ramsar wetlands, heritage sites, and sacred groves enhances facilitation of in situ conservation.

12.4.1.1 National Parks

It is a protected area needed for the purpose of protecting and propagating or developing wildlife therein or its environment whether it lies within a sanctuary or not. Inside the national park, no human activity is permitted except for the ones permitted by the Chief Wildlife Warden of the state under specified conditions as described in Wildlife protection Act, 1972. A total of 6555 national parks have been reported in the world till 2015 (http://www.basicplanet.com/national-parks. Accessed in September 2015). There are 103 existing national parks in India covering an area of 40,500 km^2, which is 1.23% of the total geographical area of the country (http://www.wiienvis.nic.in. Accessed in September 2015).

12.4.1.2 Wildlife Sanctuaries

A wildlife sanctuary is a declared protected area, where very limited human activity is allowed. The ownership of this type of protected area could lie in the hands of either a government or in any private organization or person, provided the regulations are governed by the government. Inside a wildlife sanctuary, the hunting of animals is completely prohibited. Additionally, the trees cannot be cut down for any purpose, especially the clearing of the forest for agriculture is completely banned. A total of 531 wildlife sanctuaries have been reported in India covering about 3.58% of the total geographical area of the country (http://www.wiienvis.nic.in). Although the populations of many wild species are under heavy pressure because of overexploitation, habitat degradation, and invasive species, wild nurseries can provide an effective approach for in situ conservation of medicinal plants that are endemic, endangered, and in-demand (Liu et al. 2011; Li and Chen 2007).

12.4.1.3 Biosphere Reserves

The term Biosphere Reserve for conservation of natural areas and of the genetic material they contain was accorded in Man and Biosphere project by the International Co-ordinating Council (ICC) of UNESCO in its first meeting held in 1971 at Paris Biosphere Reserve network and was formally launched in 1976. The Indian Biosphere Reserve Programme was initiated in 1986 keeping in view the following objectives:

1. To conserve the diversity and integrity of plants and animals within natural ecosystems
2. To safeguard genetic diversity of species on which their continuing evolution depends

3. To provide areas for multifaceted research and monitoring
4. To provide facilities for training and education
5. To ensure sustainable use of natural resources

Biosphere Reserves generally includes one or more national parks. Flora and fauna are protected in regions along with the tribal communities inhabiting these regions (MOEFCC 2007). At present there are about 631 Biosphere Reserves reported in 119 countries across the world with the support of the World Bank, with 18 reserves in India (http://www.wiienvis.nic.in).

12.4.1.4 Mangrove Conservation Programme

In India, the MOEFCC, Government of India, has launched a Mangrove Conservation Programme in 1987 under which 39 mangrove areas have been identified on the recommendations of National Committee on Wetlands, Mangroves, and Coral Reefs on the basis of their unique ecosystems and biodiversity. The 39 mangrove areas harbour 39 unique plant species.

12.4.1.5 Medicinal Plant Conservation Areas

India has 6000–7000 species estimated to have medicinal usage in folk and documented systems of medicine like Ayurveda, Siddha, Unani, and Homoeopathy. It is estimated that about 960 medicinal plant species are in trade of which 178 species have annual consumption levels in excess of 100 metric tons (Rajpurohit and Jhang 2015). With the financial support of national and international agencies, several civil societies are engaged in in situ conservation of medicinal plant species. Presently in India, there are 108 MPCAs which are established in various states like Karnataka (3), Kerala (9), Tamil Nadu (12), Andhra Pradesh (8), Maharashtra (13), Rajasthan (7), Orissa (5), West Bengal (7), Madhya Pradesh (13), Arunachal Pradesh (7), Uttarakhand (7), and Chhattisgarh (7).

12.4.1.6 Biodiversity Hotspots

Biodiversity hotspots are the regions which contain at least 1500 species of vascular plants as endemic. The biodiversity hotspots must meet two strict criteria: it must contain at least 0.5% or 1500 species of vascular plants as endemic, and it has to have lost at least 70% of its primary vegetation (Myers et al. 2000). There are about 34 biodiversity hotspots throughout the world of which 4, namely, Himalayas (Western and Eastern Himalayas), Western Ghats and parts of Sri Lanka, the Northeast and Indo-Burma, and the Nicobar Sundalands, are present in India (MOEFCC 2014).

12.4.1.7 World Heritage Sites

A World Heritage Site is a landmark or area which is selected by the United Nations Educational, Scientific and Cultural Organization (UNESCO) as having cultural, historical, scientific, or other forms of significance and is legally protected by international treaties. To be selected, a World Heritage Site must be an already classified landmark, unique in some respect as a geographically and historically identifiable place having special cultural or physical significance (such as an ancient ruin or historical structure, building, city, complex, desert, forest, island, lake, monument, mountain, or wilderness area (James et al. 2018). Presently there are 911 world heritage sites in 152 countries of which 39 world heritage sites are present in India in the states of Kerala, Karnataka, Tamil Nadu, and Maharashtra (http://whc.unesco.org/en/list/).

12.4.1.8 Ramsar Wetland Areas

Wetland conservation is aimed at protecting and preserving areas were water exists at or near the Earth's surface, such as swamps, marshes, and bogs. Wetlands cover at least 6% of the Earth and have become a focal issue for conservation due to the ecosystem services they provide. More than three billion people, around half the world's population, obtain their basic water needs from inland freshwater wetlands. The same number of people relies on rice as their staple food, a crop grown largely in natural and artificial wetlands. In some parts of the world such as the Kilombero wetland in Tanzania, almost the entire local population relies on wetland cultivation for their livelihoods (Water Issue Brief 2010).

12.4.1.9 Sacred Grooves

Sacred groves are the fine example of in situ conservation. Sacred groves are forest fragments, size varying between 0.5 and 500 ha, (some groves are more than 500 ha in size) which are protected by religious communities, and have a significant religious connotation for the protecting community. Sacred groves are the mini forests with rich diversity. Around 15,000 sacred groves have been reported from different parts of India. Above 100 scared groups are present in Rajasthan (only few in documented form). Hunting and logging are usually prohibited in sacred groves. Developmental activities are also restricted within these patches. Sacred groves are mostly associated with temples/monasteries/shrines or with cremation grounds. Sacred groves occur in many parts of India, particularly where the indigenous communities live. Sacred groves are also culturally important; various cultural and religious festivals are often arranged by local people within these patches. Sacred groves contain various ecosystems, various food chains, and food webs. Sacred groves are found in Khasi and Jaintia Hills in Meghalaya, Aravalli Hills of Rajasthan, Western Ghats regions of Karnataka and Maharashtra, and the Sarguja, Chanda, and Bastar areas of Madhya Pradesh. In Meghalaya, the sacred groves are the last refuges for a large number of rare and threatened pants.

12.5 Obstacles to In Situ Conservation of Wild Genetic Resources

Two main obstacles to in situ conservation of wild genetic resources are sectoralism and lack of knowledge (Prescott 1982). The conservation focus of protected areas is typically on the level of ecosystems and species, not on the maintenance of crop genetic resources. The agencies responsible for protected areas are aware of the

Table 12.1 Crops for which in situ conservation and ecogeographical surveys are of high priorities

Crop	Wild-type relative	Location
Groundnut	Perennial *Arachis* spp.	Latin America
Oil palm	*Elaeis* spp.	Africa, Latin America
Banana	Wild-type diploid *Musa* spp.	Asia
Rubber	*Hevea* spp.	Amazonia
Coffee	Coffee, *Arabica* spp.	Africa
Cocoa	*Theobroma* spp.	Latin America
Onion family	Selected wild-type *Allium* spp.	Worldwide
Citrus	Wild-type *Citrus* spp.	Asia
Mango	Wild-type *Mangifera* spp.	Southeast Asia
Cherries	Wild-type *Prunus* spp.	Europe, Asia
Apples	Wild-type *Malus* spp.	Europe, Asia
Pears	Wild-type *Pyrus* spp.	Europe, Asia
Forages	Hundreds of species	Worldwide

need for conserving genetic resources; they tend to regard it as an additional responsibility for which additional resources are generally not forthcoming. Ministries of agriculture or their equivalents have a direct interest in conserving wild relatives of crops, but they may be ambivalent about the importance of in situ conservation. In part, this may be because they often lack authority over the appropriate lands. Thus, difficulties in establishing a protected area may quickly outweigh the benefits of doing so, especially if the goal is protection for only one or two wild relatives of a single crop.

Lack of knowledge of the degree and distribution of interpopulation genetic variation in the wild relatives of crops is another obstacle (Noy-Meir et al. 1989). This information is needed for answering questions such as were in situ conservation areas should be established, how large should they be, and what ways should they be managed. Ecogeographical surveys that assess the genetic variation of a species across its entire geographical and ecological range are needed (Hoyt 1988). It can take years to obtain a complete ecogeographical survey. Although such information is of potential value to all crops, related wild species with high priority have been identified (Table 12.1).

12.6 Ex Situ Conservation

Ex situ conservation literally means "offsite" conservation, i.e. conservation of biological diversity components (seed, clone, live saplings, pollen, or their DNA) outside their natural habitats in storage infrastructures termed as "gene banks". Although species conserved in their natural habitats (in situ) have the potential for continued evolution of a particular trait within the species and are subjected to natural selection, there are indeed many problems in establishing this type of reserve, for example, cost, size, and maintenance aspects, political and social issues, natural disasters, fire, etc. Ex situ conservation aims to cultivate and naturalize threatened species to ensure their continued survival and sometimes produce large quantities of planting material used in the creation of drugs and is often an immediate action taken to sustain natural plant resources (Swarts and Dixon 2009; Pulliam 2000). Many plant species, e.g. wild medicinal plants, can not only retain high potency when grown in ex situ conditions far away from the habitats but can have their reproductive materials selected and stored in seed banks for future replanting (Hamilton 2004). For ex situ conservation, germplasm collection is the first and prerequisite step (Kiambi et al. 2018). Various methods of ex situ conservation of genetic resources are discussed in brief below.

12.6.1 Botanic Gardens

Botanic gardens play an important role in ex situ conservation (Havens et al. 2006), and they can maintain the ecosystems to enhance the survival of rare and endangered plant species (Huang et al. 2002). Botanic gardens involve a wide variety of plant species grown together under common conditions and often contain taxonomically and ecologically diverse flora (Primack and Miller-Rushing 2009). However, botanical gardens have the potential risk of germplasm being lost due to disease, stress, or disaster, and a large amount of space and labour are required to maintain a small portion of diversity. There are more than 1700 botanical gardens worldwide holding plant collections that serve both conservation and educational purposes. Preservation of rare and threatened wild plants are the mandates of various botanical gardens. The Millennium Seed Bank Project (MSBP) at the Royal Botanical Garden, Kew, England, is one of the largest conservation projects (Schoen and Brown 2001). In India, Botanical Garden at Calcutta is the largest one established in 1787 which spreads over an area of 110 ha and has around 15,000 plants belonging to 2500 species.

12.6.2 Gene Banks

Gene banks are a type of biorepository which preserve genetic material. Gene banks use low temperatures to stop chemical and biological activity that might break down cells. Some banks freeze material in liquid nitrogen at −196 °C. This freezing process replaces water in cells with another fluid, such as glycerol which minimizes the development of ice crystals. Gene banks hold different types of materials including seeds (seed banks), frozen plant cuttings, or in vitro storage. Currently, there are approximately 1500 gene banks maintaining 5.5 million samples (FAO 1996). These banks maintain primarily genetic resources of basic food crops and forages, but generally no vegetable or forest germplasm.

12.6.3 Seed Banks

Seed banks offer a better way of storing the genetic diversity of many plants ex situ than through botanic gardens and are recommended to help preserve the biological and genetic diversity of wild plant species (Li and Pritchard 2009; Schoen and Brown 2001). In 90% of the cases, genetic resources are conserved as seeds in cold storage. The most noteworthy seed bank is the Millennium Seed Bank Project at the Royal Botanic Gardens in Britain (Schoen and Brown 2001). Seed banks allow relatively rapid access to plant samples for the evaluation of their properties, providing helpful information for conserving the remaining natural populations (Li and Pritchard 2009; Schoen and Brown 2000). There are significant advantages to seed banking, including "ease of storage, economy of space, relatively low labour demands and consequently, the capacity to maintain large samples at an economically viable cost" (BGCI 2016). In addition, "seeds are a convenient means of long-term storage of genetic diversity, as the samples are small in size, are easily handled, require low maintenance and frequently remain viable for long periods" (BGCI 2016).

12.6.4 Pollen Banks

Pollen is a useful source of diverse alleles within a genepool and so may be an effective propagule for gene banks. The ease of pollen storage, shipment, and the potential for its immediate use provide researchers with increased options when designing their breeding programmes. Methods for pollen collection, desiccation, viability testing, and longevity assessment have been developed for many species of interest and have revealed the critical importance for increased longevity by using high-quality pollen desiccation sufficiently in a rapid manner and subsequently storing it at very low temperatures. Reliable viability assessments are dependent upon adequate rehydration and the use of reliable stains, in vitro germination assays or in vivo pollination experiments.

12.6.4.1 Storage Temperature

It is possible to store pollen of many species at temperatures between 4 °C and −20 °C for the short-term. Dry pollen that is kept at between 4 °C and −20 °C remains viable for a few days to a year, which may be adequate for use in breeding programmes (Hanna and Towill 1995). Long-term viability can be maintained by storing pollen at −80 °C or LN temperatures (−196 °C) (Hanna and Towill 1995). Once desiccated, pollen can be dispensed into cryovials for long-term storage in LN or LN vapour (Table 12.2). Precise labelling of vials and storage locations is recommended to aid in future retrieval of samples. Vials can then be placed in boxes or cryocanes and directly immersed in the liquid or vapour phase of liquid nitrogen (Barnabás and Kovacs 1996; Ganeshan et al. 2008; Connor and Towill 1993).

Table 12.2 A selection of species for which pollen can be successfully stored at −80 °C or liquid nitrogen (LN)

Species	Storage duration	Temperature	References
Actinidia	1 year	LN	Abreu and Oliveira (2004)
Aechmea	15 min	LN	Parton et al. (2002)
Allium	1 year	LN	Ganeshan (1986a)
Beta	17 years	LN	Panella et al. (2009)
Beta	1 year	LN	Hecker et al. (1986)
Carica	485 days	LN	Ganeshan (1986b)
Carya	13 years	LN	Sparks and Yates (2002)
Carya	1 year	LN	Yates and Sparks (1990)
Carya	3 years	−80	Yates and Sparks (1990)
Citrus	3.5 years	LN	Ganeshan and Alexander (1991)
Clianthus	3 h	LN	Hughes et al. (1991)
Dioscorea	2 years	−80	Ng and Daniel (2000)
Elaeis	8 years	LN	Tandon et al. (2007)
Gladiolus	10 years	LN	Rajasekharan et al. (1994)
Glycine	7 days	LN	Tyagi and Hymowitz (2003)
Guzmania	15 min	LN	Parton et al. (2002)
Humulus	2 years	LN	Haunold and Stanwood (1985)
Juglans	2 years	LN	Farmer and Barnett (1974)
Juglans	1 year	LN	Luza and Polito (1987)
Lycopersicon	5 weeks	−80	Sacks and St. Clair (1996)
Lycopersicon	22 months	LN	Karipidis et al. (2007)
Olea	1 h	LN	Parfitt and Almehdi (1984a)
Panax	11 months	LN	Zhang et al. (1993)
Persea	1 year	LN	Sedgley (1981)
Phoenix	435 days	LN	Tisserat et al. (1983)
Protea	1 year	LN	Van der Walt and Littlejohn (1996)
Prunus	12 months	−80	Martinez-Gómez et al. (2002)
Prunus	1 h	LN	Parfitt and Almehdi (1984b)
Pyrus	3 years	LN	Akihama and Omura (1986)
Rosa	8 weeks	LN	Marchant et al. (1993)
Rosa	1 year	LN	Rajasekharan and Ganeshan (1994)
Solanum	10 min	LN	Towill (1981)
Tillandsia	15 min	LN	Parton et al. (2002)
Vitis	64 weeks	LN	Ganeshan (1985)
Vitis	5 years	LN	Ganeshan and Alexander (1990)
Vitis	1 h	LN	Parfitt and Almehdi (1983)
Vriesea	15 min	LN	Parton et al. (2002)
Zea	120 days	LN	Barnabás and Rajki (1976)

12.7 In Vitro Germplasm Conservation

In vitro methods employing shoots, meristems, and embryos are ideally suited for the conservation of germplasm of vegetatively propagated plants. The plants with recalcitrant seeds and genetically engineered materials can also be preserved by this in vitro approach.

There are mainly three approaches for the in vitro conservation of genetic resources:

1. Cryopreservation (freeze preservation)
2. Cold storage
3. Low-pressure and low-oxygen storage

12.7.1 Cryopreservation

Cryopreservation is the storage of biological materials (seeds, plant embryos, shoot tips/meristems, and pollen) at ultralow temperatures, usually that of liquid nitrogen at −196 °C (Engelmann and Takagi 2000; Reed 2010). Under these conditions, biochemical and most physical processes are halted, and materials can be conserved over the long term. The technique of cryopreservation of biological materials includes the storage at different temperatures:

(i) Over solid carbon dioxide (at −79 °C)
(ii) Low temperature deep freezers (at −80 °C)
(iii) In vapour phase nitrogen (at −150 °C)
(iv) In liquid nitrogen (at −196 °C)

Among these, the most commonly used cryopreservation is by employing liquid nitrogen. At the temperature of liquid nitrogen (−196 °C), the cells stay in a completely inactive state and thus can be conserved for long periods. The technique of freeze preservation is based on the transfer of water present in the cells from a liquid to a solid state. Due to the presence of salts and organic molecules in the cells, the cell water requires much more lower temperature to freeze (even up to −68 °C) compared to the freezing point of pure water (around 0 °C). When stored at low temperature, the metabolic processes and biological deteriorations in the cells/tissues almost come to a standstill. In fact, cryopreservation has been successfully applied for germplasm conservation of a wide range of plant species, e.g. rice, wheat, peanut, cassava, sugarcane, strawberry, coconut, etc. Several plants can be regenerated from cells, meristems, and embryos stored in cryopreservation. Table 12.3 represents the plant materials that are being successfully cryopreserved.

Table 12.3 List of plants in various forms that are successfully cryopreserved

Plant material	Plant species
Cell suspensions	*Oryza sativa*
	Glycine max
	Zea mays
	Nicotiana tabacum
	Capsicum annum
Callus	*Oryza sativa*
	Capsicum annum
	Saccharum spp.
Protoplast	*Zea mays*
	Nicotiana tabacum
Meristems	*Solanum tuberosum*
	Cicer aietinum
Zygotic embryos	*Zea mays*
	Hordeum vulgare
	Manihot esculenta
Somatic embryos	*Citrus sinensis*
	Daucus carota
	Coffea arabica
Pollen embryos	*Nicotiana tabacum*
	Citrus spp.
	Atropa belladonna

12.7.2 Cold Storage

Cold storage basically involves germplasm conservation at a low and nonfreezing temperature (1–9 °C). The growth of the plant material is slowed down in cold storage in contrast to complete stoppage in cryopreservation. Hence, cold storage is regarded as a slow growth germplasm conservation method. The major advantage of this approach is that the plant material (cells/tissues) is not subjected to cryogenic injuries. Long-term cold storage is simple and cost-effective and yields germplasm with good survival rate. Many in vitro developed shoots/plants of fruit tree species have been successfully stored by this approach, e.g. grape plants and strawberry plants. Virus-free strawberry plants could be preserved at 10 °C for about 6 years, with the addition of a few drops of medium periodically (once in 2–3 months). Several grape plants have been stored for over 15 years by cold storage (at around 9 °C) by transferring them yearly to a fresh medium.

12.7.3 Low-Pressure and Low-Oxygen Storage

As alternatives to cryopreservation and cold storage, low-pressure storage (LPS) and low-oxygen storage (LOS) have been developed for germplasm conservation.

12.7.3.1 Low-Pressure Storage (LPS)

In low-pressure storage, the atmospheric pressure surrounding the plant material is reduced. This results in a partial decrease of the pressure exerted by the gases around the germplasm. The lowered partial pressure reduces the in vitro growth of plants (of organized or unorganized tissues). Low-pressure storage systems are useful for short-term and long-term storage of plant materials. The short-term storage is particularly useful to increase the shelf life of many plant materials, e.g. fruits, vegetables, cut flowers, and plant cuttings. The germplasm grown in cultures can be stored for long term under low pressure. Besides germplasm preservation, LPS reduces the activity of pathogenic organisms and inhibits spore germination in the plant culture systems.

12.7.3.2 Low-Oxygen Storage (LOS)

In the low-oxygen storage, the oxygen concentration is reduced, but the atmospheric pressure (260 mm Hg) is maintained by the addition of inert gases (particularly nitrogen). The partial pressure of oxygen below 50 mm Hg reduces plant tissue growth (organized or unorganized tissue). This is due to the fact that with reduced availability of O_2, the production of CO_2 is low. As a consequence, the photosynthetic activity is reduced, thereby inhibiting the plant tissue growth and dimension.

12.7.3.3 Animal Translocation

Release of animals in a new locality which have been transported from other habitats is called animal translocation. Animal translocation provides the best method for conservation of animals. Translocation is carried in following cases:

1. When a species on which an animal is dependent becomes rare
2. When a species is endemic or restricted to a particular area
3. Due to habit destruction and unfavourable environment conditions
4. Increase in population in an area

12.8 Lacunae and Future Prospects

A concise assessment of the status and trends of plant genetic resources is needed to identify the most significant gaps and needs in order to provide the basis to update the rolling GPA (Global Plan of Action). Modern agricultural practices strongly favour reduction of crop diversity by providing the subsidies for cultivating high-yielding varieties and reducing weed/wild plant diversity by using crop protection measures. Plant Genetic Resource for Food and Agriculture (PGRFA) diversity should be assessed at regular intervals to record changes in species population as well as monitoring genetic erosion, if any. Regular surveys and upgradation to assess status of PGRFA with reference to the number of crops cultivated in an area and number of varieties of each crop being cultivated in relation to diversity available and genetic erosion, to add unexplored areas as well as areas explored more than 20 years back, are recommended. International linkages, project formulation, and sufficient funding for such activities needed to be explored. Training on the study of genetic erosion needs to be given to at least one professional plant breeder from each crop-based agricultural university. An awareness campaigns need to be launched at gross root level among actual stakeholders.

Expanding agriculture leads to habitat loss and fragmentation, drainage of wetlands, and impact of freshwater and marine ecosystems through sedimentation and pollution. The important threats include introduction of high-yielding and improved varieties, urbanization, deforestation, shifting cultivation, overexploitation and lack of regeneration, human interference, modernization of agriculture, biotic and abiotic stresses, natural disasters, lack of availability of seeds of local varieties, lack of awareness, etc. The main constraints indicate lack of focussed and coordinated approach by the concerned organizations, insufficient financial support, and need for strengthening skills and technical staff. Although in most of the crops, a large number of accessions have been assembled, their management for effective use has become difficult. Core collections have been developed only in few crops. Therefore, there is a need for strengthening application of techniques like development of gene pools, core collections, trait specific core sets, etc.

12.9 Conclusion

Conservation of Plant Genetic Resource for Food and Agriculture (PGRFA) ensures the future adaptability of cultivars and wild populations; preserved traits ensure sustainable agriculture to promote the use of genetic resources commercially and to conserve genetic diversity for multiple reasons. Food and agriculture production have relied collectively on worldwide efforts, like genetic resources domesticated elsewhere and subsequently developed in other countries and regions. Agriculture itself is one of the major threats to biodiversity worldwide. Today's international breeds are based on a narrow genetic base; food security is threatened when these breeds are affected by pests and diseases and climate changes. The drastic reduction in the diversity of genetic resources is due to changes in consumer preferences, population growth, and changes in land cover and land use. The combination of a well-designed, well-monitored, and well-managed system of protected areas, with ex situ conservation in seed banks and, where necessary, living collections and cryogenic storage, should be enough to protect all land plant species through the next few decades of rapid global change. It is now recognized that an appropriate conservation strategy for a particular plant genepool requires a holistic approach, combining the different ex situ and in situ conservation techniques available in a complementary manner. Decisions related to selection of the most efficient ways of conservation should not be based only on genetic strategies but should also take into account the economical aspect, i.e. costs, benefits, and risks of these strategies. Realizing that ex situ conservation is not the unique way of conserving genetic material for the improvement of future varieties but in situ and on-farm conservation have a role to play in the process of conservation introduces new relationships between farmers and breeders and between users and managers of genetic resource.

It is a field wide open for new investigations and experiments, for new relations between scientists and users, and a different distribution of responsibility in conservation of genetic resources. For sustainability of these resources, the conservation, diversification, adaptation, improvement, and delivery to farmers through seed systems are needed.

References

Abreu I, Oliveira M (2004) Fruit production in kiwifruit (*Actinidia deliciosa*) using preserved pollen. Aust J Agric Res 55:565–569

Akihama T, Omura M (1986) Preservation of fruit tree pollen. In: Bajaj YPS (ed) Biotechnology in agriculture and forestry 1: Trees. Springer, Berlin, pp 101–112

Barnabas B, Kovacs G (1996) Storage of pollen. In: Shivanna KR, Sawhney VK (eds) Pollen biotechnology for crop production and improvement. Cambridge University Press, Cambridge, pp 293–314

Barnabas B, Rajki E (1976) Storage of maize (*Zea mays* L.) pollen at −196°C in liquid nitrogen. Euphytica 25:747–752

BGCI (Botanic Gardens Conservation International) (2016) Seed Banks. Https://www.bgci.org/resources/seedbanks/. Accessed 4 Nov 2016

Birnbaum K, DeSalle R, Peters CM, Benfey PN (2003) Integrating gene flow, crop biology, and farm management in onfarm conservation of avocado (*Persea americana*, Lauraceae). Am J Bot 90:1619–1627

Brush SB (1992) Ethnoecology, biodiversity, and modernization in Andean potato. Agric J Ethnobiol 12:161–185

Brush SB (1995) In situ conservation of landraces in centers of crop diversity. Crop Sci 35:346–354

Coley PD, Heller MV, Aizprua R, Arauz B, Flores N, Correa M, Gupta M, Solis PN, Ortega-Barria E, Romero LI, Gómez B, Ramos M, Cubilla-Rios L, Capson TL, Kursar TA (2003) Using ecological criteria to design plant collection strategies for drug discovery. Front Ecol Environ 1:421–428

Connor KF, Towill LE (1993) Pollen-handling protocol and hydration/dehydration characteristics of pollen for application to long-term storage. Euphytica 68:77–84

Engelmann F, Takagi H (2000) Cryopreservation of tropical plant germplasm: current research progress and application. Japan International Research Centre for Agricultural Science/IPGRI, Tsukuba

Engels JMM, Arora RK, Guarino L (1995) An introduction to germplasm exploration and collecting: planning, methods and procedures, follow-up. In: Guarino L, Rao VR, Reid R (eds) Collecting plant genetic diversity: technical guidelines. CAB International, Wallingford

FAO (1996) Report on the state of the World's plant genetic resources, Rome Italy

Farmer RE, Barnett PE (1974) Low-temperature storage of black walnut pollen. Cryobiology 11:366–367

Fehr WR (1984) Genetic contributions to yield gains of five major crop plants. CSSA, Madison

Figueiredo MSL, Grelle CEV (2009) Predicting global abundance of a threatened species from its occurrence: implications for conservation planning. Divers Distrib 15:117–121

Forest F, Grenyer R, Rouget M, Davies TJ, Cowling RM, Faith DP, Balmford A, Manning JC, Proche S, Bank M, Reeves G, Terry AJ, Savolainen V (2007) Preserving the evolutionary potential of floras in biodiversity hotspots. Nature 445:757–760

Ganeshan S (1985) Cryogenic preservation of grape (*Vitis vinifera* L.) pollen. Vitis 24:169–173

Ganeshan S (1986a) Viability and fertilizing capacity of onion pollen (*Allium cepa* L.) stored in liquid nitrogen (−196°C). Trop Agric (Trinidad) 63:46–48

Ganeshan S (1986b) Cryogenic preservation of papaya pollen. Sci Hortic 28:65–70

Ganeshan S, Alexander MP (1990) Fertilizing ability of cryopreserved grape (*Vitis vinifera* L.) pollen. Vitis 29:145–150

Ganeshan S, Alexander M (1991) Cryogenic preservation of lemon (*Citrus limon* Burm.) pollen. Gartenbauwissenschaft 56:228–230

Ganeshan S, Rajasekharan PE, Shashikumar S, Decruze W (2008) Cryopreservation of pollen. In: Reed BM (ed) Plant cryopreservation: a practical guide. Springer, New York, pp 443–464

Gepts (2006) Plant genetic resources conservation and utilization: the accomplishments and future of a societal insurance policy. Crop Sci 46:2278–2292

Green RE, Cornell SJ, Scharlemann JPW, Balmford A (2005) Farming and the fate of wild nature. Science 307:550–555

Hamilton AC (2004) Medicinal plants, conservation and livelihoods. Biodivers Conserv 13:1477–1517

Hanna WW, Towill LE (1995) Long-term pollen storage. Plant Breed Rev 13:179–207

Haunold A, Stanwood PC (1985) Long-term preservation of hop pollen in liquid nitrogen. Crop Sci 25:194–196

Havens K, Vitt P, Maunder M, Guerrant EO, Dixon K (2006) *Ex situ* plant conservation and beyond. Bioscience 56:525–531

Hecker RJ, Stanwood PC, Soulis CA (1986) Storage of sugarbeet pollen. Euphytica 35:777–783

Hoyt E (1988) Conserving the wild relatives of crops. IPGRI/IUCN/WWF, Rome

Huang H, Han X, Kang L, Raven P, Jackson PW, Chen Y (2002) Conserving native plants in China. Science 297:935

Hughes HG, Lee CW, Towill LE (1991) Low-temperature preservation of *Clianthus formosus* pollen. Hortic Sci 26:1411–1412

James AR, Kormos C, Jaeger T, Venter O, Bertzky B, Shi Y, MacKey B, Van Merm R, Osipova E, Watson JEM (2018) Gaps and opportunities for the world heritage convention to contribute to global wilderness conservation. Conserv Biol 32(1):116–126

Karipidis C, Olympios C, Passam HC, Savvas D (2007) Effect of moisture content of tomato pollen stored cryogenically on *in vitro* germination, fecundity and respiration during pollen tube growth. J Hortic Sci Biotech 82:29–31

Kiambi DK, Ford-Lloyd BV, Jackson MT, Guarino L, Maxted N, Newbury HJ (2018) Collection of wild rice (*Oryza sativa* L.) in east and southern Africa in response to genetic erosion. PGR Newsl 142:10–20

Li XW, Chen SL (2007) Conspectus of ecophysiological study on medicinal plant in wild nursery. Zhong Guo Zhong Yao Za Zhi 32:1388–1392

Li DZ, Pritchard HW (2009) The science and economics of ex situ plant conservation. Trends Plant Sci 14:614–621

Liu C, Yu H, Chen SL (2011) Framework for sustainable use of medicinal plants in China. Zhi Wu Fen Lei Yu Zi Yuan Xue Bao 33:65–68

Long CL, Li H, Ouyang ZQ, Yang XY, Li Q, Trangmar B (2003) Strategies for agrobiodiversity conservation and promotion: a case from Yunnan, China. Biodivers Conserv 12:1145–1156

Luza JG, Polito VS (1987) Effects of desiccation and controlled rehydration on germination in vitro of pollen of walnut (*Juglans* spp.). Plant Cell Environ 10:487–492

Ma J, Rong K, Cheng K (2012) Research and practice on biodiversity in situ conservation in China: progress and prospect. Sheng Wu Duo Yang Xing 20:551–558

Marchant DR, Denton GH, Swisher CC (1993) MiocenePliocene-Pleistocene glacial history of Arena Valley, Quartermain Mountains, Antarctica Geogr Ann Ser A 75:269–302

Martínez-Gómez P, Gradiel TM, Ortega E, Dicenta F (2002) Low temperature storage of almond pollen. HortSci 37:691–692

MoEFCC (2007) Annual report 2006–07. Ministry of Environment, Forests and Climate Change, Govt. of India, India

MoEFCC (2014) Annual report 2013–14. Ministry of Environment, Forest and Climate change Govt. of India

Myers N, Mittermeier RA, Mittermeier CG, da Fonseca GAB, Kents J (2000) Biodiversity hotspots for conservation priorities. Nature 403:855–858

Ng NQ, Daniel IO (2000) Storage of pollens for long-term conservation of yam genetic resources. In: Engelmann F, Takagi H (eds) Cryopreservation of tropical plant germplasm. IPGRI, Rome, pp 136–139
Noy-Meir I, Gutman M, Kaplan Y (1989) Responses of Mediterranean grassland plants to grazing and protection. J Ecol 77:290–310
Panella L, Wheeler L, McClintock ME (2009) Long-term survival of cryopreserved sugarbeet pollen. J Sugar Beet Res 46:1–9
Parfitt DE, Almehdi AA (1983) Cryogenic storage of grape pollen. Am J Enol Viticult 34:227–228
Parfitt DE, Almehdi AA (1984a) Cryogenic storage of olive pollen. Fruit Var J 38:14–16
Parfitt DE, Almehdi AA (1984b) Liquid nitrogen storage of pollen from five cultivated *Prunus* species. Hortic Sci 19:69–70
Parton E, Vervaeke I, Delen R, Vandenbussche B, Deroose R, De Proft M (2002) Viability and storage of bromeliad pollen. Euphytica 125:155–161
Perales H, Brush SB, Qualset CO (2003) Dynamic management of maize landraces in Central Mexico. Econ Bot 57:21–34
Pistorius R (1997) Scientists, plants and politics. A history of the plant genetic resources movement. Bioversityinternational.org, Rome
Prescott RA (1982) Resource policy analyst, personal communication, June. News
Primack RB, Miller-Rushing AJ (2009) The role of botanical gardens in climate change research. New Phytol 182:303–313
Pulliam HR (2000) On the relationship between niche and distribution. Ecol Lett 3:349–361
Rajasekharan PE, Ganeshan S (1994) Freeze preservation of rose pollen in liquid nitrogen: feasibility, viability and fertility status after long-term storage. J Hortic Sci 69:565–569
Rajasekharan PE, Rao TM, Janakiram T, Ganeshan S (1994) Freeze preservation of gladiolus pollen. Euphytica 80:105–109
Rajpurohit D, Jhang T (2015) *In-situ* and *Ex-situ* conservation of plant genetic resource and traditional knowledge. In: Salgotra RK, Gupta BB (eds) Plant genetic resource and traditional knowledge. Springer, Rome, pp 137–162
Reed BM (2010) Plant cryopreservation. A practical guide. Springer, New York
Rubenstein KD, Heisey P, Shoemaker R, Sullivan J, Frisvold G (2005) Crop genetic resources: an economic appraisal. Available at www.ers.usda.gov/publications/eib2/ (June 2 2006). USDA-ERS, Washington, DC
Sacks EJ, Clair DA (1996) Cryogenic storage of tomato pollen: effect on fecundity. HortSci 31:447–448
Schoen DJ, Brown AHD (2001) The conservation of wild plant species in seed banks. Bioscience 960–6(60):51
Sedgley M (1981) Storage of avocado pollen. Euphytica 30:595–599
Sparks D, Yates IE (2002) Pecan pollen stored over a decade retains viability. HortSci 37:176–177
Swarts ND, Dixon KW (2009) Terrestrial orchid conservation in the age of extinction. Ann Bot 104:543–556
Tandon R, Chaudhury R, Shivanna KR (2007) Cryopreservation of oil palm pollen. Curr Sci 92:182–183
Thayer and Alan (2005) A new look at conservation as a means of maintaining wild type genetic variation among economically important crop species. Intl J Agric Res 20(16):117–132
Tisserat B, Ulrich JM, Finkle BJ (1983) Survival of phoenix pollen grains under cryogenic conditions. Crop Sci 23:254–256
Towill LE (1981) Liquid nitrogen preservation of pollen from tuber-bearing *Solanum* species. HortSci 16:177–179
Tyagi RK, Hymowitz T (2003) Pollen from *Glycine* species survive cryogenic exposure. Cryo Lett 24:119–124
Ulukan H (2011) Plant genetic resources and breeding: current scenario and future prospects. Intl J Agric Biol 13:447–454

United Nations Food and Agriculture Organization (1996) Report on the state of the World's plant genetic resources for food and agriculture, prepared for the international technical conference on plant genetic resources, Leipzig, June17–23. FAO, Rome
Van der Walt ID, Littlejohn GM (1996) Storage and viability testing of *Protea* pollen. J Am Soc Hort Sci 121:804–809
Water Issue Brief (2010) Using wetlands sustainably. International Water Management Institute
Yates IE, Sparks D (1990) Three-year-old pecan pollen retains fertility. J Am Soc Hort Sci 115:359–363
Zhang LX, Chang WC, Wei YJ, Liu L, Wang YP (1993) Cryopreservation of ginseng pollen. HortSci 28:742–743
Zizumbo-Villarreal DP, Colunga-Garcia M, Payro de la Cruz E, Delgado-Valerio P, Gepts P (2005) Population structure and evolutionary dynamics of wild-weedy-domesticated complexes of common bean in a Mesoamerican region. Crop Sci 35:1073–1083

Crop Landraces: Present Threats and Opportunities for Conservation

13

Rakeeb Ahmad Mir, Arjun Sharma, and Reetika Mahajan

Contents

Abstract

Crop landraces are important source of novel alleles which can be utilized for improvement of desired crops. They have variable phenology and moderate edible yield. Landraces provide traits for more efficient nutrient uptake and utilization, as well as useful genes for adaptation to stressful environments such as water stress, salinity, and high temperatures for development of improved cultivars. However, since last few decades, modern agricultural practices have resulted in decline of diversity in crop landraces. Various environmental factors like genetic erosion and local cultivation practices have threaten the landrace diversity. To overcome these threats, certain conservation methods have been adapted, and these methods have been reported to play critical role in conserving

R. A. Mir (✉) · A. Sharma
Department of Biotechnology, School of Biosciences and Biotechnology, BGSB University, Rajouri, Jammu and Kashmir, India

R. Mahajan
Division of Plant Biotechnology, Sher-e-Kashmir University of Agricultural Sciences and Technology of Kashmir, Srinagar, Jammu and Kashmir, India

R. K. Salgotra, S. M. Zargar (eds.), *Rediscovery of Genetic and Genomic Resources for Future Food Security*,
https://doi.org/10.1007/978-981-15-0156-2_13

crop landrace diversity. Furthermore, there is a need for proper documentation of the information available on remedial measures to cope up with the stress mediated by gene flow to crop landraces. Overall information generated may provide a framework to initiate different approaches for the crop improvement.

Acronyms

CRISPR	Clustered regularly interspaced short palindromic repeats
CMT3	Chromomethylase 3
DCL	DICER-like enzymes
DNA	Deoxyribonucleic acid
GM	Genetically modified
GR	Green Revolution
HYVs	High-yielding varieties
IBPGR	International Board for Plant Genetic Resources
LR	Landraces
NGOs	Nongovernmental organizations
RNA	Ribonucleic acid
RNAi	RNA interference
RISC	RNA-induced silencing complex
siRNA	Small interfering RNA
TALENs	Transcription activator-like effector nucleases
ZNFs	Zinc finger nucleases

13.1 Introduction

Agriculture is one of the oldest livelihood sources for mankind. History revealed that with the onset of civilization, agriculture had played a key role in sustainable development of mankind. From centuries, sowing of seeds saved from superior parents in next season by the farmers had led to the identification of various important traits of crops which can be used for crop improvement programme (Zeven 1998). Crop landraces are the locally adapted varieties with important traits but lack proper knowledge. Landraces have an important role in crop improvement and agricultural production, and it is for these former reasons they have been found to exist since the origin of agriculture (Zeven 1998). Modernization in agriculture and lack of information regarding the landraces possess great threat to crop landraces. Here in this chapter, we have discussed various threats and opportunities faced by crop landraces.

13.2 Crop Landraces and Their Classification

From the period since 1909 to 1974, numerous attempts have been made by scientists to define the term 'landraces' properly, but till date a well-defined definition of landraces based on knowledge of their traits, utilization, ecogeographic adaptation, cultivation and management procedures is yet to be established (Harlan 1975; IBPGR 1980; Brush 1995; Tsegaye et al. 1996; Pistorius 1997; Zeven 1998; Louette 1999; Friis-Hansen and Sthapit 2000; Saxena and Singh 2006; Berg 2009; Newton et al. 2010). Primitive cultivars, traditional varieties or conservation varieties are some of the synonyms of landraces used in literature (Camacho Villa et al. 2005). Specifically defining the seed-propagated landraces, they are the crops which have been identified and given a vernacular name. Their evolution and adaptation are restricted to the habitat they have been grown since centuries. Landraces (LR) are usually adapted to one specific geographical location, whereas cultivars are bred in remote areas being cultivated in diverse locations (Hawkes 1983). Each crop landrace has a specific local name assigned to it, highlighting its features and importance to the particular habitat and representing the class of humans inhabiting that area (Von Rünker 1908). They have been so closely associated with the particular habitat that indigenous farmers have developed a data set revealing their traditional use, knowledge of their habitat, utilization pattern and importance in several religious celebrations. Thus crop landraces can be defined as a 'dynamic population(s) of a cultivated plants that have historical origin, distinct identity and lacks formal crop improvement, as well as often being genetically diverse, locally adapted and associated with traditional farming systems' (Camacho Villa et al. 2005). In 1890, for the first time, landraces were thought to be genetic resources (Zeven 1998). Landraces are not only important for maintaining biodiversity but are also important source of superior nutritional and medicinal values. Before the invention of modern breeding technologies, every year farmers utilize their seasonally saved seeds from grown crops such that they can be used in following year for cultivation. The seeds were selected from the parent plants having best traits which evolved due to natural and non-orientated anthropogenic selections (Carvalho et al. 2012). Thus by using this selective breeding approach, various desirable traits have been developed over generations. Seed-saving method is used for the development of crops resistant to local diseases, and this method maintained the genetic diversity of the crops grown in that particular habitat, leading to the evolution of these landraces as valuable genetic resources for future generations. Crop landraces can be used to study variation in various desirable traits and to develop improved crop varieties (Table 13.1). Crop landraces has been classified into various categories as mentioned in Table 13.2. Mayr's (1934) classified landraces into five categories: autochthonous (a landrace cultivated for more than a century in the same region), autochthogenous (a landrace derived from a new genotype due to spontaneous mutations or derivative of a natural cross originating from an autochthonous landrace), allochthonous (an autochthonous landrace from one region introduced into another region and adapting itself in

Table 13.1 Showing different crop landraces used for the development of desirable trait

Crop landrace	Desired trait	Studied for	References
Barley	Plant height and crown rot disease	QTL identification	Li et al. (2009)
Triticum turgidum (*turgidum* convar. *durum*) durum wheat	Glutenin protein subunits	Genetic diversity	Moragues et al. (2006)
Durum wheat	Morphological and agronomical traits and protein	Genetic variation	Pecetti et al. (2001)
	Composition		
Ethiopian tetraploid wheat germplasm	Grain yield potential and quality traits	Genetic diversity	Teklu and Hammer (2009)
Triticum turgidum L. (tetraploid wheat)	Agronomic traits	Genetic diversity	Tsegaye et al. (1996)
Syrian durum wheat landraces	Glutenin content	Diversity	van Hintum and Ellings (1991)
Hexaploid wheat	Abiotic stress	Identification of novel germplasm resource	Trethowan and Mujeeb-Kazi (2008)
Wheat wild relatives and landraces	Drought-adaptive traits		Reynolds et al. (2007)
Barley (*Hordeum vulgare*) from Egypt	–	Genetic diversity	Sarker et al. (2008)
Rice	Drought tolerant	QTL identification	Kumar et al. (2014)
Rice	Salt tolerant	QTL identification	Ren et al. (2005), Bonilla et al. (2002), Thompson et al. (2010) and Kumar et al. (2015)

Table 13.2 Different classifications of landraces

Classification	Basis of classification	Types	References
Christiansen-Weniger's	–	Primary landrace	Christiansen-Weniger (1931)
		Secondary landrace	
Mayr's	Breeding history	Autochthonous, Autochthogenous, Allochthonous, Allochthogenous, Zucht-Landsorte'	Mayr (1934)
Mayr's	Breeding values	Primitive landrace	Mayr (1937)
		Secondary landrace	
Zeven's	Based on Christiansen-Weniger's classification	Clean multiline landrace	Zeven (1975)
		Dirty multiline landrace	

new environment), allochthogenous (a landrace being grown for a longer period in a non-native region and has been changed by this new environment although the original type is still recognizable) and Zucht-Landsorte' (improved landrace derived form a 'reversed' cultivar).

13.3 Insight into Threat Assessment of Crop Landraces

Threat is the ultimate indicator of species extinction rate, and it is the basis of this relative threat we can establish the conservation priorities. The higher the rate of threat, the higher will be the priorities for conservation. Assessment of threat, i.e. the higher probability of genetic erosion, generates the realistic data to conserve the landraces and their extinction. The loss of landraces can be assessed through 'local cultural erosion' and 'genetic erosion'. Genetic erosion can be analysed through crop loss and their varieties or allelic diversity, decrease in richness of a species and loss of genetic diversity. The local cultural erosion refers to the unending use of landraces in different cultural activities. In addition, the intervention of modern technology has drastically transformed the traditional agricultural developments into modernized agricultural practices, resulting in great impact on crop yield and diversity. This revolution has led to the global development of various stress-tolerant crops. These factors have negatively affected the landraces and in turn resulted in their extinction, whereas local cultural erosion caused loss of biodiversity by replacing local varieties by crops having desired traits, genetically uniform hybrids and improved cultivars by practicing monocropping (Ceccarelli and Grando 2000; Sarker and Erskine 2006; Rodriguez et al. 2008; Abay and Bjørnstad 2009; Frison et al. 2011). Presently, most of the population feeds on few improved cultivars of wheat, rice, maize and potato which account only 60% of diets (Esquinas-Alcazar 2010). According to World Conservation Monitoring Centre (1992), 74% rice cultivars (staple crop) of Indonesia are mainly derived from a single stock. Also in the USA, 50% wheat is derived from 9 cultivars, 75% potato is derived from 4 cultivars, and 50% soybean is derived from 6 cultivars. Genetic erosion had decreased the landrace diversity in southern Italy by 72.8% and in Albania by 72.4% (Hammer et al. 1996), and in Greece, 95% wheat landraces were lost after utilization of modern practices (Lopez 1994).

Since the advent of Green Revolution (GR) locally adapted populations of plants or 'landraces' have been replaced by HYVs (high-yielding varieties) or modern varieties, forcing farmers to leave behind the historically adapted mechanisms to conserve the landraces through seeds and other traditional conservation strategies. Out of this revolutionary process, i.e. GR, diversity of rice cultivars and other landraces decreased drastically in India and at global level. These multiple reasons lead to the threatening of landraces and it has became pretinent to asses the reasons behind erosion of landraces. Threats to landraces can be assessed (Fig. 13.1) either at individual level or at genetic level by a three-stage method which includes defining the different indicators of threat and then identifying threats to LR diversity and evaluation of the relative degree of threat (Negri 2003). Based on different

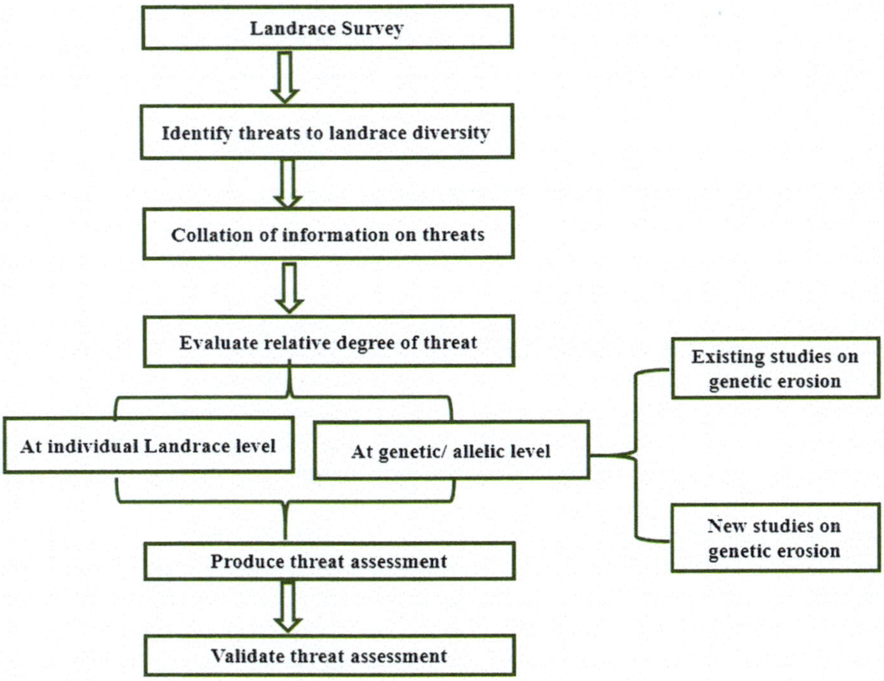

Fig. 13.1 Landrace diversity threat assessment methodology. (http://www.fao.org/fileadmin/templates/agphome/documents/POR)

categories, various alternative methods for assessing threats to landrace diversity have been developed (Joshi et al. 2004; Antofie et al. 2010; Porfiri et al. 2009). Besides a number of methods available for threat assessment not even a single method is standardized for threat assessment of erosion of landraces. Certain methodologies which rely on assessment of threat indicators include simple analysis like farmer's wealth, access to seed planting material, farming area, system of cultivation, ability of a plant material to multiply, use of plant material by local farmers, historical indicators include first time development of a landraces temporally and spatially, socio-economic indicators, conservation status of landraces, uniqueness to the habitat, familiarity of genetic diversity and data generated regarding the cataloguing of landraces. From centuries, plant breeding approaches were used by farmers for selecting superior varieties which later resulted in evolution of important landraces. But with the advancement in plant breeding approaches, there is decline in the diversity of landraces as these approaches are shifting the landraces towards a model of agriculture based on uniformity (Van de Wouw et al. 2010; Frison et al. 2011; Ceccarelli 2012). Threat to landrace diversity has direct effect on global food security. Thus, to minimize the negative effects on food security, threat identification and evaluation are of dire importance.

13.4 Approaches to Understand Impact of GM Crops on Crop Landraces

Generally speaking, gene flow is the natural incorporation of genes from one population to the other (Futuyma 1998). This unidirectional flow of genes from cultivated crops to wild species/landraces has been reported since thousands of years (Ellstrand et al. 1999). Irony to this flow of genes is that illegitimate gene flow has been further stimulated by introduction of GM crops (Snow and Moran-Palma 1997; Hall et al. 2000; Ellstrand 2001). The term genetically modified (GM) refers to the transfer of genes between organisms using a series of laboratory techniques for cloning genes, modifying DNA segments together and inserting genes into cells. The 'genetically modified' is a vague term and a potentially confusing one, in that virtually everything we eat has been modified genetically through domestication from wild species and many generations of selection by humans for desirable traits. Crop plants have been improved for different applications, and few of them are worth to mention, like to enhance the plant productivity and production of disease-resistant plants and pest-resistant plant and also to improve the quality of the plant products. Apart from being important for increasing the crop yield through the introduction of different classes of desired genes from different origins, GM crops have been found to be negatively effecting the landraces. The contamination of landraces has been reported by several studies. In Mexico, maize landraces have been found to contain genes from transgenic crop plants causing lot of controversy globally (Carpentier and Herrman 2003; Christou 2002; Kaplinsky et al. 2002; Metz and Fütterer 2002; Quist and Chapela 2001, 2002). It has raised the issue of whether the commercial introduction of transgenic maize varieties may have a deleterious effect on the diversity of maize landraces. This issue is significant because Mexico is a centre of maize domestication and maize diversity. Highlighting the negative side of GM crops, the cross contamination of local landraces of Mexican maize may serve purpose to highlight the negative effect of GM crops. Ignacio Chapela and his student David Quist collected corncobs (maize landraces) for cross-checking the contamination of Mexican maize landraces by GM maize imported from the USA where at least 40% of crops are GM based (Quist and Chapela 2001). This series of troubleshooting experimentations was famously being highlighted as 'Chapela Affair'. The duo checked the contamination through repeated amplification of CaMV promoter (a marker DNA sequence to check the transgenic plants) from crop landraces by PCR and inverse PCR (iPCR). These results showed that four maize landraces were found to be positive out of six tested maize landraces (Quist and Chapela 2001) further compounding the safety of GM crops. The maize landraces in Mexico have been conserved and bred by local farmers since centuries in small patches of agricultural fields. Out of these findings, both the investigators finally concluded that local landraces of maize have been contaminated by GM maize somehow imported from the USA. This potential of GM crops to contaminate the local landraces is alarming to eradicate the allelic diversity of wild plants. Against this backdrop of technological intervention of transgenic plants, Mexican government doesn't allow the cultivation of GM crops.

13.5 Opportunities for Conservation and Remedial Measures to Protect Diversity of Crop Landraces

Landraces play a key role in crop improvement programme as they are important source of novel alleles. Farmers from underdeveloped nations and rural areas in developing nations depend on the landraces diversity for food and seeds for next season (Joshi and Bauer 2007). The main aim of landraces conservation is to conserve the full range of genetic diversity within the LR from the threat (Negri et al. 2009). Thus, conservation of these landraces is an important task in today's world. There are various strategies for protecting the landrace diversity from the threats they face time and again mostly due to anthropogenic activities. These opportunities are divided into short- and long-term priorities. Sutherland and Woodroof (2009) reported that new threats to biological diversity and new opportunities for landrace conservation could be identified by horizon scanning. Landraces can be conserved by in situ or on-farm conservation methodologies. Various programmes have been started for conservation of landraces around the globe. These conservation methods of landraces result in establishment of biodiversity links, highlighting the need for conserving specific populations, and provide full range of ecogeographic data and genetic diversity of crop landraces. These conservation methods provide opportunities to the farmers to identify improved cultivars which can be employed for resolving food security issues globally. For conserving crop landraces, genetic reserves in secondary ecosystems (human disturbed, e.g. roadside and railroad banks) could be established. However, this approach is stringent to infrastructure and can be used as an approach to mitigate other biodiversity losses. There is a need for the development of improved national landraces inventories and prioritization of inventory on economic value, breeding demands and threat and biogeographic responsibility which are one of the significant steps towards landrace biodiversity information system. Another approach is to develop participatory management and monitoring models for landrace conservation so that it can increase emphasis on holistic approach to conservation strategies and methodologies and integration of genetic resource conservation into mainstream biodiversity conservation. Promotion of biodiversity friendly agriculture systems through NGOs could help the mankind for recognizing value of landrace biodiversity. There is an awful need of professionals and famers of traditional knowledge to timely intervene and identify the problem to chart down the strategies for protection of landraces existing in the ecosystem/s. A number of strategies have been developed to improve the crops and simultaneously protect the diversity of crop landraces. Few of them are mentioned below.

13.5.1 Strategy I: Mutational Breeding Systems

To circumvent the gene flow effect of GM crops for crop improvement, researchers recommend different alternative technologies to raise plants with desired trait. One such strategy which is beneficial to produce plants with desired traits is mutational breeding systems. These changes may be permanent or temporary. As far as

Table 13.3 Number of officially released mutant varieties in the top six countries (total 2252)

Country	Number of released mutant cultivars	Percent of total
China P.R.	605	26.8
India	259	11.5
USSR + Russia	210	9.3
Netherlands	176	7.8
USA	128	5.7
Japan	120	5.3

spontaneous mutations are concerned, they occur naturally with very low frequencies of 10^{-6} due to transposable elements which move into genome and cause alteration in DNA sequence (Wessler 2006), whereas induced mutations are caused by either chemical mutagens or other agents like X-rays, UV radiation, α-particles and β-particles. The main purpose of mutation breeding technology is the development of new and desired variation(s) through breeding programmes for crop improvement. Induced mutations can play an important role in the conservation and preservation of crop biodiversity. Induced mutations and related advanced technologies are important not only for increasing the genetic diversity of crops but also as a source for additional biodiversity enhancement of neglected and local crops/landraces (Hussain et al. 2012; Roychowdhury and Tah 2013). In this approach, mutants with desired traits were selected in the M_1 or M_2 generation after treatment with mutagens and then released as new variety for cultivation after evaluation and trials. Those mutants which are not selected as cultivars are rather used in cross-breeding programmes for tracing desired alleles (Roychowdhury and Tah 2013). More than 2000 (Table 13.3) plant varieties that contain induced mutations have been officially either released for cultivation directly as new varieties or used as parents to derive new varieties without the regulatory restrictions faced by genetically modified material (Maluszynski et al. 2000; Waugh et al. 2006). The number of mutant varieties released in China and India places Asia at the top of the list. This approach is the best alternative to transgenic biology to prevent gene flow within populations.

13.5.2 Strategy II: RNA Interference Systems

The phenomena of RNA interference is employed to produce crops having desirable traits. The process of RNAi can be triggered by the entry of small siRNA into a cell by several different ways, such as by *Agrobacterium*-mediated gene transfer, viral-mediated dsRNA transfer and particle bombardment method (Sijen and Kooter 2000). An RNAi vector is used to transform cell and produce stable dsRNA in vivo and further mediate silencing of target gene. RNA interference is an emerging tool in biotechnology for crop improvement. It has been widely used for increasing crop yield, quality and resistance against biotic and abiotic stresses. RNAi includes the sequence-specific gene silencing at post-transcription level (Kamthan et al. 2015).

Two major players of RNA interference are (endogenous) microRNA and exogenous, such as transgene and small interfering RNA (siRNA). They are produced by the breakdown of dsRNA by the ribonuclease enzyme DICER or DICER-like enzymes (DCL) (Bernstein et al. 2001; Hutvagner et al. 2001). Then a RNA-induced silencing complex (RISC) is activated by the incorporation of these single-stranded RNAs. RISC contains protein which has ribonuclease activity to degrade the mRNA- and RNA-binding domains (Hammond et al. 2000). RISC contains another important protein, Argonaute, that has been reported in *Arabidopsis thaliana*, which makes the catalytic core of RISC be involved in silencing (Vaucheret 2008). Activated RISC-RNA (antisense strand) then binds to complementary sequence and degrades the mRNA (Williams et al. 2004). siRNAs can also regulate gene expression at transcription level by regulating the chromatin modelling. siRNA maintains the transcription rate at minimal level by controlling histone modification including the cytosine methyltransferase; chromomethylase 3 (CMT3) keeps DNA into transcriptionally inactive state (Ossowski et al. 2008). Major threat of the transgenics is gene flow, which may lead to the genetic erosion. RNAi technology-based suppression of targeted expression of a gene evades this possibility and has been employed to conserve the parental crops/landraces. It can be employed to generate total sterility resulting in restriction of gene flow.

13.5.3 Strategy III: Somaclonal Variations

Australian scientists were the leaders in the field of somaclonal variations (SVs), demonstrating the efficiency in improvement of sugar cane, wheat and other crops. Somaclonal variations are genetic or epigenetic changes which are induced in plant cell and tissue culture (Fig. 13.2). The induction of somaclonal variation is an alternate approach to conventional breeding and transgenic approaches to introduce desirable genetic variability in the gene pool, thus protecting the crop landraces from selection pressure and extinction. The efficiency of developing disease-resistant SVs is accomplished with the imposition of an appropriate in vitro selection pressure. Selection agents that have been applied include pathogen elicitors, pathogen culture filtrates and purified pathotoxins. This method of SV selection has been successful in enhancing disease resistance in several crops, and it is an accepted biotechnological approach with tremendous potential for crop improvement. The Biotechnology Centre at the Indian Agricultural Research Institute (IARI) has standardized protocols for plant regeneration of *Brassica carinata* and is isolating somaclonal variants. Useful somaclonal variants for earliness, maturity, plant height, etc. have been induced in *B. juncea* and *B. napus*.

13.5.4 Strategy IV: Total Sterility

Total sterility involves the deletion of a portion of the gene involved in the production of pollen or flower or ovule. So this strategy forces the farmers to propagate the plant vegetatively, thus preventing the possibility of gene flow (Sharma et al. 2013).

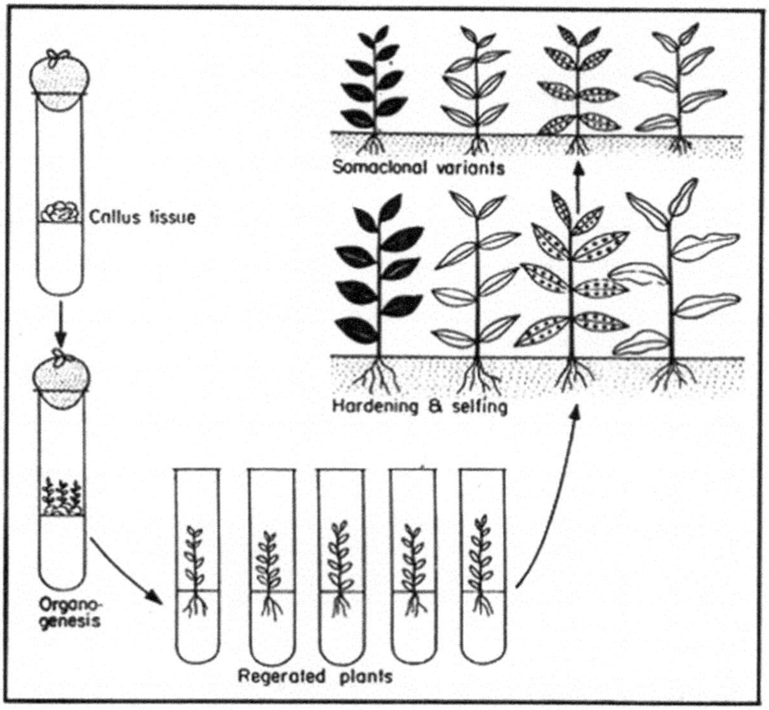

Fig. 13.2 Steps involve in induction and selection of somaclonal variation

This technique would be restricted to the plants which can be propagated through bulb propagules, leafy vegetable crops and forest plants. Due to high metabolic energy cost of sexual processes in plants, there will be higher yield of vegetative part of the plant, increasing the biomass production.

In addition to above strategy for preventing gene flow, the total sterility can be achieved by different genome-editing technologies like zinc finger nucleases (ZNFs), transcription activator-like effector nucleases (TALENs) (Christian et al. 2013) and CRISPR-Cas systems (Shan et al. 2013) to easily remove the target gene.

13.6 Conclusion

In fact, most of the threats to crop landraces need to be explored and documented for further fortification and conservation purpose. But it is evident from the above-mentioned literature that crop landraces are important to be focused for food security and diversity. In addition, crop landraces provide genetic resources that can be used for meeting current and new challenges of farming in stressful environments. However, proper knowledge of the crop landrace threats and opportunities are yet to be explored. Thus, we need to study these threats and opportunities so that crop landraces could be protected from extinction and could be used as a source for improvement of crops and to enhance production to meet the increased food demand throughout the globe. Research programmes need to be initiated to develop a stringent regulatory system to disseminate the knowledge regarding the prime importance of crop landraces and their threats and systemic application of specific transgenic approaches to prevent gene flow.

References

Abay F, Bjørnstad A (2009) Specific adaptation of barley varieties in different locations in Ethiopia. Euphytica 167:181–195

Antofie MM, Sand MPC, Ciotea G, Iagrăru P (2010) Data sheet model for developing a red list regarding crop landraces in Romania. Ann Food Sci Technol 11(1):45–49

Berg T (2009) Landraces and folk varieties: a conceptual reappraisal of terminology. Euphytica 166:423–430

Bernstein E, Caudy AA, Hammond SM, Hannon GJ (2001) Role for a bidentate ribonuclease in the initiation step of RNA interference. Nature 409(6818):363–366

Bonilla P et al (2002) RFLP and SSLP mapping of salinity tolerance genes in chromosome 1 of rice (*Oryza sativa* L.) using recombinant inbred lines. Philipp Agric Sci 85:68–76

Brush SP (1995) In situ conservation of landraces in centres of crop diversity. Crop Sci 35:346–354

Camacho Villa TC, Maxted N, Scholten MA, Ford-Lloyd BV (2005) Defining and identifying crop landraces. Plant Genet Res Char Util 3:373–384

Carvalho M, Bebeli P, Bettencourt E, Costa G, Dias S et al (2012) Cereal landraces genetic resources in worldwide gene banks. A review. Agronomy for sustainable development. Springer/EDP Sciences/INRA 33(1):177–203

Carpentier CL, Herrmann H (2003) Maize and biodiversity: the effects of transgenic maize in Mexico. Issues summary. Part of the Article 13 initiative on Maize and biodiversity: the effects of transgenic maize in Mexico. Accessed 19 April 2005 at http://www.cec.org/files/PDF// Issue_summary-e.pdf

Ceccarelli S, Grando S (2000) Barley landraces from the Fertile Crescent: a lesson for plant breeders. In: Brush SB (ed) Genes in the field: on-farm conservation of crop diversity. International Development Res Center, Boca Raton, pp 51–76

Ceccarelli S (2012) Landraces: importance and use inbreeding and environmentally friendly agronomic systems. In: Maxted N, et al (eds) Agrobiodiversity conservation: securing the diversity of crop wild relatives and landraces. CAB International, pp 103–117

Christian M, Qi Y, Zhang Y, Voytas D (2013) Targeted mutagenesis of *Arabidopsis thaliana* using engineered TAL effector nucleases. Genes Genome Genet 3:1697–1705

Christou P (2002) No credible scientific evidence is presented to support claims that transgenic DNA was introgressed into traditional maize landraces in Oaxaca, Mexico. Transgenic Res 11(1):3–5

Christiansen-Weniger F (1931) Bericht über eine Studienreise durch das ostanatolische Hochland. Zeitschr Züchtung A Pflanzenzüchtung 18:73–108

Ellstrand NC, Prentice HC, Hancock JF (1999) Gene flow and introgression from domesticated plants into their wild relatives. Annu Rev Ecol Syst 30:539–563

Ellstrand NC (2001) Crop transgenes in natural populations. Abstr Pap Am Chem Soc 221(1–2):AGFD 37

Esquinas-Alcázar J (2010) Protecting crop genetic diversity for food security: political, ethical and technical challenges. Nature 6:946–953

Friis-Hansen E, Sthapit B (2000) Participatory approaches to the conservation and use of plant genetic resources. Intl Plant Gen Res Inst (IPGRI), Rome

Frison AE et al (2011) Agricultural biodiversity is essential for a sustainable improvement in food and nutrition security. Sustainability 3:238–253

Futuyma D (1998) Evolutionary biology, 3rd edn. Sinauer, Sunderland

Hall L, Topinka K, Huffman J, Davis L, Good A (2000) Pollen flow between herbicide-resistant *Brassica napus* is the cause of multiple-resistant *B. napus* volunteers. Weed Sci 48:688–694

Hammer K et al (1996) Estimating genetic erosion in landraces –two case studies. Genet Res Crop Evol 43:329–336

Hammond SM, Bernstein E, Beach D, Hannon GJ (2000) An RNA-directed nuclease mediates post-transcriptional gene silencing in Drosophila cells. Nature 404(6775):293–296

Harlan J (1975) Our vanishing genetic resources. Science 188:618–621

Hawkes JG (1983) The diversity of crop plants. Harvard University Press, Cambridge, MA, p 102

Hussain B, Khan MA, Ali Q, Shaukat S (2012) Double haploid production is the best method for genetic improvement and genetic studies of wheat. Int J Agro Vet Med Sci 6(4):216–228

Hutvagner G, McLachlan J, Pasquinelli AE, Balint E, Tuschl T, Zamore PD (2001) A cellular function for the RNA-interference enzyme Dicer in the maturation of the let-7 small temporal RNA. Science 293(5531):834–838

IBPGR (1980) A glossary of plant genetic resources terms. IBPGR Secretariat, Rome

Joshi BK, Upadhyay MP, Gauchan D, Sthapit BR, Joshi KD (2004) Red listing of agricultural crop species, varieties and landraces. Nepal Agric Res J 5:73–80

Kamthan A, Chaudhuri A, Kamthan M, Datta A (2015) Small RNAs in plants: Recent development and application for crop improvement. Front in Plant Sci 6 https://doi.org/10.3389/fpls.2015.00208

Kaplinsky N, Braun D, Lisch D, Hay A, Hake S, Freeling M (2002) Maize transgene results in Mexico are artifacts. Nature 416:601–602

Kumar A et al (2014) Breeding high-yielding drought-tolerant rice: genetic variations and conventional and molecular approaches. J Exp Bot 65:6265–6278

Kumar V et al (2015) Genome-wide association mapping of salinity tolerance in rice (*Oryza sativa*). DNA Res. Published online January 27, 2015 https://doi.org/10.1093/dnares/dsu046

Li HB, Zhou MX, Liu CJ (2009) A major QTL conferring crown rot resistance in barley and its association with plant height. Theor Appl Genet 118:903–910

Lopez PB (1994) A new plant disease: uniformity. CERES 26:41–47

Louette D (1999) Traditional management of seed and genetic diversity: what is a landrace? In: Brush SB (ed) Genes in the field: onfarm conservation of crop diversity. Lewis Publishers, CRDI/IPGRI, Boca Raton, pp 109–142

Maluszynski M et al (2000) Officially released mutant varieties – the FAO/IAEA Database. Mutat Breed Rev 12:1–84

Mayr E (1934) Die Bedeutung der alpinen Getreidelandsorten für die Pflanzenzüchtung and Stammesforschung mit besonderer Beschreibung der Landsorten in Nordtirol und Vorarlberg. Zeitsch f Züchtung A: Pflanzenzüchtung 19:195–228

Metz M, Fütterer J (2002) Suspect evidence of transgenic contamination. Nature 416:600–601

Mayr E (1937) Alpine Landsorten in ihrer Bedeutung für die praktische Züchtung. Forschungsdienst 4:162–166

Moragues M, Zarco-Hernandez J, Moralejo MA, Royo C (2006) Genetic diversity of glutenin protein subunits composition in durum wheat landraces [*Triticum turgidum ssp. turgidum convar. durum (Desf.) MacKey*] from the Mediterranean basin. Genet Res Crop Evol 53:993–1002

Negri V (2003) Landraces in Central Italy: where and why they are conserved and perspectives for their on-farm conservation. Genet Res Crop Evol 50:871–885

Negri V, Maxted N, Vetelainen M (2009) European landrace conservation: an introduction. In: Vetelainen M, Negri V, Maxted N (eds) European landrace: on-farm conservation, management and use. Bioversity Technical Bulletin 15. Bioversity International, Rome, pp 1–22

Newton AC, Akar T, Baresel JP, Bebeli PJ, Bettencourt E, Bladenopoulos KV, Czembor JH, Fasoula DA, Katsiotis A, Koutis K, Koutsika-Sotiriou M, Kovacs G, Larsson H, Pinheiro de Carvalho MAA, Rubiales D, Russell J, dos Santos TMM, Vaz Patto MC (2010) Cereal landraces for sustainable agriculture. A review. Agron Sustain Dev 30: 237–269

Ossowski S, Schwab R, Weigel D (2008) Gene silencing in plants using artificial microRNAs and other small RNAs. Plant J 53(4):674–690
Pecetti L, Doust MA, Calcagno L, Raciti CN, Boggini G (2001) Variation of morphological and agronomical traits, and protein composition in durum wheat germplasm from Eastern Europe. Genet Resour Crop Evol 48:609–620
Pistorius R (1997) Scientists, plants and politics. A history of the plant genetic resources movement. IPGRI, Rome
Porfiri O, Costanza MT, Negri V (2009) Landrace inventories in Italy and the Lazio region case study. In: Veteläinen M, Negri V, Maxted N (eds) European landraces: on-farm conservation, management and use. Bioversity technical bulletin 15. Bioversity International, Rome, pp 117–123
Quist D, Chapela IH (2001) Transgenic DNA introgressed into traditional maize landraces in Oaxaca, Mexico. Nature 414:541–543
Quist D, Chapela IH (2002) Reply. Nature 416:602
Ren ZH et al (2005) A rice quantitative trait locus for salt tolerance encodes a sodium transporter. Nat Genet 37:1141–1146
Reynolds M, Dreccer F, Trethowan R (2007) Drought-adaptive traits derived from wheat wild relatives and landraces. Integrated approaches to sustain and improve plant production under drought stress. J Exp Bot 58(2):177–186
Rodriguez M et al (2008) Genotype by environment interactions in barley (*Hordeum vulgare* L): different responses of landraces, recombinant inbred lines and varieties to Mediterranean environment. Euphytica 163:231–247
Roychowdhury R, Tah J (2013) Mutagenesis—a potential approach for crop improvement. In: Hakeem KR, Ahmad P, Ozturk M (eds) Crop improvement. Springer, USA, pp 149–187. https://doi.org/10.1007/978-1-4614-7028-1_4
Sarker A, Erskine W (2006) Recent progress in the ancient lentil. J Agric Sci 144:19–29
Sarker M, Adawy S, Smith CM (2008) Entomological and genetic variation of cultivated barley (*Hordeum vulgare*) from Egypt. Arch Phytopathol Plant Prot 41:526–536
Saxena S, Singh AK (2006) Revisit to definitions and need for inventorization or registration of landrace, folk, farmers' and traditional varieties. Curr Sci 91:1451–1454
Shan Q, Wanp Y, Li J, Zhang Y et al (2013) Targeted genome modification of crop plants using a CRISPR–Cas system. Nat Biotech 31:686–688
Sharma S, Shahzad A, da Silva JAT (2013) Synseed technology – a complete synthesis. Biotech Adv 31:186–207
Sijen T, Kooter JM (2000) Post-transcriptional gene-silencing: RNAs on the attack or on the defense. BioEssays 22(6):520–531
Snow AA, Moran-Palma P (1997) Commercialization of transgenic plants: potential ecological risks. BioScience 47:86–96
Sutherland WJ, Woodroof HJ (2009) The need for environmental horizon scanning. Trends Ecol Evol 24:523–527
Teklu Y, Hammer K (2009) Diversity of Ethiopian tetraploid wheat germplasm: breeding opportunities for improving grain yield potential and quality traits. Plant Genet Resour 7:1–8
Thompson MJ et al (2010) Characterizing the Saltol quantitative trait locus for salinity tolerance in rice. Rice 3:148–160
Trethowan RM, Mujeeb-Kazi A (2008) Novel germplasm resources for improving environmental stress tolerance of hexaploid wheat. Crop Sci 48:1255–1265
Tsegaye S, Tesemma T, Belay G (1996) Relationships among tetraploid wheat (*Triticum turgidum* L.) landrace populations revealed by isozyme markers and agronomic traits. Theor Appl Genet 93:600–605
Van de Wouw M et al (2010) Genetic diversity trends in twentieth century crop cultivars: a meta-analysis. Theor Appl Genet 120:1241–1252
van Hintum TJL, Ellings A (1991) Assessment of glutenin and phenotypic diversity of Syrian durum wheat landraces in relation to their geographical regions. Euphytica 55:209–215
Vaucheret H (2008) Plant Argonautes. Trends Plant Sci 13(7):350–358

Von Rünker K (1908) Die Systematischeeinteilung und Benen-ung der Getreidesortenfu¨rpr aktische Zwecke. Jahrbuch der Deutschenlandwirtschafts-Gesellschaft 23:137–167

Waugh R, Leader DJ, McCallum N, Caldwell D (2006) Harvesting the potential of induced biological diversity. Trends Plant Sci 11(2):71–79

Wessler SR (2006) Transposable elements and the evolution of eukaryotic genomes. Proc Natl Acad Sci 103(47):17600–11760

Williams M, Clark G, Sathasivan K, Islam AS (2004) RNA interference and its application in crop improvement. Plant Tissue Cult Biotechnol 1:18. https://extension.colostate.edu/topic-areas/agriculture/genetically-modified-gm-crops-techniques-and-applications-0-710

World Conservation Monitoring Centre (1992) In: Groombridge B (ed) In global biodiversity: status of the Earth's living resources. Chapman & Hall, London

Zeven AC (1975) Domesticatie en evolutie van de kultuurplant. Wageningen Agricultural University, Dept of Plant Breeding. Mimeographed Lecture Notes pp 177

Zeven AC (1998) Landraces: a review of definitions and classifications. Euphytica 104:127–139

Future Threats and Opportunities Facing Crop Wild Relatives and Landrace Diversity

14

Chainika Gupta, Romesh Kumar Salgotra, and Gulshan Mahajan

Contents

Abstract

Crop wild relatives and landraces possess novel alleles for biotic and abiotic stress resistance which can be used to develop varieties with superior traits. They can survive in different agro-environmental conditions as they have not undergone through genetic bottlenecks of domestication. They have broad genetic base with a wide range of allele diversity, thus contributing towards food safety and livelihood security. Due to natural calamities and anthropogenic activities, CWR and landraces are under threat of the risk of extinction as a result of deforestation, genetic erosion, industrialized agriculture, dryland destruction and desertification, urbanization and climate change. Thus, a coordinated global approach is needed for conservation of CWR and landraces as they play a major role in providing ecosystem services which are beneficial to humans. It is also

C. Gupta (✉) · R. K. Salgotra
School of Biotechnology, Sher-e-Kashmir University of Agricultural Sciences and Technology of Jammu, Jammu, Jammu and Kashmir, India

G. Mahajan
Queensland Alliance for Agriculture and Food Innovation (QAAFI), The University of Queensland, Gatton, QLD, Australia
e-mail: g.mahajan@uq.edu.au

R. K. Salgotra, S. M. Zargar (eds.), *Rediscovery of Genetic and Genomic Resources for Future Food Security*,
https://doi.org/10.1007/978-981-15-0156-2_14

necessary to screen novel genes in both wild plants that can be used in crop improvement programmes and threatened wild plants and landraces which need to be conserved. Plant genetic resources can be conserved either in situ or ex situ. Also, CWR and landraces have been utilized in crop improvement programmes through modern approaches like tissue culture, genetic engineering, AB-QTL and alien transfer of genes from wild relatives to chromosomes.

Acronyms

AB-QTL	Advanced backcross-QTL
BC	Backcross
CWR	Crop wild relatives
DNA	Deoxyribose nucleic acid
FAO	Food and Agriculture Organization
GE	Genetic engineering
IPCC	Intergovernmental Panel on Climate Change
PGR	Plant genetic resources
PGRFA	Plant genetic resources for food and agriculture
QTL	Quantitative trait loci
T-DNA	Transfer DNA

14.1 Introduction

With the rapid increase in human population and impact of natural calamities and anthropogenic activities on agriculture, food security is a global concern (Schmidhuber and Tubiello 2007; FAO 2008). It is necessary that crop production should be sufficient enough to meet the needs of population. Since the origin of agriculture, plant genetic resources for food and agriculture (PGRFA) including landraces, crop wild relatives (CWR), breeding lines and modern cultivars are valuable resources of our ecosystem and play a vital role in contributing to food security (Maxted et al. 2006). They contain useful genes for pest and disease resistance; resistance to abiotic stresses like drought, salinity and heat; increase in yield; cytoplasmic male sterility and fertility restorers; and enhanced quality. To feed the growing population, crops with resistance to biotic and abiotic stresses, high yield, and improved quality need to be developed. Hence, plant breeders can utilize plant genetic resources to develop novel varieties with superior traits. Domestication of crops is carried out by repeated selection of desirable trait from wild species crop contributing to uniformity and high productivity at the cost of reduction in genetic variation and allele diversity, making them more susceptible to biotic and abiotic stresses (Tanksley and McCouch 1997). For example, reduction in genetic diversity has been seen in rice, soybean and maize (Xu et al. 2012; Wright et al. 2005; Zhou et al. 2015).

The wild species closely related to crop and found naturally in the environment are known as crop wild relatives (CWR). Crop wild relatives (CWR) were initially used by plant breeders in the 1940s and 1950s for improvement of major crops (Plucknett et al. 1987). By the 1970s and 1980s, the use of wild relatives in crop improvement programmes has been increased tremendously (Hoyt 1988). For example, Norin 10, a dwarf variety of wheat that contains a gene for dwarfism, has been introduced in wheat, resulting in increased uptake of nitrogen and increased production (Kihara 1983). They grow in different habitats and adapt to a wide range of environmental conditions (FAO 2008). These are rich sources of important genes for yield and quality traits as they have not been undergone through genetic bottleneck of domestication (Vollbrecht and Sigmon 2005; Heywood et al. 2007). The significance of preserving a broad range of genetic diversity in crops as well as in their wild relatives was first realized by Vavilov (1922). About 10,000 of the 60,000 crop wild species act as donors of gene to cultivated variety, thus contributing to food security (Maxted and Kell 2009).

Crop wild relatives are the ancestors of crops possessing broad genetic base with a wide range of allele diversity, resulting in traits like high yield and resistance to biotic and abiotic stresses (Dempewolf et al. 2014). A few examples showing CWR as source of novel alleles for biotic and abiotic stress resistance are shown in Table 14.1. CWR possess potential useful genes which enables crop species to survive in harsh climatic conditions enabling breeders to produce new improved variety by crossing them with domesticated crops. CWR also contains genes for cytoplasmic male sterility and fertility restorer. For example, cytoplasmic male sterility has been reported in pigeon pea where cytoplasm from cultivated pigeon pea has been crossed

Table 14.1 Crop wild relatives as source of novel alleles for biotic and abiotic stress resistance

Crop wild relative	Gene transferred to cultivated variety	Resistant to	References
Wheat	*TmHKT1;5-A*	Salt tolerance	James et al. (2006)
Soybean	*GmCHX1*	Salt tolerance	Guan et al. (2014)
	GmSALT3		Qi et al. (2014)
Barley	QTLs identified	Drought tolerance	Diab et al. (2004), Suprunova et al. (2004), Suprunova et al. (2007), Fischer et al. (2013), and Placido et al. (2013)
Wheat			
Tomato			
Mexican maize	Blight-resistant alleles	Corn blight	Maxted and Kell (2009)
Rice	*Xa3, Xa4*	Bacterial blight	Song et al. (1995)
	Xa23, Xa21		Zhou et al. (2011)
Barley	*RYM16*^{Hb}	Barley mild mosaic virus (BaMMV); barley yellow mosaic virus (BaYMV)	Ruge-Wehling et al. (2006)
T. tauschii	*Lr41, Lr 42 and Lr 43*	Rust resistance	Cox et al. (1994)

with wild relative, *C. acutifolius* (donor of nuclear genes) (Mallikarjuna and Saxena 2005). There were two concepts given to define CWR. The first one is gene pool concept (Harlan and de Wet 1971). On the basis of genetic diversity and ease with which genetic material is transferred across species, CWR have been categorized into primary gene pool, secondary gene pool and tertiary gene pool. Crops belonging to primary gene pool are closely related to crop; hence, genes can be easily transferred to crops by traditional breeding methods. In secondary gene pool, the crops are less closely related, but genetic material can be transferred to crops. Tertiary gene pool includes distant related species; as a result, genetic material can be transferred through approaches like genetic engineering. The second concept is taxon groups which takes taxonomic hierarchy into account (Maxted et al. 2006). Whether the taxa belong to similar tribe, genus, species and subgenus as crop, five taxon groups have been allotted. By these two concepts, we can assume relatedness of crop wild species. Maxted et al. (2006) proposed a working definition of CWR as 'a wild plant taxon with indirect use, which is genetically related to crop and this interrelation can be assumed as CWR associated with primary or secondary Gene pools or taxon groups of the crop'. Many wild species of genus *Lycopersicon* possess disease-resistant genes like *L. hirsutum* and *L. peruvianum* contain fungus-resistant genes, *L. peruvianum* contains nematode-resistant genes and *L. hirsutum* contains insect-resistant genes. Besides this, *L. chmielewskii* contains quality improvement genes, and *L. cheesmaniae* contains genes for adaptation in harsh conditions. Thus, by crossing wild species with cultivated species (*L. esculentum*), desirable genes can be introgressed, resulting in crop improvement (Esquinas-Alcazar 2005). Many of the improved cultivars contain genes derived from wild relatives which plays a critical role in food security (Meilleur and Hodgkin 2004; Stolton et al. 2006). It was impossible to cultivate sugar cane, tomato and tobacco commercially on a global scale if disease-resistant genes were not transferred from wild relatives (FAO 1998). Hence, CWR have been employed in crop improvement programmes of certain crops depending on the compatibility and how much crop and its wild relative are taxonomically related, fertility in F_1 and next generations as well as availability of CWR (Zhang and Mittal 2016). A wide range of novel alleles have been transferred from wild relatives to cultivated crops through map-based cloning and backcrossing. Through biotechnological approaches like next-generation sequencing and high-throughput phenotyping, many QTLs associated with agronomic traits in CWR have been identified (Honsdorf et al. 2014; Qi et al. 2014).

Along with artificial selection of desirable phenotype, natural selection also plays a vital role in selection of phenotypes at specific geographical locations. These are known as landraces. Landraces are variable genotypes, locally adapted in particular geographical region, maintained by farmer and have been grown for a thousand years (Hawkes 1983; Tudge 1988; Harlan 1975). They are well adapted to the location where they grow, and no artificial selection is done. They retain their original attributes even when cultivated away from their place of origin (Zeven 1998). They have been named on the basis of their origin (Von Rumker 1908). For example, Tuxpeno maize got its name from its origin: Tuxpan region in Mexico (Maxted et al. 2006). They are very stable and produce high yield even under various environmental stresses, thus containing genes for disease and pest resistance, nutritional quality, heat stress, drought stress, salinity stress and marginal environment tolerance contributing to food security (Brush 1995; Frankel et al. 1998). Landraces act as source of novel alleles for abiotic stress tolerance as depicted in Table 14.2. Another reason for adaptability in stressful conditions is that there are segments of genes with little recombination (Gepts 2006). They are not uniform and possess genetic diversity (FAO 1998). Primitive cultivar, primitive variety, primitive form, farmers' variety, traditional variety, local variety and folk variety are synonyms of landrace (Maxted et al. 2005). Landraces are classified as autochthonous, allochthonous and Creole. Autochthonous is a landrace which has been cultivated in a particular area for more than a century (Zeven 1998). Allochthonous is a landrace

Table 14.2 Landraces as source of novel alleles for abiotic stress tolerance

Crop	Landrace	Resistant to	Gene	Released variety	References
Lowland rice	FR13A	Submergence tolerance	SUBMERGENCE 1 (*SUB1A-1*) on chromosome 9	Swarna-Sub1, BR11-Sub1, IR64-Sub1	Xu and Mackill (1996), Bailey-Serres et al. (2010), and Xu et al. (2006)
Upland rice	Aus 257, Aus Bak Tulsi, Azucena, basmati 370, Dular, Kalia, Kali Aus, Lal Aus and N22	Drought tolerance	QTL identification	Swarna, IR64, Sabitri, TDK1 and BR11	Dwivedi et al. (2016)
Rice	Kasalath	Phosphate tolerance	Phosphate uptake 1 (*Pup1*)	Phosphate-tolerant cultivars	Wissuwa and Ae (2001) and Wissuwa et al. (2002)
Rice	Nona Bokra and Pokkali	Salt tolerance	QTL identification	Salt-tolerant cultivars	Ren et al. (2005) and Bonilla et al. (2002)
Barley	TX9425	Drought tolerance and salt tolerance	QTL identification	Drought- and salt-tolerant cultivars	Fan et al. (2015)
Barley	Sahara 3763	Boron toxicity tolerance	*Bot1*	Cultivars with low boron concentrations	Yao and Ryan (2008) and Nable (1988)

which is originated in a particular area but locally adapted in another region (Zeven 1998), for example, Tuxpeno maize. Creole is a landrace which requires successive rounds of seed selection in a particular area (Wood and Lenne 1997).

Landraces and CWR play a vital role in ensuring food security. CWR have also contributed economically by enhancing global crop production with profit of approximately 115 billion USD worldwide (Pimentel et al. 1997). The life span of commercial cultivars is around 5–10 years which forces breeders to produce new cultivars with novel genetic variants, resistance to marginal environments and high productivity, of which these traits are found in landraces and CWR (Maxted et al. 2011). Due to land degradation by human activities, agriculture is carried out in adverse conditions. So, crops which can survive in these conditions can be developed from CWR and landraces. The improved varieties contain alleles from CWR, thus promoting livelihood security.

In marginal areas with poor agronomic conditions and subsistence farming system, landraces are preferred by farmers due to their high adaptability in a wide range of agroecological conditions and high stability (Frankel et al. 1998; Brown 1999). Landraces can sense even minor changes in surroundings and respond accordingly which might be due to their inherent population structure (Bennett 1970; Harlan 1975). Yield stability is a fundamental characteristic in landraces (Zeven 1998). Landraces have good nutrition value and are healthy and hence play a major role in food safety (Azeez et al. 2018).

14.2 Threats Faced by CWR and Landraces

It has been estimated that global population will tremendously rise by 2050 to 9.3 billion demanding food production from agriculture (U.S. Census Bureau 2014; Kastner et al. 2012). CWR and landraces are under threat of the risk of extinction (Maxted et al. 1997; Stolton et al. 2006). At such an alarming situation, conservation of biodiversity is of great concern. The challenges faced by biodiversity are deforestation, genetic erosion, industrialized agriculture, dryland destruction and desertification, urbanization, invasive species and climate change. One of the major challenges faced by them is climate change. According to Intergovernmental Panel on Climate Change (2007), global rise in temperature by 2–4 °C till 2050 causes changes in precipitation pattern (IPCC 2007). In Europe, 27–42% of species will be lost as a result of change in climate by 2080 (Thuiller et al. 2005). It has also been predicted the extinction of CWR species (16–22%) by 2055 due to genetic erosion (Jarvis et al. 2008). As a result, environment in which crop grows is continuously changing; hence, crops needed to be replaced by new varieties which can be well adapted to altering environmental conditions (Deryng et al. 2011; Li et al. 2011; Luck et al. 2011). The crop wild relatives ranged from 15 to 264 for citrus fruits and root, bulb or tuber vegetables. By 2070, more than 50% of CWR are going to be lost as a result of climate change. Among them, the most vulnerable are spices, followed by sugar CWR, cereal CWR and beverages (Vincent et al. 2019). The increase in temperature due to global warming will have an adverse effect on crops. At high temperature, sterility will be induced in rice, and senescence in wheat and yield loss in maize will be the major problem (Lobell et al. 2012; Dempewolf et al. 2014). Jarvis et al. (2008) reported 75% of CWR have been threatened. The preference of cultivars produced by breeding over landraces can result in extinction as well as genetic erosion of gene pool of landraces.

Another threat to CWR and landraces is genetic erosion. Genetic erosion is the change in allele number and evenness of potential agronomic traits in a defined geographical area, resulting in loss of genetic diversity (FAO 1999). It occurs mainly due to deforestation, replacement of landraces and CWR with modern cultivars, urbanization and degradation of environment (Bettencourt et al. 2008). Genetic erosion had many disastrous impacts. According to the National Research Council (1972), more than 50% of the standing maize crop had been eradicated in the USA due to narrow genetic base of seed, thereby increasing its susceptibility to fungus *Helminthosporium maydis*. There were reports of loss of landraces from Bara District and Seti River Valley (Sherchand et al. 1998; Rijal et al. 1998).

Modern agriculture is considered as the major threat to biodiversity as farmers have shifted towards cultivation of high-yielding, genetically uniform homozygous crop (monoculture) instead of locally adapted landraces leading to decrease in genetic diversity (Dwivedi et al. 2016; Sarker and Erskine 2003). In the USA, farmers cultivated genetically uniform homozygous varieties of corn and potato, as a result threatened with Irish potato famine and Southern corn leaf blight epidemic (Govindaraj et al. 2015). Majority of the world population is dependent on only 12 crops, among which 60% of their diet is constituted by rice, wheat, maize and potatoes (Esquinas-Alcazar 2005). It was estimated by the World Conservation Monitoring Centre (1992) that in Indonesia, 74% of rice cultivars are developed from the same stock. On the other hand, in the USA, 50% wheat is developed from nine cultivars, 75% potato is derived from four cultivars and 50% soybean is derived from six cultivars. Reduction in variability of landraces has been seen due to genetic erosion in Albania and Southern Italy to an extent of 72.4% and 72.8%, respectively (Hammer et al. 1996). After the Green Revolution, the replacement of landraces with modern cultivars led to great loss in their number from 400,000 to 30,000 in case of rice (Dwivedi et al. 2016), while 95% of wheat landraces have been lost in Greece (Lopez 1994).

14.3 Opportunities for Conservation of Crop Wild Relatives and Landraces

Plant genetic resources (PGR) play a major role in providing ecosystem services which are beneficial to humans in many ways, majorly in providing food security globally (Naidoo et al. 2008). Food insecurity is a major threat that we are facing, and a coordinated global approach is needed for conservation of CWR (Maxted and Kell 2009). The natural calamities like change in climatic conditions and anthropogenic activities like degradation of habitats enforce development of novel varieties well adapted to marginal environments by looking for genes in CWR and landraces imparting resistance to pest/disease, drought, flood and heat (Heywood et al. 2007; Negri and Tiranti 2010). It is necessary to screen novel genes in wild plants that can be used in crop improvement programmes and threatened wild plants and landraces which need to be conserved (Maxted et al. 2008). The Green Revolution relied on novel genes from CWR and landraces that resulted in enhanced crop yield and thus production. The conservation of plant genetic resources can be done either in situ or ex situ. In situ conservation means conservation of genetic resources in natural environment, while ex situ conservation means conservation of genetic resources as seeds in gene bank or explants in tissue culture. To meet the demands of present as well as future generations, in situ conservation of crop wild relatives is needed as the resultant genetic diversity within CWR is due to environment (Maxted and Kell 2009). Though a crop wild relative might seem agriculturally unimportant but as natural selection is also acting on it, with passage of time, they acquire desirable traits. By conserving plant genetic resources ex situ as seeds, decrease in genetic diversity, increased homozygosity, inbreeding depression and no natural selection have been observed (Schoen and Brown 2001).

14.4 Use of CWR and Landraces in Crop Improvement Through Modern Approaches

14.4.1 Alien Gene Transfer from Wild Relatives Through Chromosomes

Alien genes are those genes which are not closely related and are present in distant relatives. The phylogenetic relationship has been established by the ease with which genetic material will be transferred. The major challenge in transfer of alien genes from wild relatives to cultivated species is cross-incompatibility and limited recombination (Brar and Khush 1986; Khush and Brar 1992; Sitch 1990). Through chromosome engineering, *Ph1* gene, inhibitor of chromosome pairing, has been suppressed, and alien genes have been transferred in wheat (Jauhar 2006). Using cytogenetic techniques, leaf rust-resistant gene from *Aegilops umbellulata* has been transferred to wheat chromosome 6B (Sears 1956).

14.4.2 Advanced Backcross-Quantitative Trait Loci Analysis (AB-QTL)

Another problem encountered in transfer of alien gene is of linkage drag. Along with desirable gene, many undesirable genes also get transferred, thereby affecting yield and quality traits (The et al. 1988; Lukaszewski 2000; Labuschagne et al. 2002). AB-QTL is a combination of QTL analysis and variety development. It reduces the major problem associated with linkage drag and has been successfully done in crops like tomato, wheat and rice (Tanksley and Nelson 1996; Kumar and Pratap 2016). In this technique, molecular markers linked with improved background are utilized in genotyping BC2/ BC3 progenies, followed by selection of improved genotypes having recovered genome. These progenies possess minimum linkage drag (Kumar and Pratap 2016). Linh et al. (2012) used AB-QTL technique to remove linkage drag in rice cultivar 'BT7' by incorporating salt-tolerant allele 'Saltol' from 'FL478'.

14.4.3 Tissue Culture

Hybridization in cross-incompatible species has been made possible due to tissue culture techniques like embryo rescue, ovule culture and in vitro hormonal treatments (Gupta and Sharma 2005; Mallikarjuna et al. 2006). Cultivated lentil has been successfully hybridized with *L. ervoides* and *L. nigricans* to obtain distant hybrids using embryo rescue (Cohen et al. 1984). Clarke et al. (2006) reported viable embryos in chickpea by crossing *C. arietinum* with *C. bijugum* and *C. arietinum* with *C. pinnatifidum*. By somatic hybridization using protoplast fusion, genes can be successfully transferred from wild to cultivated species in *Pisum* (Davey et al. 2005; Ochatt et al. 2000).

14.4.4 Genetic Engineering

GE is defined as manipulation of plant's genetic material by insertion of foreign DNA sequence controlling specific trait which can be from related species or unrelated species. It is also known as recombinant DNA technology. The most versatile method for transfer of gene is '*Agrobacterium*-mediated' gene transfer. *Agrobacterium tumefaciens* is gram-negative, plant pathogenic bacteria. It has ability to incorporate T-DNA of Ti plasmid into plant's genetic material causing crown gall disease. This has been utilized by molecular biologists to transfer desirable gene in plants by replacing gall formation genes with suitable gene to produce desirable trait, thus acting as natural genetic engineer. Its disadvantage is it is not able to transform monocots. To overcome this limitation, microprojectile bombardment method (also known as gene gun) is used. The DNA-coated tungsten or gold particles are accelerated under high pressure and shot into plant tissue (Gan 1989). The DNA gets integrated into chromosomal DNA of plant. Selection of transformed plants is done using selectable markers which are subsequently regenerated by tissue culture. Some other methods are electroporation and microinjection. In electroporation, plant protoplast takes foreign DNA from surrounding when electric impulse is applied. In microinjection, foreign DNA is directly injected into plant cells.

14.5 Conclusion

The crop wild relatives and landraces have rich source of genetic variation and novel alleles which can be used in crop improvement programmes to make crops resistant to biotic and abiotic stresses. Since we are under constant threat of climate change and genetic erosion, it is very important to preserve these PGR to meet the needs of present as well as future generations. In situ conservation of genetic diversity should be given priority so as to maintain allelic variation in natural environment. Globally, lots of initiatives have been taken to conserve biodiversity due to their potential in contributing towards food security. Modern techniques for gene transfer from wild relatives would further reinforce their use in crop improvement.

References

Azeez MA, Adubi AO, Durodola FA (2018) Landraces and Crop Genetic Improvement. https://doi.org/10.5772/intechopen.75944

Bailey-Serres J, Fukao T, Ronald P, Ismail A, Heuer S, Mackill D (2010) Submergence tolerant rice: SUB1's journey from landrace to modern cultivar. Rice 3:138–147

Bennett E (1970) Adaptation in wild and cultivated plant populations. In: Frankel OH, Bennett E (eds) Genetic Resources in Plants—Their Exploration and Conservation. International Biological Programme Handbook No. 11. Oxford: Blackwell, pp. 115–129

Bettencourt E, Ford- Lloyd BV, Dias S (2008) Genetic erosion and genetic pollution of Crop wild relatives: the PGRForum perspective and achievements. CAB International, Crop wild relative conservation and use (eds N Maxted et al)

Bonilla P, Dvorak J, Mackill D, Deal K, Gregorio G (2002) RFLP and SSLP mapping of salinity tolerance genes in chromosome 1 of rice (*Oryza sativa* L.) using recombinant inbred lines. Philipp Agric Sci 85:68–76

Brar DS, Khush GS (1986) Wide hybridization and chromosome manipulation in cereals. In: Evans DH, Sharp WR, Ammirato PV (eds) Handbook of plant cell culture, vol 4, Techniques and applications. MacMillan Publish Co, New York, pp 221–263

Brown AHD (1999) The genetic structure of crop landraces and the challenge to conserve them in situ on farms. In: Brush S (ed.) Genes in the Field. International Plant Genetic Resources Institute, Rome, pp 29–48

Brush SB (1995) In situ conservation of landraces in centers of crop diversity. Crop Sci 35:346–354

Clarke HJ, Wilson JG, Kuo I, Lulsdorf MM, Mallikarjuna N, Kuo J, Siddique KHM (2006) Embryo rescue and plant regeneration in vitro of selfed chickpea (*Cicer arietinum* L.) and its wild annual relatives. Plant Cell Tissue Organ Cult 85:197–204

Cohen D, Ladizinsky G, Ziv M, Muehlbauer FJ (1984) Rescue of interspecific Lens hybrids by means of embryo culture. Plant Cell Tissue Organ Cult 3:343–347

Cox TS, Raup WJ, Gill BS (1994) Leaf rust-resistance genes *Lr41, Lr42,* and *Lr43* transferred from *Triticum tauschii* to common wheat. Crop Sci 34:339–343

Davey MR, Anthony P, Power JB, Lowe KC (2005) Plant protoplasts: status and biotechnological perspectives. Biotechnol Adv 23:131–171

Dempewolf H, Eastwood RJ, Guarino L, Khoury C, Müller JV, Toll J (2014) Adapting agriculture to climate change: a global initiative to collect, conserve, and use crop wild relatives. Agrocecol Sust Food Syst 38:369–377. https://doi.org/10.1080/21683565.870629

Deryng D, Sacks WJ, Barford CC, Ramanicutty N (2011) Simulating the effects of climate and agricultural management practices on global crop yield. Global Biogeocheni Cycles 25:GB2006. https://doi.org/10.1029/2009GB003765

Diab AA, Teulat-Merah B, This D, Ozturk NZ, Benscher D, Sorrells ME (2004) Identification of drought-inducible genes and differentially expressed sequence tags in barley. Theor Appl Genet 109:1417–1425
Dwivedi SL, Ceccarelli S, Blair MW, Upadhyaya HD, Are AK, Ortiz R (2016) Landrace germplasm for improving yield and abiotic stress adaptation. Trends Plant Sci 21(1):31–42
Esquinas-Alcazar J (2005) Protecting crop genetic diversity for food security: political, ethical and technical challenges. Nature 6:946–953
Fan Y, Shabala S, Ma Y, Xu R, Zhou M (2015) Using QTL mapping to investigate the relationships between abiotic stress tolerance (drought and salinity) and agronomic and physiological traits. BMC Genomics 16:43
FAO (1998) The State of the World's Plant Genetic Resources for Food and Agriculture. FAO, Rome
FAO (1999) Report of the Technical Meeting on the Methodology of the World Information and Early Warning System on Plant Genetic Resources. Research Institute of Crop Production, Prague, Czech Republic. 21–23 June 1999. FAO, Rome, Italy
FAO (2008) Climate change and Biodiversity for Food and Agriculture. Food and Agriculture organization of the United Nations, Rome
Fischer I, Steige KA, Stephan W, Mboup M (2013) Sequence evolution and expression regulation of stress-responsive genes in natural populations of wild tomato. PLoS One 8(10):e78182
Frankel OH, Brown AHD, Burdon JJ (1998) The Conservation of Plant Biodiversity, 2nd edn Cambridge: Cambridge University Press, pp 56–78
Gan C (1989) Gene Gun Accelerates DNA-Coated Particles To Transform Intact Cells. The Scientist, 3(18):25
Gepts P (2006) Plant genetic resources conservation and utilization: the accomplishments and future of a societal insurance policy. Crop Sci 46:2278–2292
Govindaraj M, Vetriventhan M, Srinivasan M (2015) Importance of Genetic Diversity Assessment in Crop Plants and Its Recent Advances: An Overview of Its Analytical Perspectives. Genetic Research International. https://doi.org/10.1155/2015/431487
Guan RX, Qu Y, Guo Y, Yu LL, Liu Y, Jiang JH, Chen JG et al (2014) Salinity tolerance in soybean is modulated by natural variation in *GmSALT3*. Plant J 80:937–950
Gupta D, Sharma SK (2005) Embryo-ovule rescue technique for overcoming post-fertilization barriers in interspecific crosses of Lens. J Lentil Res 2:27–30
Hammer K, Knupffer H, Xhuveli L, Perrino P (1996) Estimating genetic erosion in landraces – two case studies. Genet Resour Crop Evol 43:329–336
Harlan JR (1975) Our vanishing genetic resources. Science 188:618–621
Harlan JR, de JMJ W (1971) Towards a rational classification of cultivated plants. Taxon 20:509–517
Hawkes JG (1983) The Diversity of Crop Plants. Cambridge, MA: Harvard University Press, p. 102
Heywood V, Casas A, Ford-Lloyd B, Kell S, Maxted N (2007) Conservation and sustainable use of crop wild relatives. Agric Ecosyst Environ 121(3):245–255
Honsdorf N, March TJ, Berger B, Tester M, Pillen K (2014) High-throughput phenotyping to detect drought tolerance QTL in wild barley introgression lines. PLoS One 9(5):e97047
Hoyt E (1988) Conserving the Wild Relatives of Crops. IPGRI/IUCN/WWF, Rome
Intergovernmental Panel on Climate Change (IPCC) (2007) Fourth assessment report. Climate change 2007 synthesis report. Intergovernmental Panel on Climate Change, Geneva
James RA, Davenport RJ, Munns R (2006) Physiological characterization of two genes for Na^+ exclusion in durum wheat, *Nax1* and *Nax2*. Plant Physiol 142:1537–1547
Jarvis A, Lane A, Hijmans RJ (2008) The effect of climate change on crop wild relatives. Agric Ecosyst Environ 126:13–23
Jauhar PP (2006) Cytogenetic architecture of cereal crops and their manipulation to fit human needs: opportunities and challenges. In: Singh RJ, Jauhar PP (eds) Genetic resources, chromosome engineering, and crop improvement, vol 2, Cereals. CRC Taylor & Francis Press, Boca Raton, FL, pp 1–25

Kastner T, Rivas MJI, Koch W, Nonhebel S (2012) Global changes in diets and the consequences for land requirements for food. Proc Natl Acad Sci 109:6868–6872

Khush GS, Brar DS (1992) Overcoming the barriers in hybridization. Theor Appl Genet Monograph No.16:47–61

Kihara H (1983) Origin and history of 'Daruma', a parental variety of Norin 10. Proc Sixth Int Wheat Genet Symp 6:13–19

Kumar J, Pratap A (2016) Alien gene transfer: Challenges and opportunities. Alien Gene transfer in crop plants, Volume 1: Innovations, Methods and Risk assessment. Springer Science+ Business Media, New York. https://doi.org/10.1007/978-1-4614-8585-8_12. 2013

Labuschagne MT, Pretorius ZA, Grobbelaar B (2002) The influence of leaf rust resistance genes Lr29, Lr34, Lr35 and Lr37 on breadmaking quality in wheat. Euphytica 124:65–70

Li X, Takahashi T, Suzuki N, Kaiser HM (2011) The impact of climate change on maize yields in the United States and China. Agrie Syst 104:348–353

Linh LH, Linh TH, Xuan TD, Ham LH, Ismail AM, Khanh TD (2012) Molecular breeding to improve salt tolerance of rice (Oryza sativa L.) in the red river delta of Vietnam. Int J Plant Genomics 2012:9. https://doi.org/10.1155/2012/949038

Lobell DB, Sibley A, Ortiz-Monasterio JI (2012) Extreme heat effects on wheat senescence in India. Nat Clim Chang 2:186–189

Lopez PB (1994) A new plant disease: uniformity. Ceres 26:41–47

Luck J, Spackmand M, Freemand A, Trçbicki P, Griffiths W, Finlay K, Chakraborty S (2011) Climate change and diseases of food crops. Plant Pathol 60:113–121

Lukaszewski AJ (2000) Manipulation of the 1RS.1BL translocation in wheat by induced homoeologous recombination. Crop Sci 40:216–225

Mallikarjuna N, Saxena KB (2005) A new cytoplasmic nuclear male-sterility system derived from cultivated pigeonpea cytoplasm. Euphytica 142:143–148

Mallikarjuna N, Jadhav D, Reddy P (2006) Introgression of *Cajanus platycarpus* genome into cultivated pigeonpea, *C. Cajan*. Euphytica 149:161–167

Maxted N, Kell SP (2009) Establishment of a global network for the *in situ* conservation of crop wild relatives: Status and needs. Commission on Genetic Resources for Food and Agriculture

Maxted N, Hawkes JG, Ford-Lloyd BV, Williams JT 1997. A practical model for *in situ* genetic conservation. In: Plant genetic conservation: the in situ approach (eds. Maxted, N., Ford-Lloyd, B.V. and Hawkes, J.G.), Pp. 339–367. Chapman & Hall, London

Maxted N, Villa TCC, Scholten M, Ford-Lloyd B (2005) Defining and identifying crop landraces. Plant genetic resources 3:373–384

Maxted N, Ford-Lloyd BV, Jury S, Kell S, Scholten M (2006) Towards a definition of a crop wild relative. Biodivers Conserv 15:2673–2685

Maxted N, Ford-Lloyd BV, Kell SP (2008) Crop wild relatives: Establishing the context. Pages 3–30 in Maxted N, Ford-Lloyd BV, Kell SP, Iriondo JM, Dulloo ME, Turok J, eds. Crop Wild Relative Conservation and Use. CAB International Publishing

Maxted N, Kell S, Brehm JM (2011) Options to promote food security: on-farm management and in situ conservation of plant genetic resources for food and agriculture. FAO

Meilleur BA, Hodgkin T (2004) In situ conservation of crop wild relatives. Biodivers Conserv 13:663–684

Nable RO (1988) Resistance to boron toxicity amongst several barley and wheat cultivars: a preliminary examination of the resistance mechanism. Plant Soil 112:45–52

Naidoo R, Balmford A, Costanza R, Fisher B, Green RE, Lehner B, Malcolm TR, Ricketts TH (2008) Global mapping of ecosystem services and conservation priorities. Proc Natl Acad Sci 105:9495–9500

National Research Council (1972) Genetic Vulnerability of Major Crops. (National Academy of Sciences, Washington, DC, 1972)

Negri V, Tiranti B (2010) Effectiveness of *in situ* and *ex situ* conservation of crop diversity. What a *Phaseolus vulgaris* L. landrace case study can tell us. Genetica 138:985–998

Ochatt SJ, Mousset-Declas C, Rancillac M (2000) Fertile pea plants regenerate from protoplast when calluses have not undergone endoreduplication. Plant Sci 156:177–183

Pimentel D, Wilson C, McCullum C, Huang R, Owen P, Flack J, Tran Q, Saltman T, Cliff B (1997) Economic and environmental benefits of biodiversity. Bioscience 47:747–757
Placido DF, Campbell MT, Folsom JJ, Cui X, Kruger GR, Baenziger PS, Walia H (2013) Introgression of novel traits from a wild wheat relative improves drought adaptation in wheat. Plant Physiol 161:1806–1819
Plucknett D, Smith N, Williams J, Murthi Anishetty N (1987) Gene Banks and the World's Food. Princeton University Press, Princeton, NJ
Qi XP, Li MW, Xie M, Liu X, Ni M, Shao GH, Song C et al (2014) Identification of a novel salt tolerance gene in wild soybean by whole-genome sequencing. Nat Commun 5(1):920
Ren ZH, Gao JP, Li LG, Cai XL, Huang W, Chao DY, Zhu MZ, Wang ZY, Luan S, Lin HX (2005) A rice quantitative trait locus for salt tolerance encodes a sodium transporter. Nat Genet 37:1141–1146
Rijal DK, Kadayat KB, Joshi KD, Sthapit BR (1998) Inventory of indigenous rainfed and aromatic rice landraces in Seti river valley, Pokhara, Nepal. LI-BIRD Technical Paper No. 2, ISSN 1561–1558
Ruge-Wehling B, Linz A, Habeku A, Wehling P (2006) Mapping of *RYMl6Hb,* the second soilborne virus resistance gene introgressed from *Hordeum bulbosum*. Theor Appl Genet 113:867–673
Sarker A, Erskine W (2003) Recent progress in the ancient lentil. J Agric Sci 144:19–29
Schmidhuber J, Tubiello FN (2007) Global food security under climate change. Proc Natl Acad Sci 104(50):19703–19708. https://doi.org/10.1073/pnas.0701976104
Schoen DJ, Brown AHD (2001) The conservation of wild plant species in seed banks. Bioscience 51(11):960. https://doi.org/10.1641/0006
Sears ER (1956) The transfer of leaf-rust resistance from *Aegilops umbellulata* to wheat. Brookhaven Symp Biol 9:1–22
Sherchand KK, Adhikari NP, Khatiwada SP, Shrivastav AC, Bajracharya J, Joshi KD, Kadayat KB, Chaudhary M, Vhaudhary P, Vishwakarma SS, Yadav S (1998) Strengthening the scientific basis for in situ conservation of agrobiodiversity: Findings of site selection in Bara, Nepal, NP Working Paper No 2/98. NARC/LIBIRD/IPGRI
Sitch LA (1990) Incompatibility barriers operating in crosses of *Oryza sativa* with related species and genera. In: Gustafson JP (ed) Genetic manipulation in plant improvement II. Plenum Press, New York, pp 77–94
Song WY, Wang GL, Chen LL, Kim HS, Pi LY, Holsten T, Gardner J et al (1995) A receptor kinase-like protein encoded by the rice disease resistance gene, *Xa21*. Science 270:1804–1806
Stolton S, Maxted N, Ford-Lloyd B, Kell SP, Dudley N (2006) Food Stores: Using protected areas to secure crop genetic diversity. WWF Arguments for protection series. WWF, Gland Switzerland
Suprunova T, Krugman T, Fahima T, Chen G, Shams I, Korol A, Nevo E (2004) Differential expression of dehydrin genes in wild barley, *Hordeum spontaneum*, associated with resistance to water deficit. Plant Cell and Environment, 27:1297–1308
Suprunova T, Krugman T, Distelfeld A, Fahima T, Nevo E, Korol A (2007) Identification of a novel gene (*Hsdr4*) involved in water-stress tolerance in wild barley. Plant Mol Biol 64:17–34
Tanksley SD, McCouch SR (1997) Seed banks and molecular maps: Unlocking genetic potential from the wild. Science 277:1063–1066.https://doi.org/10.1126/science.277.5329.1063
Tanksley SD, Nelson JC (1996) Advanced back cross QTL analysis, a method for the simultaneous discovery and transfer of valuable QTLs from unadapted germplasm into elite breeding lines. Theor Appl Genet 92:191–203
The TT, Latter BDH, McIntosh RA, Ellison FW, Brennan PS et al (1988) Grain yields of near – isogenic lines with added genes for stem rust resistance. In: Miller TE, Koebner RMD (eds) Proceedings of the 7th international wheat genetics symposium, vol 2, Cambridge, UK, pp 901–906
Thuiller T, Lavorel S, Araujo MB, Sykes MT, Prentice IC (2005) Climate change threats to plant diversity in Europe. Proc Nati Acad Sci USA 102(23):8245–8250
Tudge C (1988) Food Crops for the Future. Oxford:Basil Blackwell, p. 83

U.S. Census Bureau. 2014. International database, June 2011 update. http://www.census.gov/population/international/data/idb/worldpoptotal.php. Accessed 15 Jan 2014

Vavilov N (1922) The law of homologous series in variation. J Genet 12:47–89

Vincent H, Amri A, Castaneda-Alvarez NP, Dempewolf H, Dulloo E, Guarino L, Hole D, Mba C, Toledo A, Maxted N (2019) Modeling of crop wild relative species identifies areas globally for in situ conservation. Communications Biology, 2:136

Vollbrecht E, Sigmon B (2005) Amazing grass: developmental genetics of maize domestication. Biochem Soc Trans 33:1502–1506

Wissuwa M, Ae N (2001) Genotypic variation for tolerance to phosphorus deficiency in rice and the potential for exploitation in rice improvement. Plant Breed 120:43–48

Wissuwa M, Wegner J, Ae N, Yano M (2002) Substitution mapping of the Pup1: a major QTL increasing phosphorus uptake of rice from a phosphorus deficient soil. Theor Appl Genet 105:890–897

Wood D, Lenne ´ JM (1997) The conservation of agrobiodiversity on-farm: questioning the emerging paradigm. Biodivers Conserv 6:109–129

World Conservation Monitoring Centre (1992) In global biodiversity: Status of the Earth's Living Resources (Groombridge, B., ed.), Chapman & Hall, London

Wright SI, Bi IV, Schroeder SG, Yamasaki M, Doebley JF, McMullen MD, Gaut BS (2005) The effects of artificial selection of the maize genome. Science 308:1310–1314

Xu K, Mackill DJ (1996) A major locus for submergence tolerance mapped on rice chromosome 9. Mol Breed 2:219–224

Xu K, Xu X, Fukao T, Canlas P, Maghirang- Rodriguez R, Heuer S, Ismail AM, Bailey-Serres J, Ronald PC, Mackill DJ (2006) Sub1A is an ethylene responsive-factor-like gene that confers submergence tolerance to rice. Nature 442:705–708

Xu X, Liu X, Ge S, Jensen JD, Hu F, Li X, Dong Y, Gutenkunst RN, Fang L et al (2012) Resequencing 50 accessions of cultivated and wild rice yields markers for identifying agronomically important genes. Nat Biotechnol 30:105–111

Yau SK, Ryan J (2008) Boron toxicity tolerance in crops: a viable alternative to soil amelioration. Crop Sci 48:854–865

Zeven AC (1998) Landraces: a review of definitions and classifications. Euphytica 104:127–139

Zhang H, Mittal N, Leamy1 LJ, Barazani O, Song BH (2016) Back into the Wild – Apply Untapped Genetic Diversity of Wild Relatives for Crop Improvement. Evol Appl, 10(1): 5–24. https://doi.org/10.1111/eva.12434

Zhou YL, Uzokwe VNE, Zhang CH, Cheng LR, Wang L, Chen K, Gao XQ et al (2011) Improvement of bacterial blight resistance of hybrid rice in China using the *Xa23* gene derived from wild rice (*Oryza rufipogon*). Crop Prot 30:637–644

Zhou Z, Jiang Y, Wang Z, Gou Z, Lyu J, Li W, Yu Y, Shu L, Zhao Y, Ma Y et al (2015) Resequencing 302 wild and cultivated accessions identifies genes related to domestication and improvement in soybean. Nat Biotechnol 33:408–414

The manufacturer's authorised representative in the EU is Springer Nature Customer Service Centre GmbH, Europaplatz 3, 69115 Heidelberg, Germany. If you have any concerns regarding our products, please contact ProductSafety@springernature.com

Printed and bound by CPI Group (UK) Ltd, Croydon, CR0 4YY
30/09/2026
02225247-0010